Customer Choice:
Finding Value in Retail Electricity Markets

AHMAD FARUQUI AND J. ROBERT MALKO, EDS.

Public Utilities Reports, Inc.
Vienna, Virginia
1999

First printing, March 1999

Library of Congress Cataloging-in-Publication Data
Customer choice : finding value in retail electricity markets / Ahmad
    Faruqui and J. Robert Malko, eds.
        p.   cm.
    Includes bibliographical references and index.
    ISBN 0-910325-73-1 (alk. paper)
    1. Electric utilities--United States. 2. Electric utilities-
-Deregulation--United States. 3. Electric utilities--Government
policy--United States. I. Faruqui, Ahmad. II. Malko, J. Robert
(John Robert), 1943–   .
    HD9685.U5C88   1999
    333.793 ' 2 ' 0973--dc21   98-50438
                                                                CIP

Printed in the United States of America

TABLE OF CONTENTS

LIST OF TABLES

LIST OF FIGURES

PREFACE

This book is written for individuals interested in the future of customer choice in the American electric utility industry. Our hope is that those who read this book will acquire a richer understanding of the factors causing change in this industry, and a better appreciation for the different outlooks of the various affected groups concerning customer choice. We believe that the industry's transformation over the next decade will be smoother for all concerned if the participants seriously consider not just their own needs, but also the needs of other affected groups. In that spirit, we sought out a diverse group of experts who could articulate their thoughts and opinions concerning customer choice issues in a manner that would be understandable to persons who may not possess their expertise and perspective.

This book would not have been possible without the efforts of the authors who have contributed chapters to it. We express our sincere thanks for the work they have done. We are indebted to our book editor, Lori M. Rodgers of the PUR Press, for her meticulous reading and editing of the entire manuscript. We also thank her colleagues, Susan M. Johnson and Paul C. Lasky of Public Utilities Reports, Inc., for their devotion to moving this project along for the past year. We would like to thank the individuals who assisted Susan and Paul in making this book possible—most of whom we have not met. We would also like to thank several members of the support staff at EPRI who worked with us in assembling the manuscript. These include Pamela Arauz, Dee Autio, Davina Atwal and Peggy Prater.

In preparing this book, we were inspired by the contributions to the industry made by our friend and colleague, Gregory B. Enholm. His innovations at the Public Service Commission of Wisconsin and Salomon Brothers Inc. continue to influence utilities and regulatory agencies nationwide. Finally, and most importantly, we would not have been able to produce this book without the encouragement and understanding of our spouses, Nuzhat Faruqui and Sandra Malko.

We dedicate this book to our children, Furah and Saba Faruqui and Heather and Jeffrey Malko.

A. F.
Danville, California
J. R. M.
Salt Lake City, Utah

PROLOGUE

In the more than one-hundred-year history of the electric utility industry there have been few cataclysmic changes in institutional organization and none so potentially revolutionary as the ongoing deregulation of the industry.

One could argue that the horizontal and vertical integration that marked the first four decades of the industry was a gradual change of equal importance. Or, one could argue that the industry's submission to regulation in a trade-off for monopoly power, a position fostered by Samuel Insull after World War I, was a faster change of equal magnitude. Certainly, the Federal Power Act of 1935 creating the Federal Power Commission (now the Federal Energy Regulatory Commission) was an instantaneous change, which affected utility behavior then and now. The more recent momentous events such as the Public Utility Regulatory Policies Act of 1979 had great impact on utility operations but little effect on institutional organization.

Two decades ago, the idea of instituting competition within the electric utility industry seemed impossible except to a few prescient industry observers. Even today, there remain leaders in the industry who think that deregulation leading to full-scale competition will not happen, at least, not in their state, not in their lifetime and never to their utility.

Yet, the seeds for this momentous change were sown long ago and the concept tested in many venues. Within the United States, deregulation of the telephone industry was the clearest portent of what might happen in the electric utility industry. It was, however, the successful deregulation of the gas industry, the airlines, the trucking industry and others that made the change in the electric industry inevitable.

Perhaps of equal importance in changing intellectual perception on the need for competition was the growing world wide view that centrally controlled systems, whether political or economic, do not function efficiently and never can compete effectively with democratic, free market systems. As this wave of understanding swept the planet, we saw the breakdown of centralized planned economies replaced by trials at free market economies. Similarly, we are witnessing the break-up and sale of centralized electric utilities. This has happened most notably in the United Kingdom, Chile, and Argentina and, to a more limited extent, in many other countries, even in Russia and China.

What is driving this trend? A major force is the perception, and probably the reality in many cases, that utility costs are higher than they need to be to serve the customers' desires. One reason for this is that under a regulated environment, investments and expenditures that appear reasonable will be included by regulators for recovery under rates. It is not too hard to imagine that some utility spending decision under a regulated framework would not have been undertaken in a competitive environment. This does imply a nefarious plot to over-invest or incur unneeded expenditures, but the protection offered by the regulatory process did lead some to invest to avoid other risks (falling short of capacity) or to provide a quality of service that the customer does not want to pay for, if given a choice.

Other factors played an important role in the resulting over-capacity and over-investments. The oil shocks of the 1970s caused by the embargo and two conspired vaults in price slowed electric demand well beyond any utility planned future. Also, those commission- or federal-mandated conservation and renewable resource programs that were not economically competitive have led to the bloating of utility investments. Some utility executives became enamored with the potential of "clean" nuclear power without realizing the risks of this new technology.

Of course, the cozy system of assured return for approved investments came to an abrupt and unexpected (by utility executives) end when inflation and sky-rocketing fuel prices incited public revolt to ever-increasing rates. The traditional regulatory compact of adequate returns for prudently taken investments suddenly evaporated. This change in treatment of investments frequently was done under the guise that investments had not been prudent even if, in some cases, they had been blessed previously by regulators.

But now comes the time when utilities must face the test of real market forces, real competition. The Department of Energy reports that as of April 1998, 10 states have enacted restructuring legislation, six have issued regulatory orders; 20 have legislation pending; 12 are holding legislative or commission investigations; only three have no ongoing significant activity. To believe that a deregulated industry in some form and a fully competitive electricity market will not exist in the near future is to ignore the trends both here and abroad, both in this industry as well as in others.

Experience with such change in other industries and other countries tells us that there will be winners and there will be losers. One set of winners will be most customers who will see more choice and lower effective cost for their electric energy needs. The winners among the utilities will be those who can retool themselves and their organizations for this new environment. Since the drive for such change is to push institutions toward more optimum results for the consumer, utility managers must reinvent themselves into market-oriented, consumer-focused experts. They must learn the tools of successful combat to meet the onslaught of more experienced competitors. They must structure their organizations to deliver what consumers desire, at an acceptable price, in an attractive package, with sensitivity to the consumer decision-maker.

René H. Malès
Retired President
IES Utilities

FOREWORD

It is not surprising that a regime of retail competition in electricity involves major changes in policy and in law. One important area in which these changes already have taken place is in access to transmission. Under statutory authority, administratively implemented with a broad construction of anti-discrimination principles and as an extension of essential facilities doctrine, suppliers of electricity generally have acquired the right to use the transmission facilities of their competitors, for a price. This in itself is a revolutionary development, climaxing many years of efforts to endow the transmission system with common carrier status. A provision to that effect was originally proposed to be included in Part II of the Federal Power Act, enacted in 1935, but was dropped in the final version of that legislation. In 1978, the Federal Power Act was amended to give the Federal Energy Regulatory Commission authority to order owners of transmission facilities to provide transmission service to other electric suppliers, but the language of the new provision was so opaque that it could not be put to practical use. Finally, in 1992, new legislation (the Energy Policy Act) conferred broad powers on the FERC to order wheeling (transmission service) if the public interest required.

Originally, the drive to impose common carrier obligations on transmission owners was primarily carried on by public power agencies seeking to have cheap (usually public) power brought to them over the transmission lines of intervening investor-owned utilities. The 1992 provisions, however, were intended to accommodate a more general sort of wholesale competition among electricity providers. In fact, the 1992 provisions expressly prohibited their invocation to facilitate retail competition. More recently, however, state law, applied in conjunction with federal regulation, has authorized retail competition and presumably made it possible to order transmission service for the purpose of serving retail customers.

As indicated, these expansive provisions for transmission access have been revolutionary in purpose and effort. For all practical purposes, property in transmission assets has been dedicated to public use (although the owner is entitled, of course, to fair compensation for use by others). In this respect, perhaps ironically, retail (and other) competition, an individualist concept if there ever was one, has led to the effective socialization of the transmission system. Certainly, transmission assets, although they may have a nominal owner, are readily available at a price for the benefit of whoever finds it useful to use them. Among other things, this makes it unclear who has an economic incentive to expand or improve the transmission system. Nonetheless, transmission access is clearly an idea whose time has come. It is unlikely to be restricted in the years ahead.

Another, and more vexing, problem involved in retail competition implicates a traditional and fundamental principle of public utility law. This is the principle that a public utility has an obligation to provide service on demand. In considerable measure this principle is a corollary of the grant of a monopoly franchise. If there is only one provider, that provider must have an obligation to respond to all the demands for service that it receives. If there were no such obligation, some potential users of electricity could be deprived of this essential service. As indicated, the obligation to serve is primarily a corollary of a utility's monopoly status. However, there may be providers of certain services, that may not be monopolies—such as innkeepers—who have a traditional obligation to serve all comers. These are occupations generally under the rubric of "businesses affected with a public interest," which are treated in many ways like monopoly utilities even though they may not possess the economic characteristics of monopolies. See *Munn v. Illinois* (94 U.S. 113 [1877]).

The obligation to serve is, of course, a status-based duty arising from the nature of the provider and imposed as a matter of law. With retail competition, on the other hand, the law of electricity moves more toward a thoroughgoing basis in contract rather than retaining its traditional status-based elements. Consumer choice necessarily involves a buyer "shopping" to make a purchase of power and free to enter into a contract with a supplier of its choice. The terms of the contract represent the obligations that the parties bargain for and agree to impose on each other. This is in contrast to obligations imposed by the law without particular regard to the special requirements of the parties. This transition from status to contract is in line with the capitalist and individualist orientations that are otherwise dominant in Western economies.

However, the question for the law of electricity may be whether the traditional obligation to serve can survive the transition to competition. Can such an obligation be thrust on a "supplier of last resort" if, for some reason, contractual arrangements for supply entered into with another provider break down and cannot be replaced? Supplier of last resort obligations, where they have been recognized, usually have been imposed on the territorial utility directly connected to the customer in question (the "local utility")—that is the utility which, before competition, would have had the exclusive franchise to serve the customer in question. Presumably, under this kind of arrangement, service provided by the local utility would be at regulated rates. After all, an obligation to serve without any prescription of rate would be no obligation at all.

Under another scenario, the obligation to serve may evolve into an obligation to deliver (that is, an obligation of the infrastructure utility to provide transmission service to electric suppliers to enable them to reach the load). The utility that operates the infrastructure would owe a legal duty to the seller of the product. The local utility would be required to aver that it would hold its system open for marketers to deliver their product to the marketers' customers. Under this arrangement, the product marketer becomes the local utilities' customer. Nevertheless, a retail customer could block this scenario by choosing to take bundled service from the local utility, presumably at the standard offer.

A retail customer, therefore, would have a potential for service either from the local utility or from a distant source (the market). If a customer were permitted to divide its power requirements, it would be free to buy its requirements for on-peak hours when the temperature was above 90 degrees from the local utility at regulated rates while buying everything else in the market during all other hours. It would be extraordinarily difficult for the local utility and its regulators to design partial requirements rates to fit these circumstances. A less taxing but probably much more common problem would revolve around what ancillary or control area services (for example, load following service and voltage support) were necessary and whether the local utility ought to provide those services to make the power supply fully adequate.

There are, of course, other cases than the ones just examined where the customer deliberately splits its demand between the market and the local utility. There is the case, for example, where the customer for whatever reason fails to contract for all the electricity that it needs and has to make up the shortfall from the local utility. This is essentially similar to the first case, except that the split of requirements is no longer "deliberate." Thus, the state of mind of the customer is different but the economics may be indistinguishable.

Finally, there is a third case where sufficient power is under contract, but, due to equipment failure, external events or whatever, the electrical requirements cannot be met by the original contract partner, and, for some reason, the market is otherwise unable or unwilling to make up the deficiency. Here there may be breaches of contract requiring the contracted-for supplier to buy power to fulfill the contract, or a variety of other contract remedies may be available. Still, in the end, and all else having failed, the customer may have to fall back on the local utility as the supplier of last resort.

There are, of course, arguments to be made for eliminating the supplier of last resort function— for directing parties to electric transactions to provide by contract for the fulfillment of their requirements, no matter what the circumstances, and for leaving them ultimately to recourse in the market. In the world of contract there seems no good reason to allow users of electricity to "mix and match" competitive rates with regulated rates. If electricity is to be treated as a competitive commodity like many others, why aren't the remedies of the Uniform Commercial Code enough? It would certainly simplify matters to require retail customers entering the market to make their own contractual arrangements for electric power, including back-up power, emergency power, supplementary power, etc.*

But before adopting such a seemingly simple solution, do not forget that electric power is an infrastructure industry, providing the backbone of the economy. Electricity is rich in external benefits—benefits to third parties with no contractual relations with the electric supplier. Tenants in an apartment house or shoppers in a store all benefit from an electric supply procured by someone else. If the power supply to a factory fails, many more than the factory owner suffer. The chances of third parties like these recovering their damages from a power failure do not seem totally promising.

---

*I owe a great deal to Joe Pace and his paper, "Is Retail Wheeling Right for the Electric Utility Industry?" (Federal Energy Bar Association Meeting, May 17, 1995), for much of this analysis.

Therefore, to keep the lights on, someone may have to serve as the supplier of last resort, and that someone will have to be fairly compensated. This is a difficult concept to reduce to workable form, but this challenge may have to be met if the future is to live up to its billing.

Judge Richard Cudahy
U.S. Circuit Court of Appeals
Chicago, Illinois

INTRODUCTION

CHAPTER 1

The Brave New World
of Customer Choice

Ahmad Faruqui, EPRI
J. Robert Malko, Utah State University

As this book goes to press, millions of electricity customers in California and Pennsylvania have an option—but not an obligation—to choose their electricity supplier. As Milton Friedman might express it, they are "free to choose." Millions more in New York, New England, and Illinois are likely to obtain the same freedom by the year 2000. The inexorable movement towards customer choice is shown in Figure 1–1.

By exercising their right to choose, customers can reduce their energy bills. In addition, due to the competitive pressures created by market forces for innovation, they also can derive additional benefits from value added services that are bundled around the core product, commodity electricity.

In California and elsewhere, scores of energy service providers (ESPs) are engaged in a massive campaign to lure customers to their particular brand of electricity, while incumbent utilities are seeking to ensure that they can keep a large fraction of their historical customer base. Many incumbents, faced with unprecedented change in their business environment, are re-engineering their business processes. Some have let these change management practices, and the consequent cycles of reorganizations that they embody, dominate their day-to-day business activities, to an almost pathological extent. Others are creating nimble new corporate cultures, bringing in executives and managers from recently deregulated industries such as telecommunications and natural gas. Still others are doing all of the above, and still keeping a sharp eye on the market.

This book surveys the dynamics of this brave new world of customer choice, and identifies a wide range of strategies and tactics that can help ESPs create lasting economic value for both their shareholders and their customers.

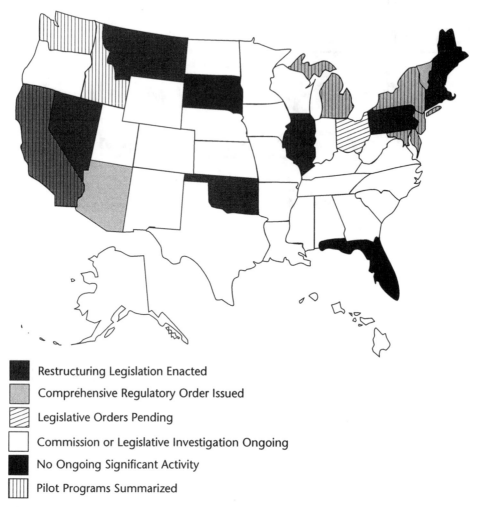

Restructuring Legislation Enacted

Comprehensive Regulatory Order Issued

Legislative Orders Pending

Commission or Legislative Investigation Ongoing

No Ongoing Significant Activity

Pilot Programs Summarized

Figure 1–1   *Restructuring of the U.S. Electricity Industry*
Source: U.S. Department of Energy, Energy Information Administration, as of Sept. 1, 1998

It is written for all the participants in this "great game" of choice that has all the elements of high drama that one might find in one of Rudyard Kipling's essays about the turn-of-the-century conflict between the British Lion and the Russian Bear. Participants include ESPs, incumbent utilities, regulators, consultants, researchers and academics. It is our hope that electricity customers also will use it to understand their choices, and to realize the benefits that they deserve.

This book represents a continuation of the tradition established in a book co-edited by one of us with Greg Enholm, *Electric Utilities Moving into the 21st Century*.[1] That book provided readers with a broad overview of industry trends. This book focuses on what is arguably the most significant of these trends, the provision of choice to electricity customers.

This book is organized into six topic areas: (1) What Is Customer Choice?; (2) What Is Driving Customer Choice?; (3) What Opportunities Are Created by Customer Choice?; (4) What Strategies Should Energy Service Providers Pursue?; (5) What Enabling Technologies and Infrastructure Are Needed?; and (6) What Are the Key Market Structure Issues? These topics, shown in Figure 1–2, provide a road map to how this book is organized.

Each of these topic areas is addressed by "thought leaders" drawn from a cross-section of the industry. To establish a backdrop against which to view the key points made by our distinguished contributors, we provide a "sneak preview" of what has already happened in three states: California, Pennsylvania, and Rhode Island. This sneak preview is based on a recent study of 150 customers and 12 energy service providers with customer choice.[2] The study finds that the key driver for switching electricity supplier is the desire to save money.

(1) In markets where ESPs can offer significant savings, customers are switching electricity suppliers.

- In California, 17.6 percent of industrial accounts (> 500 kilowatts) have switched, representing 27.2 percent of utility distribution company (UDC) load prior to switching (as of November 30, 1998).
- In Pennsylvania, approximately 227,000 customers are participating in the state's pilot program.

(2) Without a price incentive, customers are reluctant to switch.

- In California, few ESPs are offering price discounts to residential and small business customers because they can't beat the mandated 10 percent rate reduction. Without a price incentive, only 0.8 percent of residential customers and 2.3 percent of commercial customers (under 20 kW) have switched (as of September 30, 1998).

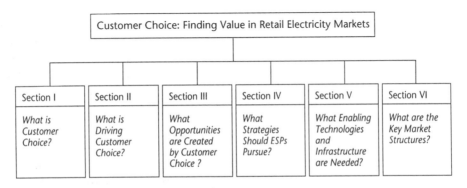

Figure 1–2 *Road Map for the Book*

- In Rhode Island, ESPs have not been able to compete with the standard offer rates; thus, few customers have switched suppliers.

(3) If customers perceive that they can save money, they will switch.

- In California, many of the residential and small business customers who switched think (perhaps incorrectly) that they will save more than 10 percent by switching.

(4) Other drivers of choice include green power, customer service and reliability, and value-added services.

- To date, most activity in the California residential market is from customers switching to purchase more expensive green power.
- Customer service and reliability are important considerations for some customers. Some customers are switching because they are not happy with their current supplier and hope to get improved service/reliability from their new supplier.
- For some commercial and industrial customers, value-added services, in combination with price, is the driving factor.

(5) Switchers may be switching to unregulated affiliates of UDCs.

- In California, industry sources speculate that many switchers are not leaving the parent corporation.
- In several of the pilot programs, many customers switched to unregulated affiliates of UDCs. In the New Hampshire pilot, for example, 70 percent of switchers switched to the affiliate of their local distribution company.

With this perspective in mind, let us now take a quick tour of the major sections of the book, defined earlier in Figure 1–2.

Section I: What Is Customer Choice?

Clark Gellings discusses the increasing choices in technologies, in quality, in bundling of services, and in suppliers facing the "Open Market Customer." Raymond Lawton, Francine Sevel, and David Wirick explore the two sides of the consumer communication transaction, including the receipt and use of consumer information and communications from utilities and regulatory agencies to consumers. Michael Rufo and Kenneth Train analyze empirical choice modeling methods with two case studies using revealed preference data from Pennsylvania, and two case studies using stated preference data from California. Lisa Wood, Sheryl Cates, and Sue Winemiller examine a qualitative approach, Voice of the Customer, and a quantitative approach, choice-based conjoint analysis, for identifying customer needs and preferences.

Section II: What Is Driving Customer Choice?

Karl Stahlkopf discusses technological challenges and opportunities associated with the convergence of utility services. Robert Michaels examines economic issues concerning convergence mergers in the natural gas and electricity markets. Michael Peevey argues that California already is demonstrating that a competitive—as opposed to regulated—market will stimulate construction of new power plants, bringing both lower energy costs and greater system reliability.

Section III: What Opportunities Are Created by Customer Choice?

Robert Crandall and Jerry Ellig examine the potential implications for electricity customers of the value of consumer benefits and related price reductions associated with deregulation and regulatory reform in other industries, including long-distance telecommunications, airlines, trucking, and railroads. Dean Jones provides insights from the competitive movement in the wholesale natural gas industry concerning opportunities for customers.

Section IV: What Strategies Should Energy Service Providers Pursue?

To prosper in a world of customer choice, energy service providers should develop and pursue new strategies and tactics that include listening, pricing, marketing, and modeling. These will enable them to differentiate their offerings from other suppliers, and enhance profitability. At the same time, as demonstrated in other industries that have been deregulated over the past two decades, such differentiated offerings will permit customers to fulfill needs previously unmet by the simple undifferentiated offerings available in a regulated marketplace, thereby enhancing their economic well-being.

Ahmad Faruqui creates a framework for strategic listening. Joseph Ewing provides several examples demonstrating the necessity of engaging in listening to customers.

Doug Caves, Stefan Brown, and Ahmad Faruqui introduce the concept of risk-based pricing, and demonstrate how different types of risk-based pricing can enhance profitability and economic value to energy suppliers while providing customers more choices and enhanced well-being.

Kenneth Bartkus presents general principles of market-orientation and examines how these principles relate to the rapidly changing electric utility industry. David Lineweber, Kerry Diehl, and Patricia Garber provide a "how-to" guide or framework to assist energy providers to acquire and retain market share and create customer loyalty.

Daniel Violette and Michael King present viable business models for electricity generation for conditions of competitive markets and retail choice. Robert Wayland and David Jones examine and apply a new framework called the Value Compass^{sm} that emphasizes different dimensions of shareholder and customer value creation.

Section V: What Enabling Technologies and Infrastructure Are Needed?

Eric Cody discusses the various mysteries of competitive infrastructure in order to define approaches and activities that moderate the costs of implementation. Melanie Mauldin examines technologies and related issues concerning the markets for metering and information services.

Section VI: What Are the Key Market Structure Issues?

Kenneth Lay argues that the most troubling aspect of the current market structure in California is that it guarantees market share for incumbents. He wants to give all customers the right to choose immediately, and to discontinue the utility's exclusive franchise for customers who neglect to make a choice. James Malachowski examines changing and new responsibilities and considerations for regulatory agencies including the creation and maintenance of meaningful choice for consumers and reasonably efficient markets. Becky Kilbourne analyzes the roles, including a reliable clearing function, of commodity exchanges and discusses the development of electricity power exchanges in several countries that have introduced competition in electricity. Kenneth Costello proposes a comprehensive regulatory strategy associated with mitigating the vertical-control problem in retail electricity markets, and this proposed strategy combines command-and-control options with market-based options.

NOTES

1. Gregory B. Enholm and J. Robert Malko, eds., *Electric Utilities Moving into the 21st Century* (Arlington, VA: Public Utilities Reports Inc., 1994).

2. For details, see "What Drives Customer Choice in Competitive Power Markets?" (Electric Power Research Institute report, [October 1998] TR-111806).

Section I

What is Customer Choice?

CHAPTER 2

The Open Market Customer

Clark W. Gellings

Today's revolution in the energy industry is not about deregulation, restructuring, or the creation of new institutions, like Independent System Operators (ISOs). It is about consumerism. Through the forces brought on by the emergence of new technologies and a changing regulatory framework, a new wave of consumerism is evolving, embodied in a concept I have chosen to call the "Open Market Customer." The term is intended to describe a complex energy and energy service market in which customers increasingly have choice and begin to make energy purchase decisions in a more integrated way. This integration captures not just the purchase of energy, but also the purchase of the related energy-consuming devices, appliances, infrastructure, financing, and management. Today, the Open Market Customer in the U.S. spends an estimated $780 billion annually on non-transportation energy and on the devices, appliances, and apparatus that help to consume it. Only about one-third of that sum is viewed today as being part of the energy business.

Over the last 20 years, Americans have continued to prosper through their dependence on a safe, reliable, and economical energy supply system. This dependence has been paralleled by underlying changes in regulation initiated in the electric business by the Public Utility Regulatory Policies Act (PURPA) of 1978.

Despite the fact that energy costs in the U.S. in the 1990s were relatively low and stable, the Open Market Customer was conceived through the public's desire for choice of supplier and desire for control of energy costs. It was the regulators' interpretation of the public desire that laid the groundwork for the changes that now are well under way.

PERIODS OF GROWTH AND EVOLUTION

The electric energy business can be viewed as having transitioned through four periods. It started a little more than 100 years ago, and has gone through phases of development triggered by regulatory changes, technological progress, and external pressure.

The first period spanned from the birth and initial growth of the electric energy industry until 1910. Thomas Edison did not envision the creation of utilities, but found himself engaged in launching a business to create a demand for his technologies and to enable financing. Edison started with a light bulb, and went on to invent a variety of electrical apparatus necessary to support the infrastructure for light bulbs and, soon afterward, motors. Competition ensued involving alternating versus direct current, ever larger generators, various prime energy sources, and competition for customers with parallel distribution systems. Business interests began to dominate, and various sized competitors began to bundle power generation with delivery in order to compete.

The second period witnessed the creation of ordered processes for solving much of the confusion created in the first period. Growth and demand for capital pressured smaller enterprises to combine or to be assimilated into larger enterprises. Public concern stimulated policymakers to begin constructing a regulatory framework to watch over the new industry, as had recently been constructed for railroads. Policymakers also sponsored the development of municipal and federal utilities.

The second period electrified America in all but the rural areas. The lack of profitability in serving rural America prompted intervention by the federal government and the formation of rural electric cooperatives. Regulation formed franchised service areas in exchange for various types of gross receipts and franchise taxes, and obligated utilities to serve anyone in their areas. A basis for calculating how much utilities were allowed to earn also was established. This period extended from 1910 through 1970. Technology continued to make utilities more profitable, and until the late 1960s, allowed each new power production facility to be more efficient than the last. Nuclear power promised further economies, and utility stocks continued to rise.

The third period highlighted what some would consider to be weaknesses in the U.S. energy system. The Organization of Petroleum Exporting Countries and its embargoes brought disruptions in oil supply and, subsequently, inflation. The environmental movement, born in the late 1960s, began to become a powerful force in influencing the public and its policymakers. Electricity, whose infrastructure makes it the most visible form of energy, became the target of regulation and concern. Nuclear power had lost its appeal, and the Three Mile Island incident together with the Chernobyl disaster put an end to a burgeoning segment of the industry. Regulators began to squeeze utility profits, and with the interest rates so high that utilities could not afford to build new power plants, the utilities embraced demand-side management. Incentives in some states along with federal subsidies spurred a renewable industry.

At the end of the third period, utilities that invested in nuclear power largely wished they had not; most were not building power plants, and many were pursuing demand-side management either in response to regulatory incentives or because it was cheaper to pay customers not to use the product than to build facilities to service their needs. It was the equivalent of a McDonald's restaurant that is too busy to serve all of its luncheon customers, and hires a traffic cop to wave away people who are trying to approach the driveway. Picture the cop holding up a sign that reads: "Come back after 3:00 p.m."

The fourth period is a fundamental re-envisioning of the industry. Imagine a market that creates new price structures, bundles services, and offers new products. Power plants are built based on business interests, prompted by consumer desire for energy. Customers now become the focus of energy and energy service providers. It is no longer just about utilities, or just about electricity. It is now about meeting the needs of the ultimate consumer. This new environment is the world of the Open Market Customer. In this world, suppliers of all sizes behave as reliable, aggressive companies who meet the needs of the Open Market Customer and create a synergy that will lead to more choice, lower prices, and an increased array of quality products and services.

HOW BIG IS THE MARKETPLACE?

The energy functions of traditional electric and gas utilities represent only one-third of the total marketplace. In the total marketplace, the Open Market Customer is served by a broader industry encompassing all energy users and all organizations with whom energy users must interface in order to obtain services. This includes:

- providers of electricity, natural gas, and related raw energies or fuels to residential, commercial, and industrial energy users;
- providers of equipment, technology, products, services, and systems needed to convert energy to meet the customers' needs;
- providers of design, construction, operation, maintenance, and finance of such products and systems;
- those who provide knowledge and information to the energy marketplace; and
- federal, state, and local level regulators of the marketplace and its inhabitants.

The Open Market Customer must:

1. Obtain commodity energy and ensure that it is contracted for delivery to the requisite sites.

2. Arrange for a working infrastructure of pipes, wires, breakers, controls, and energy-conserving devices and appliances to convert energy into the desired light, heat, motive power, and production.

3. Finance and administer commodity energy and infrastructure purchase, leasing, and operations.

Table 2–1 depicts one estimate of the annual expenditures and total annual investment by all residential, commercial, and industrial energy consumers. As stated above, energy is only a third of the total estimated expenditures. The overall energy infrastructure and its financing and administration amount to the remaining two-thirds. These should not be viewed as precise market measurements.

With deregulation and restructuring at hand, both the customers and their traditional suppliers, and many potential new players are looking at similar data. Several important conclusions are obvious:

- **There is a major customer position and perspective shift.** The customer is now faced with reconsidering suppliers for commodity energy, and is faced with an enormous array of questions. These questions were always there, but they were always answered by others, either as part of the commodity purchase or because they were not viewed as part of the energy and energy service purchase. Once the commodity question is on the table, the original paradigm begins to unravel.

TABLE 2–1 Functions the Open Market Customer Must Perform		
	Major Category	**Functions Included**
Obtain Commodity Energy	Energy Procurement	Electricity purchasing Gas purchasing
	Utility Distribution	Electric wires Gas pipes
	Energy Delivery and Billing	Electricity meters Gas meters
Arrange a Working Infrastructure	Energy Delivery	Substations, breakers, regulators, various distribution equipment
	Environment Systems	Lighting HVAC
	Process Systems	Process heating, motive power, magnetics, dielectrics
Finance and Administer	Financing and Investing	Project financing, leasing, financial instruments
	Administration	Personnel, accounting, compliance
Source: C. W. Gellings, *Effective Power Marketing* (Tulsa, Oklahoma: PennWell Publishing Co., 1998).		

TABLE 2–2
Business Functions in the Open Market Customer Energy Industry

Business Function	Annual Expense ($ Millions)	Total Investment ($ Millions)	Expense percent of Total
Energy Procurement	110,000	—	14 percent
Utility Distribution Company	155,000	—	20 percent
Energy Metering and Billing	5,000	—	1 percent
Energy Infrastructure	130,000	1,750,000	17 percent
Energy/Environment Systems	100,000	350,000	13 percent
Energy/Process Systems	25,000	100,000	3 percent
Energy Monitoring/ Management/Integration	5,000	25,000	1 percent
Energy Financing and Investing	225,000	—	29 percent
Energy Administration	25,000	25,000	4 percent
TOTAL	$780,000	$2,250,000	

Source: C. W. Gellings, *Effective Power Marketing* (Tulsa, Oklahoma: PennWell Publishing Co., 1998).

- **There is a major utility perspective change.** Utilities began to realize two important facts. Once there is an open market: (1) 13
percent of their customers would switch for no reason, and up to
70 percent would shift for a 20 percent reduction in price;[1] and (2)
it might cost $500 per customer to acquire new customers whose
annual contribution to the margin could be no more than $25 to
$50.[2] Both of these problems can be overcome, in whole or in
part, by reducing the price, enhancing the brand image, and
bundling services. As a result of utility interest in competing we
now find utilities interested in the same customer infrastructure,
financing, and administration issues that the Open Market
Customer has drawn its attention to.

TO COMPETE OR NOT TO COMPETE?

In response, traditional utilities are embracing one or two business areas.
Some are deciding to focus on the regulated wires business, while others are
adding the energy and energy services components. In particular, those who
have decided to compete on the unregulated side are developing new approaches to marketing and customer service that view the purchase of energy and energy services from the customer's perspective. This involves
addressing their issues, solving their problems, and helping the customers
manage their energy purchases. It is not about selling more units of energy,
but about offering a combination of products, services, and technologies to
meet the customers' needs. The engagement may reduce energy purchases,
increase them, or lengthen and/or change the bundle of services offered.

In many cases, this new view of the market forces an integration of energy use, the conversion of devices and appliances, and a balance between electricity and gas purchases. This integration provides marketing and customer service opportunities to the suppliers of the Open Market Customer, as well as increased value to the customers involved.

WHAT ARE THE REAL COSTS?

To verify the concept of the Open Market Customer, the Electric Power Research Institute commissioned case studies of four large commercial and industrial customer sites. These included the following:

- A wood products plant that manufactured fiberboard (particle board) and represented about 10 percent of total U.S. output at one location.
- A university accommodating 11,600 students on its main campus.
- A large, 850-bed, full-service general hospital.
- A major airport serving 24 million passengers per year.

Tables 2–3 and 2–4 depict the annual expenditures and total investment for energy and energy services for the case study sites.

Figure 2–1 depicts the relative weight energy procurement has in the total energy and energy service expenditure in the case studies. In this illustration, energy services include metering and billing, operations, and financing.

TABLE 2–3 Annual Expenditures for Energy/Services ($ Thousands)				
Business Function	Industrial	University	Hospital	Airport
Energy				
1) Energy Procurement	1,597	4,723	2,475	12,502
2) Utility Distribution Company	2,193	2,902	2,403	7,082
3) Energy Metering and Billing	90	472	284	638
Subtotal	3,880	8,097	5,162	20,222
Energy Operations				
4) Energy Infrastructure	142	2,272	1,951	1,976
5) Space Conditioning	105	2,251	1,948	1,768
6) Process Systems	278	3,484	3,479	4,826
7) Energy Monitoring/Management	0	344	311	210
Subtotal	525	8,351	7,689	8,780
Energy Financing				
8) Energy Financing and Investing	1,088	1,885	1,715	5,786
9) Energy Administration	85	932	889	613
Subtotal	1,173	2,817	2,604	6,399
TOTAL	$5,578	$19,265	$15,455	$35,401
Source: Electric Power Research Institute				

Business Function	Industrial	University	Hospital	Airport
TABLE 2–4 Total Investment for Energy/Services ($ Thousands)				
Energy				
3) Energy Metering and Billing	0	0	0	3,638
Subtotal	0	0	0	3,638
Energy Operations				
4) Energy Infrastructure	2,043	7,568	15,790	21,286
5) Space Conditioning	256	9,374	5,336	8,428
6) Process Systems	9,639	684	1,484	50,413
7) Energy Monitoring/Management	0	3,000	3,000	379
Subtotal	11,938	20,626	25,610	80,506
Energy Financing				
9) Energy Administration	31	100	12	2,322
Subtotal	31	100	12	2,322
TOTAL	$11,969	$20,726	$25,622	$86,466
Source: EPRI				

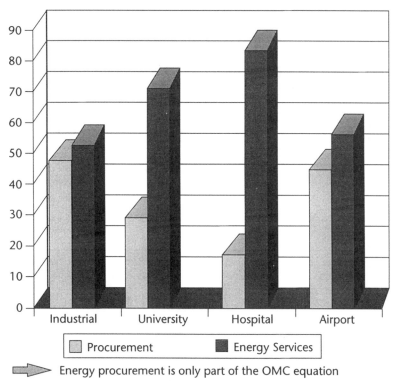

Energy procurement is only part of the OMC equation

Figure 2–1   *Open Market Customer Case Study—Observations*
Source: EPRI

Table 2–5 maps the Open Market Customer (OMC) business functions against the industry participants' functions.

In short, there is a range of issues confronting the Open Market Customer:

1. **Energy Purchase:** Price; contract terms; credibility of the supplier.

2. **Energy Delivery:** Distribution reliability; interface between customer system and the distribution entity.

3. **Metering:** Accuracy; promptness of readings; information content of bills.

4. **On-Site Energy Distribution:** Infrastructure; stand-by generation capabilities; fuel flexibility; infrastructure replacement; power quality.

5. **Energy Utilization:** Lighting, heating and cooling systems' adequacy; efficiency; cost effectiveness; maintenance.

6. **Industrial Process Utilization:** Technologies for productivity; product quality; environmental compliance.

7. **Monitoring and Measurement:** Data collection, dissemination and evaluation; efficiency versus infrastructure trade-offs; technology evaluation; telecommunication and computer technology.

8. **Financing and Investing:** Lease versus own; in-house versus outsourcing; finance versus invest; capitalization versus expensing; financial services bundling; shared savings versus performance contracts.

9. **Administration and Overhead:** Personnel competency and training; staff size; accounts receivable and payable management; regulatory compliance awareness; incentive program awareness; energy options awareness.

TABLE 2–5
Open Market Customer: Importance to Industry Participants

OMC Business Function	Open Market Customer	Traditional Vertical Utility	ESCO	Super ESCO	UDC	Power Marketer	Technology Manufacturer	Finance Industry
Energy Procurement	X	X		X		X		X
Utility Distribution Co.	X	X			X		X	
Metering/Billing	X	X		X		X	X	X
Infrastructure	X	X	X	X			X	
Space Conditioning	X	X	X	X	X		X	
Process Systems	X	X	X	X	X		X	
Integration	X	X		X			X	
Finance	X	X	X	X			X	X
Administration	X			X				

IS THERE ONE OPEN MARKET CUSTOMER?

This chapter refers to the Open Market Customer as if it were one customer; however there is not just one homogenous Open Market Customer, as the referenced studies have shown. Energy consumers are extremely diverse, not only in the ways they use energy, but also in how they make decisions. In addition, about one-third of the Open Market Customer business functions are now, and are likely to continue to be, subject to some form of regulation. The remaining two-thirds of the Open Market Customer business functions have always been subject to market forces and are "open." What is new is the increasing choice, and the opportunities for better purchase decisions through integration and heightened competition.

MORE CHOICES

With this broader view of the energy and energy service industry, it is important to understand that the Open Market Customer may be faced with increasing choice in technologies, in quality, in bundling of services, and in choice of suppliers. Here is what the Open Market Customer will see in the future:

- Many more technologies that convert energy to services will be available. These expanding portfolios will result from a resurgence of research, development, and successful commercialization. While the Open Market Customer sees lower energy prices almost immediately, what will be less apparent is that lower energy prices will significantly improve the cost-effectiveness of many energy process technologies. As sales of these technologies increase and the implications of global climate change mature, regulators will increasingly mandate technologies for environmental improvements, further reducing equipment prices.

- A renewed energy services industry will foster new energy-efficient lighting and heating/ventilating/air conditioning (HVAC) technologies, spurred by the retail affiliates of energy companies, an increasing number of public benefit programs that are developing in more and more states, and continued interest in federal and state appliance and building standards. Survivors in the energy service industry will understand customer needs and the bundles of technologies that meet those needs.

- A variety of offerings will be available with differing quality of energy attributes at significant price and service differentiation. There is an increasing recognition of the consumers' demand for a wide variety of power quality offerings and a ready market for interruptible power, firm power, standby generation, cogeneration, uninterruptible power systems, clean power technologies, and similar offerings. Integration of each of these options could well become a significant service in the Open Market Customer Energy Industry.

- Dynamic new products and services will be available. Many of these will result from synergy between the Open Energy Marketplace and the information management and tele-communications industries. Service providers will offer monitoring, management, and integration of energy and energy services. Open Market Customers with multi-site locations will be offered instant monitoring, management, and integration capabilities that will allow them a wide range of cost and service enhancements.

- The Open Market Customer will be able to select from an array of energy marketing and billing options. The right to provide energy metering is being contested heavily in the electricity industry re-structuring process. Metering's position as the primary interface to the customer, insulating commodity supply functions from the energy service functions, is critical. Multisite Open Market Customers will be able to leverage electronic billing and smart meters, and choose the best of several preselected supply options.

- The Open Market Customer will witness innovative marketing strategies from potential suppliers. Customers will be able to choose a wide variety of energy marketing options (such as a free airline mile for each kilowatt-hour used). These marketing strategies will be combined with new technologies to create a "first on the block" syndrome. A recent survey by EPRI, for example, with a sample of 45,000 households nationwide, showed that 56 percent of residential consumers already have or want to know more about home security, 53 percent already have or want to know more about on-line management, and 46 percent already have or want to know more about whole-house surge protection.

- Partially in response to the near overload of information, energy and energy service outsourcing will be available. The Open Market Customer will be presented with a greater variety of choices in all of the business areas. The customers' lack of familiarity with energy and energy service issues, and a lack of desire to take the time to fully comprehend the impacts of various options to their businesses, will create a demand for expert assistance.

- A plethora of financial offerings will be available. Financial services businesses will develop new financial offerings for energy customers. Creative financing will package energy procurement expenses with energy equipment capitalization. Financial options also will feature new forms of packaged energy and energy services such as end-use metering.

- A host of private and public agents will offer methods to control energy consumption. This will spur product offerings for design, construction, operations, maintenance, and financing services for downstream equipment systems.

CONCLUSION

Change brings opportunity, and enhances value to the discriminating buyer. The current evolution in the energy industry is no exception. The Open Market Customer will see more choices, with more options on how to manage them than ever before. As the industry that serves the Open Market Customer matures, those options will narrow, and in the end the survivors will find a market where customers examine the consequences of energy and all the devices, appliances, and apparatuses that help to consume it in an integrated fashion.

NOTES

1. Adapted from "Request II: An Investigation of Consumer Attitudes Toward Telecommunication and Electric Services," (Electric Power Research Institute Report [March 1996] TR-106166).

2. C. W. Gellings, "Is Your Refrigerator on the Internet?" Presented to the Edison Electric Institute Critical Issue Forum (Philadelphia, June 10, 1997).

ADDITIONAL REFERENCES

1. Gellings, Clark and Ken Gudger, "The Open Market Customer." Presented to the 4th Annual Utility Strategic Marketing Conference. Orlando, Florida. April 17–18, 1997.

2. Kotowski, Dr. John, "The Open Market Customer (OMC): Concept and Case Studies." Presented to EPRI Retail Market Tools and Services Business Area Council Meeting. September 25, 1997.

CHAPTER 3

Determining What Consumers Really Want

Raymond W. Lawton
Francine Sevel
David W. Wirick

In the heyday of ratebase/rate-of-return regulation, utilities and regulators were locked in a dance that focused their attention almost exclusively on one another. Consumers, who to some extent were held captive by the monopoly provider, were often overlooked. As competition slowly finds its way into utility service delivery, those forgotten customers will be invited to the dance and, in fact, must become the focus of both service providers and regulators.

In competitive markets, it is likely that utility service providers first will look to providing service at rates lower than those offered by competitors or potential competitors, as price is clearly a dominant concern of consumers. But not all utility service providers can be the lowest price provider, and all but the lowest price provider will need to find other ways to maintain or secure market share. They will be forced to compete with information about consumer needs and desires and will use that information to expand their service offerings from utility service as a commodity to focused utility service tailored to the needs of specific classes or groups of consumers. Even the low-cost provider will need to find ways to get the word out to consumers that it is the low-cost provider.

Similarly, public utility regulators, faced with increasing numbers of and a wider variety of service providers coupled with decreasing ability to reach into the operations of those providers, will need to turn their attention to the operation of utility markets and the impact of those markets on consumers. Market analysis and consumer affairs will become their stock-in-trade, replacing rate setting and command-and-control regulation.

Communicating with consumers is not a one-way street. It is not simply a matter of using a variety of media to send a message to consumers. Nor is it simply the art of assessing consumer desires through focus groups or consumer surveys. It is, instead, an integrated process that includes both transmitting communications to consumers and receiving communications from them. It is a synergistic process that builds on each communication to and from consumers and that permeates every facet of utility or regulatory operations.

This chapter explores both sides of the consumer communication transaction, first by looking at the receipt and use of consumer information and, second, by addressing communications from utilities and regulators to consumers.

GLEANING INFORMATION FROM CONSUMERS

Utilities and state public utility commissions did not used to have a need for as great a level of consumer understanding as they do in the emerging environment. In the past, they took information from the external environment and primarily used it for rate cases and to develop quality-of-service standards. (Of course, the utilities collected extensive consumer information for billing purposes.) However, in the new regulatory environment, utility service providers and regulatory commissions are placing increasing emphasis on consumer and public input as a vehicle for identifying the attributes of their services that consumers value, appropriate prices and price structures, the information needs of consumers, and incentives that both will motivate and impede the endeavors of providers to attract and retain customers.

The transition to being an organization that maximizes its use of consumer information is not easy. It requires, among other things, the development of a "customer-centered mindset." Alan Andreasen discusses the differences between the traditional organization-centered mindset and a customer-centered mindset and warns of the adopting the former.[1] The salient difference between the two approaches is that unlike the organization-centered marketer who keeps track of how the organization is doing, the customer-centered marketer keeps track of how the customer is doing. The customer-centered mindset places customers and their objectives (perceptions, needs, and attitudes) first. This mindset acknowledges that an effective transition to the new regulatory environment must encompass customer satisfaction, as defined by the customer. As an example, if a pilot program did not successfully attract consumers, the customer-centered marketer would acknowledge that the fault lies not with the consumer but with the pilot program, and the program would be adjusted accordingly.

Some of the differences between organization- and customer-centered mindsets are listed in Table 3–1.

Traditionally, state commissions and jurisdictional utilities have relied upon a number of ways of determining what consumers want in terms of the pric-

TABLE 3–1	Characteristics of Organization-Centered and Customer-Centered Mindsets	
Difference	**Organization-Centered Mindset**	**Customer-Centered Mindset**
Mission	The organization's mission is seen as inherently good.	The organization's mission is to meet the target market's needs and wants.
Customer	Customers are problems to be solved.	The customer is seen as someone with unique perceptions, needs, and wants to which the organization must adapt.
Market Research	Market research has a limited role.	Market research is vital.
Customer Groupings	Customers are treated as a mass.	Customers are grouped into segments.
Competition	Competition is ignored.	Competition is seen to be everywhere and never ending.
Source: Author's construct based on Andreasen, *Marketing Social Change.*		

ing, variety, quality, and reliability of the services sold to consumers. These methods have their origins in monopoly markets characterized by stability, with authoritatively set levels of price, quality, and reliability. As monopoly utility markets career towards markets that are more competitive, the old ways of determining what consumers really want need to be reassessed and new strategies need to be invented.

There are three basic ways to find out what consumers want. The first is to apply a standard that purports to represent what consumers want. For commissions and utilities, clean air and water standards are good examples of the standards approach. The second way is a market-test approach, where competitive firms provide services and those that are the most successful are deemed to have been the most responsive to consumer preferences. The third way is to ask the consumers directly. Each of these approaches has advantages and disadvantages that are discussed below.

Standards-Based Approach

Commissions and utilities traditionally have relied upon the standards approach, which has three parts: (1) determining what consumers want, (2) codifying preferences, and (3) enforcement. Unfortunately, the first step is the most important and the least understood. Because of the technical nature of gas, electric, water, and telecommunications services, engineers traditionally have been responsible for determining standards. They typically relied on some combination of commission, utility, public health, environmental, and professional engineering bodies to determine standards. These authoritative sources are then modified by systematic and ad hoc information from consumers, including surveys, focus groups, complaints, personal experience, and judgement. These standards are then authoritatively adopted and enforced as representing the standards, or parameters, that consumers want their services to have.

This approach suffers from the classic "client-agent" problem. Namely, how does the utility or commission know whether or not the resulting standards represent the true needs of consumers? This dilemma is akin to the old saying, "the camel is a horse designed by a committee." It is not the case that, at any one point in time, the standard writer sought to subvert the process, rather that the resulting standards represent a distillation and optimization of preferences against assorted constraints.

State commissions continually worry about the client-agent problem. When state commissioners hear an articulate complaint about a standard in a public hearing, they temper their initial response by weighting the single complaint against their assessment of how representative it is of a potentially larger, more widespread problem. They ask, "Is the complaint the odd case or is it a lead indicator of a larger, previously unrecognized problem?" Utilities also do this, but in a less public process.

Market-Test Approach

Most economists reject a standards approach and urge a market-test approach. They argue that a profit-maximizing firm will seek to be responsive to consumer wishes as the best way to grow and earn a profit. Assorted market forces, they posit, will discipline firms, and those that are not meeting consumer demand will not prosper. The main problem with this approach is that it depends upon vigorously competitive markets with well-informed consumers. Otherwise, firms are not efficiently disciplined and may use incumbency and market power to give consumers a limited menu of pricing, service offerings, quality distinctions, and reliability choices. Most utility markets can be characterized as being in some form of transition, but as yet not sufficiently competitive for market discipline to be assured or effective.

Asking the Consumer Directly

A third approach to finding out what consumers want is to ask them directly. This approach is strong on intent and input, but requires significant effort to convert survey information into service attributes valued by consumers. The Maine Public Utilities Commission asked consumers, as a part of its electric industry restructuring effort, about their price, reliability, and power generation preferences. This information was incorporated, in part, into the policies developed. The Wisconsin Public Service Commission used two surveys to determine the billing frequencies of utility customers. In this case, the utility and the commission believed that a monthly billing procedure was preferred by all residential customers, and by senior citizens living on fixed incomes in particular. Monthly billing also allowed the utility to meet its own and the state commission's financial and conservation goals. A pair of surveys indicated, however, that consumers actually preferred semi-annual billing for water service.

Several difficulties arise in asking consumers directly. One is the difficulty in translating preferences into services or standards. The second is the technical nature of utility operations. Consumers may express overwhelming preference for fewer "interruptions," but this may not provide sufficient technical guidance. A second, translation step is nearly always needed. The third difficulty is in ensuring that the preferences revealed are framed in terms of pricing information and other interrelated consumer and utility impacts.

As traditional standard-setting approaches become increasingly out-of-favor, most utilities and commissions have begun to rely increasingly on strategies that integrate market test and survey approaches to obtain information about consumer preferences. A two-prong approach allows a survey to be used "up-front" to design menus of services or service attributes. The market test is necessarily "end-loaded" and is driven by the actual consumer purchases. Generally the process is iterative, with better and newer survey data being used to design consumer-responsive services. Actual demand data then serves as input in the design of a more focused set of questions for the next survey round.

State commissions and utilities necessarily have different perspectives on this type of two-prong strategy. Commissions are not service providers; they operate more in an oversight and diagnostic mode. They acquire information to see if universal service goals are being met, rather than to specify an exact menu of services. State service quality rules do influence, however, the minimum features that all services offered for sale must have. Commissions also focus on the needs of certain kinds of wholesale customers: resellers, aggregators, and interconnection and transmission customers. The intent here is to ensure that competitors get necessary interconnection and other related underlying services on a nondiscriminatory basis. Commissions acquire only surveillance information or complaint-investigation information, although they still have other means of acquiring information. They do not design the service menu.

For utilities, the link between surveying and product design is more immediate and direct. Survey information is filtered through a utility's corporate objectives and a menu of services is offered. Depending on the competitive strength of the market area, the utility can use its sales information to revise the menu it offers in its next round. Utilities use the information for design purposes, rather than for oversight.

Surveys provide a cost-effective way of obtaining information on consumer preferences. In order to provide accurate information, however, they must meet scientific tests of validity. In order to assure validity, it may be advisable to use professional survey firms rather than in-house staff, to determine the sampling framework by the nature of the decisionmaking the survey is designed to support, and to pretest surveys before full implementation.

SENDING MESSAGES TO CONSUMERS

In this rapidly changing environment, few new requirements have captured the attention of utility service providers and utility regulators more than the need to transmit messages to consumers. In the past, utilities and regulators viewed marketing as a tangential function far removed from vital work. With the advent of competition, utilities and regulators have scrambled to tell their particular stories to the public.

Obviously, not all interactions with consumers are created equal. Sending messages to consumers can be described as falling on a spectrum of activities illustrated in Figure 3–1. At the left end of the spectrum is traditional utility and regulatory commission consumer complaint handling. These interactions with consumers are initiated by the consumer and involve highly specific content areas. Though consumer complaint handling is not a new function, it is receiving renewed attention, both as a source of usable consumer information and because of recent, significant changes in the number and complexity of complaints received.

To the right of consumer complaint handling on the spectrum is the provision of information to consumers. Utility service providers and public utility regulators are adept at providing neutral, factual information to consumers. For utilities, inquiries from consumers and the news media are likely to increase as the utility marketplace becomes more complicated, and in emerging competitive markets, the provision of information by regulatory commissions is likely to blossom as a central function. Some have suggested, in fact, that a major role of commissions in the future will be that of a *Consumer Reports*-type function, in which commissions provide information to consumers about utility service offerings in a systematic and impartial way. Recent efforts to "label" electric service offerings to identify the types of fuels used is one example of regulatory commission attempts to provide information to consumers.

Both the handling of consumer complaints and the provision of impartial information are largely reactive, requiring a response from the service

| Complaint | Provision | Consumer | Public |
| Handling | of Information | Education | Relations |

Figure 3–1 *Sending Messages to Consumers*

provider or the regulatory commission to a consumer-initiated inquiry. Though careful consideration will need to be given to the content of responses, these activities do not require long-term consideration of the needs of the target audience. For regulatory commissions, they will not substantially disrupt the traditional quasi-judicial role.

Further along the spectrum of interactions with the public are consumer education and public relations. Consumer education differs from information provision in that consumer education is a proactive attempt to change the ability of consumers to make necessary decisions in a more competitive marketplace. Consumer education requires careful consideration of the target audience and the content and delivery of messages.

Public relations is also a proactive strategy that attempts to garner public support for the utility, in the form of consumer loyalty and trust, and for the regulatory commission, in the form of public and legislative trust. Successful public relations recognizes the fact that both utility service providers and regulatory commissions are undergoing transformations that require repositioning themselves in the eyes of the public. Both consumer education and public relations may require substantial short-term investment by service providers and regulators but may diminish in importance and urgency over time as consumers develop the ability to purchase utility service in competitive markets and as they develop new conceptions of the roles of utility service providers and regulators.

Some service providers and regulators may opt to invest time and resources only in those activities on the left side of the spectrum (i.e., those activities that are more traditional). Others, however, will attempt aggressively to educate consumers and reposition themselves in the new environment.

Proponents of consumer education argue that through the dissemination of consumer information to ratepayers and the collection, analysis, and synthesis of information about ratepayers, utility service providers and state public utility commissions can create in consumers the ability to make good choices and, thereby, facilitate the development of more-competitive markets. In order to effectively impact consumers, they will need to address a variety of consumer attributes: their knowledge, skills, beliefs, attitudes, and values.[2] Each of these is discussed in turn.

Knowledge

Knowledge is defined as valuable information from the human mind that includes reflection, synthesis, and context.[3] Types of knowledge that residential ratepayers might find beneficial, especially in the early stages of electric industry restructuring, include: basic knowledge of the new regulatory environment, knowledge of how to select a service provider, knowledge of the basic standards of service quality, and knowledge of the complaint filing process.

Properly designed needs assessment surveys can provide valuable information regarding consumer information needs. Through the use of these needs assessment surveys and focus groups, utility service providers and regulatory commissions can identify residential ratepayer's information needs, as well as other factors that motivate and impede them from making choices. Utilities and commissions also need to have information regarding the demographics and psychographics of the market segments that comprise their target audiences.[4]

Skills

The term "skills" refers to the ability to successfully perform a predetermined task. A consumer educator designing a brochure on choosing an energy provider might contemplate what skills the audience needs in order to make a successful choice.

Examples of skills necessary for consumers might include the ability to weigh all variables and make appropriate decisions, the ability to compare rates, the ability to read and understand service literature. Generally speaking, four categories of skills are used in the design, delivery, and evaluation of consumer education programs: strategic planning, instructional design, external relations skills, and evaluation.[5] Through the use of consultants, temporary staff, and other creative strategies consumer educators can maximize their existing skill base.

Beliefs

Beliefs are statements that the audience holds as facts. A core element in many of the models regarding consumer behavior is the proposition that individuals act on the basis of beliefs.[6] Indeed, electric industry consumer education efforts may hinge upon the extent to which consumers believe in the efficacy of the information providers. If consumers believe that the electric utilities and the public utility commissions are honest and ethical, then they may believe in the integrity of the respective consumer education programs. In California, focus group research revealed that consumers believed an independent advisory board to be the most credible source of consumer information.

It is advantageous for utilities and commissions to believe that positive consequences will result from the application of an instructional design process to the production of consumer education materials[7] and that market segmentation is a necessary strategy.

Attitudes

According to Fishbein and Ajzen: "An attitude represents a person's general feeling of favorableness or unfavorableness towards some stimulus object. As a person forms beliefs about an object, [s/he] automatically and simultaneously acquires an attitude toward that object."[8] Through focus group interviews or phone surveys, a consumer educator can determine the attitudes that the public holds toward specific attributes of industry restructuring or the utility itself. As an example, if research indicated the presence of negative attitudes toward the utility, consumer education would need to focus on changing consumer attitudes by creating positive feelings toward the utility.

Both utilities and commissions will need to display a positive attitude toward consumers. It also will be important to adopt a "guest-relations" attitude—a term borrowed from the hotel industry. Basically, a guest relations program is a management-driven endeavor in which staff makes a concerted effort to be friendly, helpful, and courteous to customers. Organizations can communicate a guest or "consumer" relations attitude in many ways, including answering consumer lines promptly, having adequate phone lines so that consumers do not get a busy signal, avoiding placing consumers on hold, having patient and friendly representatives, or having phone representatives who are bilingual/multilingual.

Values

Consumer researchers have learned that there is a strong correlation between a person's values and the attributes they seek from products and services. In many cases, in order for consumer education to be effective, it will need to influence, perhaps even change, consumer values. Focus groups and other survey research will help utilities and commissions identify the attributes that consumers value about service providers, and education will help to "reshape" consumer values. As mentioned in the previous section, the success of both consumer education and consumer choice programs will hinge, to a great extent, on the commitment of utility service providers and regulatory commissions to placing increased value on public input.

CONCLUSION

To create the new patterns of communication described here, Brenda Dervin and Peter Shields suggest that service providers and regulators divest themselves of the "empty bucket" conception of consumers—the idea that consumers are empty receptacles who can be filled with necessary bits of information by public relations/information officers.[9] Instead, they argue, now is the time to create new "models of interaction with citizens which are genuinely dialogic, which do not capture citizens merely in the mirrors provided to them. . . ."[10]

Breaking the long-held patterns of interactions with consumers will require that utility service providers and utility regulators develop new tools, new strategies, and, indeed, new visions of themselves. Though the changes required may be daunting, creation of new, synergistic patterns of communications with consumers will open exciting vistas that may supplant moribund patterns of operation.

NOTES

1. Alan R. Andreason, *Marketing Social Change* (San Francisco: Jossey-Bass Publishers, 1995).

2. This discussion is based on the consumer behavior theories of Edward W. Maibach and David Cotton, "Moving People to Behavior Change," in *Designing Health Messages: Approaches from Communication Theory and Public Health Practice,* Edward W. Maibach and Roxanne Louiselle Parrott, eds. (London: Sage Publications, 1995), 43; and Martin Fishbein and Icek Ajzen, *Belief, Attitude, Intention, and Behavior* (Reading, MA: Addison-Wesley Publishing Co., 1975).

3. Thomas H. Davenport, *Information Ecology* (Oxford: Oxford University Press, 1997), 9.

4. For a resource on the planning and design of consumer education programs and cases studies, see: Francine Sevel, ed., *Compendium of Resources on Consumer Education* (Columbus, OH: The National Regulatory Research Institute, 1998).

5. For a more detailed discussion of these skills, as well as ways to maximize existing skills, see Raymond W. Lawton, et al, *Staffing the Consumer Education Function: Organizational Innovation, Necessary Skills, and Recommendations for Commissions* (Columbus, OH: The National Regulatory Research Institute, 1998).

6. Andreasen, *Marketing Social Change,* 151.

7. An instructional design process is a systematic decision-making process that allows educators to identify the key elements of the learning process and to make decisions about what will be the most effective way to plan and implement the learning experiences. See, Gary J. Dean, *Designing Instruction for Adult Learners* (Malabar, FL: Krieger Publishing Co., 1994), 2.

8. Fishbein and Ajzen, *Belief, Attitude, Intention, and Behavior,* 213.

9. Brenda Dervin and Peter Shields, "Some Guidelines for a Philosophy of Communicating with Citizens in the New Regulatory Environment," in Sevel, ed., *Compendium of Consumer Education Resources,* 69–85.

10. Ibid., 5.

CHAPTER 4

Using Choice Modeling to Understand Customer Preferences: A Tale of Four Studies

Michael W. Rufo
Kenneth Train

The emergence of competition in the electricity industry is resulting in increasing numbers of customers being able to choose their energy suppliers. This freedom raises a host of questions as to what exactly customers will be looking for when they choose a supplier. For example, some incumbent utilities may want to know the share of customers that will switch to other providers and how best to position themselves in the eyes of customers so as to minimize this loss. Or they may need to forecast switching simply to predict the demands of such activity on their billing and information services systems. New firms, whether they sell electricity that they generate themselves or resell electricity that they purchased from others, need to know how to attract customers in a cost-effective manner, through combinations of pricing strategies, service enhancements, relationship building, and advertising. All suppliers benefit from knowing the dollar amount that customers are willing to pay for advantageous contract terms, value-added services, and more environmentally friendly generation mixes.

In short, most energy service providers and utilities are searching for ways to more efficiently market to customers while simultaneously looking for unique market niches and attributes upon which they can differentiate their services. In addition, many organizations are trying to develop and refine their pricing levels and structures, as well as a number of other aspects of their retail strategies, including marketing campaigns, approach to sales, and contract terms. Customer choice modeling attempts to untangle the customer's decision process and determine the relative value placed on various services and other supplier attributes. It also can be used to identify segments that behave differently with respect to switching propensity. By helping to understand differences in value placed by customers on alternative service and marketing attributes, customer choice modeling can assist suppliers in spending their retail marketing dollars more efficiently and to price their products and services in the most profitable manner.

This chapter describes empirical customer choice modeling methods that the authors have used to examine customers' choices in several settings.[1] In particular, we describe four case studies that span the types of situations that analysts might face in modeling customer choice of energy suppliers. We primarily focus on the selection of the modeling approaches, issues related to concomitant survey and related research demands, and lessons learned from our experience applying these methods under a variety of circumstances. The four situations are distinguished as follows, with Figure 4–1 serving as a schematic guide. The first distinction is whether or not customers have already chosen suppliers. In situations where competition has begun, or where pilot programs have been initiated, it is possible to observe the choices that customers actually made. These situations provide "revealed preference data," so named because the preferences of customers are revealed by their actual marketplace choices. In areas where competition has not yet begun, it is necessary to use hypothetical choices to analyze customer preferences. Suppliers might want to predict the market shares that they could obtain when competition starts and, perhaps more importantly, determine how best to position themselves in the upcoming competition. In these situations, "stated preference data" are used—that is, data on how customers say they would choose in hypothetical situations. In addition, stated preference approaches often continue to have value even in circumstances when customers are faced with actual choices because of the time, expense, and difficulty associated with obtaining well-controlled revealed preference results.

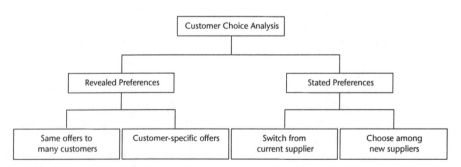

Figure 4–1   *Types of Customer Choice Analyses*

Two of the case studies presented here use revealed preference data; the other two use stated preference data. The Pennsylvania pilot program for energy competition provides the setting for the two studies with revealed preference data. One examines residential customers' choices, and the other examines commercial customers' choices. These two studies differ in an important way, as described in the next paragraph. California, prior to the advent of competition for small and medium customers, provides the setting for the two stated preference studies. Both studies address commercial customer choice behavior, but use two different methods as described below.

When the actual choices of customers are observable (that is, when revealed preference data are used) a crucial step of the analysis is to obtain data on the offers that were made by each supplier. By comparing the various offers that were available to a customer with the offer that the customer actually chose, the analyst is able to infer the aspects of offers that are important to the customer. An important distinction arises in this regard. In some settings, the same offer is made by a given supplier to all customers in a particular area and/or a particular rate class. This is generally the case for residential customers, where suppliers have a standard price and contract. In these situations, it is relatively easy to collect data on the offers made by suppliers, since the offers are public and the same for large groups of customers. In contrast, suppliers might custom-fit their offers to each customer. This is generally the case for commercial and industrial customers, particularly the larger ones but often the smaller ones as well. In these situations, it is difficult for the analyst to obtain reliable data on the offer that each supplier made to each customer. Procedures for obtaining these data become a central focus of the analysis. Our two studies of the Pennsylvania pilot program provide examples of these two situations. For residential customers in the program, the same offers were made to groups of customers. In the Pennsylvania commercial sector, offers were customer-specific.

With stated preference data, the analyst designs a survey that asks customers what they would do in hypothetical choice situations. An important decision must be made by the analyst in designing the choice questions, namely, whether to word the questions in terms of switching from the current provider or in terms of choosing a supplier from a group of available providers without reference to the current provider. An example of the first type of question is: "If an alternative supplier offered you a 10 percent discount off your current bill, would you switch to that supplier?" This type of question is relatively easy to ask, can be asked over the phone (thereby reducing survey costs), and is appropriate when the customer will stay with the incumbent utility under competition, unless the customer decides to switch. The second type of question involves the analyst describing a set of offers (one from each of several hypothetical suppliers) and asking the customer which one the customer would choose. This type of question is more complicated to convey to the customer and therefore usually involves the analyst mailing material to the customer prior to the phone survey. However, this type of question sometimes can provide more information about the customers' preferences than switching questions (since the hypothetical offers are not constrained to include the current provider's offer). Also, this type of question is appropriate when the upcoming competition will take the form of the customer actually choosing among a set of suppliers with the incumbent not in the set (though its affiliate might be). Our two analyses of commercial customers in California employed both approaches.

Examples of the specific types of questions that form the basis for our approaches in these modeling efforts are provided below.

Pennsylvania Revealed Preference Studies

- What is the value of a simple contract? A guaranteed price? Knowledgeable and professional salesmanship?
- Is calling customers effective?
- What characteristics of the suppliers and their offers influenced customers' choices the most?
- How important was advertising relative to price in attracting customers?
- Which advertising media were the most effective?
- How important was the offer of cash or a gift upon sign-up?
- Did affiliates of the host utility have an advantage, even after accounting for other differences between offers? And if so, how large of a price difference is this advantage equivalent to?

California Stated Preference Study

- What will induce customers to switch to alternative suppliers?
- How much are customers willing to pay for different goods and services?
- What is the dollar premium associated with "host" utility status? In-state status? "Green" branding?
- What is the relative value placed on customer service versus reliability?
- What customer characteristics are associated with higher propensities to switch to a new provider?

The four case studies are described in the following four sections.

RESIDENTIAL CUSTOMERS IN PENNSYLVANIA PILOT: REVEALED PREFERENCE DATA WHEN OFFERS ARE COMMON TO LARGE GROUPS OF CUSTOMERS

This section presents our analysis of residential customers' choice of supplier in Pennsylvania's pilot program. To understand the context of the revealed choice analysis described, we begin with a brief summary of the pilot background.

Pilot Background

Pennsylvania's Electricity Generation Customer Choice and Competition Act was signed into law in December 1996. This act deregulated the generation portion of the state's electric service. The legislation also required the state's eight major utilities (PECO Energy Company, PP&L, Inc., Duquesne Light Company, GPU [Metropolitan Edison and Pennsylvania Electric], Allegheny Power Company, Penn Power Company, and UGI Utilities, Inc.) to submit plans for and to run a retail access pilot. On August 29, 1997 orders from the Pennsylvania Public Utility Commission on the pilot plans were posted; electrons were delivered on November 1. Figure 4–2 displays the timeline in more detail.

Participation in the pilot was a two-step process: (1) enrollment, and (2) sign-up with an electric generation supplier (EGS). Five percent of each rate class was eligible to participate. All rate classes were overenrolled, and it was necessary to conduct lotteries to determine which customers would be eligible. Those lottery winners who did not sign-up with an EGS by November 14 (about 30 percent) had their slots taken by nonwinners. These second round winners had until January 6, 1998 to pick an EGS.

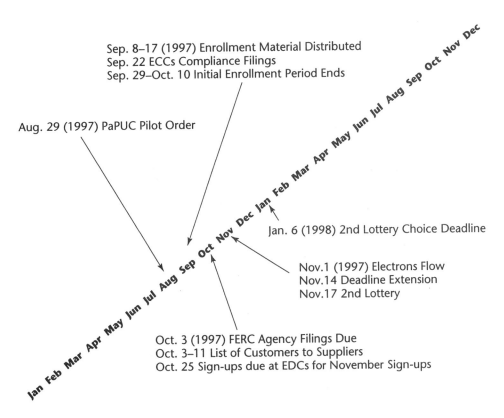

Figure 4–2 *Timeline of Pennsylvania Pilot*

Residential customers were provided with the following as part of their participation:

1. A generation (shopping) credit of 3 cents per kilowatt-hour to shop for new generation and capacity. (The PUC assumed the market price would be near this price; residential customers would save money if they could procure generation for less than 3.0¢/kWh); and

2. A participation (inducement) credit of 13 percent on the non-generation portion of their bill, which amounted to about 10 percent savings on their total bill.

The PUC mandated that each pilot would seek customer participation until 5 percent of the noncoincident peak for each customer rate class was involved. Approximately 900,00 residential customers (accounts) volunteered—17 percent of the eligible population for 235,000 slots. Of these, 221,000 signed up with an EGS. The breakdown of eligible customers by electric distribution company (i.e., EDC or host utility) is shown in Figure 4–3.

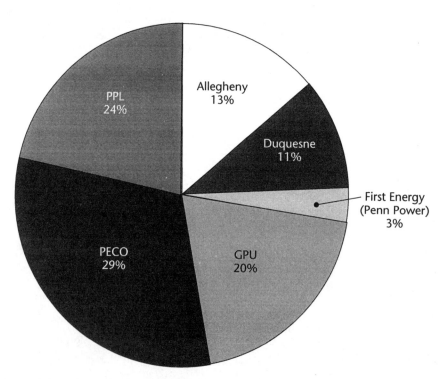

Figure 4–3 *Percent of Residential Accounts Eligible for Pilot by Incumbent Utility*

Each household that participated in the program faced a choice among various suppliers. Most importantly, each supplier made one offer to all customers in a given area and rate class. This fact made it possible to obtain data on the offers faced by each customer. Offers varied over suppliers and, for a supplier, over areas and rate classes. The offers differed in price and in whether or not the price was guaranteed for a period of time. In addition to the various offers, suppliers differed in the way that they positioned themselves competitively. Some suppliers spent more money on advertising than others and displayed their ads in multiple media (television, radio, and advertising). A few suppliers provided cash give-aways or small gifts to customers who signed with them. One supplier in each area was identified as being affiliated with the customers' previous utility, while others were not. For each respondent, we observed which suppliers were available to the customer and which of these suppliers the customer chose. Table 4–1 displays the variation in prices and selected offer characteristics across suppliers and service territories. The statistical analysis quantifies the relative impact of each attribute on customers' choices. The data and statistical procedure are described below, followed by a discussion of the challenges faced and lessons learned.

TABLE 4–1 Variation in Residential Supplier Prices, Advertising Levels, and Other Offer Characteristics

| Supplier | Average Price by Service Territory in Cents/kWh | | | | | | Incentive | Total Advertising ($ Thousands) | Incum-bency |
	1	2	3	4	5	6			
Supplier A	2.80	2.80	3.05	3.05	2.75	3.05	Free Month	$700	✔
Supplier B	—	2.98	2.94	2.94	3.00	2.93		$250	✔
Supplier C	—	—	—	—	3.00	—	Free Month	$0	
Supplier D	—	—	2.76	2.76	3.00	2.67	Free Month	$1,000	
Supplier E	—	2.95	—	—	3.00	3.00	Free Month	$500	
Supplier F	—	—	3.17	—	—	3.25		$500	✔
Supplier G	—	—	3.00	—	3.00	3.00	Gifts	$5,900	✔
Supplier H	—	—	2.78	2.78	2.68	2.80		$75	
Supplier I	—	—	—	—	3.00	—		$75	✔
Supplier J	3.09	3.00	—	—	—	—	Free Month	$200	
Supplier K	2.69	2.98	—	—	—	—		$10	
Supplier L	—	—	3.10	3.08	—	2.97	Free Month	$4,000	✔
Supplier M	3.10	3.10	3.10	3.10	3.10	3.10		$40	
Supplier N	3.00	3.00	3.00	3.00	3.00	3.00	$10 & Free Month	$0	
Supplier O	3.00	—	—	3.00	3.00	—			

Customer Data

A total of 635 residential customers in Pennsylvania were sampled from those who volunteered and were selected for participation in the pilot program. Of these, 483 selected a supplier and reported their choice in our survey. Our analysis was performed on these 483 customers. The model specification used for the analysis is discussed in the next section. Sixteen suppliers offered service to eligible residential customers in Pennsylvania. However, in each area of the state, only a subset of the 16 suppliers was available to each customer, depending on their location.

The price of an offer is obviously one of the factors in customers' choice of electricity supplier. Suppliers in the Pennsylvania pilot generally segmented their target markets by location and rate class and often offered different prices to each segment. Prices were generally advertised in mass media and presented via direct mail pieces. We were able to identify these prices for each of the market suppliers by market segment. The prices to which each sampled customer was exposed was determined on the basis of their location and rate class, and was usually quoted as a fixed amount per kilowatt-hour. One supplier offered rates "up to 12 percent off" which were based on a two-tier declining block structure. For this supplier, we assigned the mid-point price, namely, the average between the high and low price. We also included in the model a dummy variable to account for the fact that the mid-point price need not be the average price that the customer would actually pay if it chose the supplier with the two-tiered price.

For each supplier, data were obtained on the amount of money that the company spent on advertising. These expenditures were broken down by newspaper, TV, and radio. Detailed advertising data of this nature were obtained by market area. Pennsylvania contains two market areas for which such detailed expenditure data was available: Philadelphia and Pittsburgh. The Pittsburgh area expenditures were used for customers who lived in the territories of Allegheny and Duquesne, while the Philadelphia area expenditures were used for all other customers.

We also used dummy variables to indicate other aspects of the suppliers and their offers. These attributes included: whether the price was guaranteed, whether the supplier was clearly affiliated with the customers' host utility, and whether the supplier gave customers a check or other gift upon sign-up. In another similar analysis that we conducted on residential participants in the New Hampshire pilot in 1996, we included a dummy variable for green power. Note, however, that none of the suppliers offered a green power option in the Pennsylvania pilot.

Model Specification

Pennsylvania households' choices were examined with a multinomial logit model. The logit model relates the choice of the customer to the attributes of the offers that were available to the customer. In particular, the model gives the probability that the household would choose each supplier, given the attributes of the various suppliers available to the household. Mathematically, the model is expressed as:

$$P_{in} = \frac{e^{\beta x_{in}}}{\underset{j}{e^{\beta x_{jn}}}}$$

where P_{in} is the probability that customer n chose supplier i, x_{in} is a vector of characteristics of supplier i and its offer to customer n (including, for example, the price and the amount of advertising), β is a vector of coefficients that denote the relative importance of each characteristic of the supplier, and the summation in the denominator is over all suppliers that were available to customer n. The characteristics of each supplier are recorded (i.e., data on x_{in} for each supplier i is collected), as well as the choice of the customer. Estimation of the model constitutes estimation of the parameters β—that is, quantification of the relative importance of each supplier characteristic.

Lessons Learned

Our results showed that it was indeed possible to quantify the relative effect of differences in supplier prices, advertising levels, and other product attributes using a revealed preference approach. Most of the factors that we examined had coefficients that were statistically significant at the 90 percent level or better with the expected signs. In addition, by relating the coefficients of the nonprice factors to the price coefficient, all of the key parameters of interest were converted into cents per kilowatt-hour. For example, we were able to estimate the price premium associated with being an incumbent provider. Similarly, we were able to estimate the relative effectiveness of different types and levels of advertising, as well as the value of a guaranteed versus a variable rate.

One limitation of this type of revealed preference analysis of pilot data is that, in areas in which additional customers are likely to be able to choose suppliers in the future, suppliers may invest in advertising that is aimed not only at the small percentage of customers that currently have choice in the pilot, but at the larger population of future choosers as well. For example, in the Pennsylvania pilots, several suppliers spent millions of dollars on mass media advertising. These levels are unlikely to be justifiable for acquisition of a targeted share of the mere 5 percent of customers permitted to participate in the pilot; rather, it is more likely that these expenditures were also investments aimed at influencing the next two-thirds of customers that were expected to be able to choose their suppliers in the Pennsylvania phase-in in 1998 and 1999. In our revealed choice analysis, all of the expenditures to date were utilized in the model. As a result, it is possible that our estimates of the relative effectiveness of the different advertising mediums (television, radio, and newspaper) are biased against investments in advertising that were designed to meet suppliers' longer-term branding goals, rather than focusing on strategies aimed at short-term acquisition of pilot participants.

COMMERCIAL CUSTOMERS IN PENNSYLVANIA PILOT: CUSTOMER-SPECIFIC OFFERS

This section presents the model of Pennsylvania commercial customers' choice of supplier. The analysis differs in an important way from that for Pennsylvania residential customers. For both groups of customers, the actual choices of supplier were observed. However, the information that was available on the suppliers' offers differed for residential and commercial customers. For residential customers, each supplier made the same offer to customers that were in a service territory and rate class. In contrast, the offers that were made to commercial customers were often varied over customers. Suppliers could, and often did, make offers that were tailored to individual customers. The variation in prices for given suppliers across customers is shown in Figure 4–4. Note that when analyzed by customer size

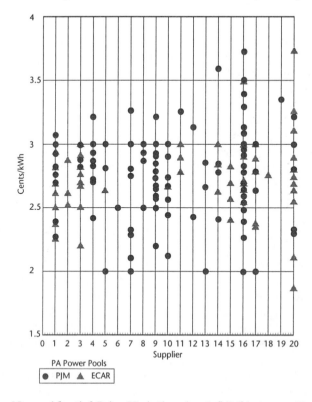

Figure 4–4   *Nonresidential Price Variation (cents/kWh) Across Customers by Supplier*

strata, the variation for the largest customers was extremely small. In addition, the mean cost estimates for the largest customers were significantly below those for the smaller ones.

We therefore collected data for the commercial customers in a way that reflected the customer-specific nature of the offers. In particular, we asked each commercial customer to tell which supplier they chose and what the characteristics of that supplier's offer were, such as the price, whether the price was guaranteed for a period of time, and the difficulty or simplicity of the terms of the offer. We then asked each customer to identify one or two other suppliers that were finalists in the customer's decision-making process and to describe the terms of these offers. Using this information, we modeled the choice that the customers made, relating their choices to the characteristics of the offers that they received.

Customer Data

We interviewed a total of 436 commercial customers that were eligible to choose an energy supplier in Pennsylvania's pilot program. Each customer was asked which supplier they chose, as well as which other suppliers were "finalists" in their decision-making process. Only 168 customers identified finalists. Our analysis was performed on these 168 customers.

For the chosen supplier and up to two nonchosen finalists identified, respondents were asked to provide information about the supplier and the supplier's offer. In particular, the customer was asked to provide:

- the price that the supplier offered;
- whether the supplier's rate was fixed or guaranteed, and would not change for the duration of the pilot;
- whether the supplier would let the customer terminate the contract with sufficient notice;
- the customer's rating (on a 1–5 scale) of the difficulty of the supplier's contract, where 1 is very simple and straightforward and 5 is difficult and complex;
- the customer's rating (on a 1–5 scale) of the knowledge of the supplier's employees, where 1 is not knowledgeable or unprofessional and 5 is very knowledgeable and professional;
- the customer's rating (on a 1–5 scale) of how comfortable the customer was with the supplier in contracting and obtaining electricity; and
- whether the supplier had called the customer.

Tables 4–2 and 4–3 show the various ratings customers gave suppliers. A summary of the variation across suppliers' contract terms and average prices is provided in Table 4–2, while noncontract characteristics are provided in Table 4–3.

TABLE 4–2 Summary of Average Nonresidential Contract Characteristics by Supplier

Supplier	Maximum Responses	Price in Cents/ kWh	Fixed Rate Guaran- teed Savings, 1=Yes, 0=No	Ter- minate Contract, 1=Yes, 0=No	Contract Difficulty, 1=Easy, 5=Complex
Supplier 1	87	2.7	0.9	1.0	1.9
Supplier 2	23	2.7	0.9	0.9	1.7
Supplier 3	55	2.9	0.9	1.0	2.1
Supplier 4	11	2.6	1.0	0.9	1.8
Supplier 5	32	2.6	0.9	0.9	2.4
Supplier 6	48	2.8	0.8	1.0	2.2
Supplier 7	33	2.8	0.9	1.0	2.1
Supplier 8	12	3.0	0.9	0.9	2.5
Supplier 9	44	2.9	1.0	0.9	2.0
Supplier 10	11	2.6	0.9	1.0	2.6
Supplier 11	110	2.9	0.9	1.0	2.1
Supplier 12	38	2.8	0.9	1.0	2.0
Refused/Don't Know	102	2.7	1.0	0.9	2.3
Total	606	2.8	0.9	1.0	2.1

TABLE 4–3 Summary of Other Nonresidential Supplier Characteristics

Supplier	Maxi- mum Re- sponses	Sales Staff Rating, 1= Unprofes- sional, 5=Profes- sional	Comfort Rating, 1=Very Uncom- fortable, 5=Very Com- fortable	Called You, 1=Yes, 0=No	Visited You, 1=Yes, 0=No	Attended Seminar, 1=Yes, 0=No
Supplier 1	87	4.1	4.1	.55	.17	.06
Supplier 2	23	3.5	3.7	.17	.04	.00
Supplier 3	55	4.0	4.1	.64	.38	.05
Supplier 4	11	4.5	4.5	.82	1.00	.00
Supplier 5	32	4.0	3.8	.84	.16	.06
Supplier 6	48	4.2	4.3	.74	.23	.10
Supplier 7	33	4.0	4.0	.78	.48	.12
Supplier 8	12	4.3	3.8	.75	.50	.00
Supplier 9	44	4.1	4.3	.68	.26	.05
Supplier 10	11	4.1	4.4	.64	.27	.00
Supplier 11	110	4.0	4.2	.61	.29	.11
Supplier 12	38	3.8	4.1	.41	.08	.00
Refused/Don't Know	102	4.0	3.9	.62	.26	.09
Total	606	4.0	4.1	.63	.32	.05

Many customers had difficulty providing price information for each supplier, particularly the "finalist" suppliers that the customer did not end up choosing. Customers were asked whether the supplier's offer was above or below that of the chosen supplier, and by how much. Alternate supplier prices were constructed to the extent possible, from the responses. A dummy variable that identified when price data could not be constructed for the supplier also was included in the model. Inclusion of this dummy allows the entire sample of 168 customers to be used rather than the subset that provided usable price data for each finalist supplier. (Usually data on other characteristics of the offers were available from the customer even when price data could not be constructed; by not eliminating customers with missing price data we were able to utilize all the available data on nonprice attributes.)

Model Specification

The commercial customers' choices were examined with a multinomial logit model, as given by the equation on page 41—the same as for the Pennsylvania residential customers. The only difference in the model specification for the two analyses is that, for commercial customers, the summation in the denominator in the equation is over the two or three firms that the customer listed as finalists (including the chosen supplier) rather than over all the available suppliers, as for the residential model.

Lessons Learned

The nonresidential revealed preference model performed reasonably well, though not as well as the residential model. Of the primary parameters investigated, about half were significant at greater than the 95 percent confidence level. All of the coefficients had the expected signs indicating, for example, that customers prefer simple contracts and knowledgeable sales staff. More importantly, we found large differences in the relative magnitude of the parameters investigated on suppliers' market share. These results were encouraging given that we were not particularly optimistic at the outset of our project design that enough customers would be able to provide accurate information on the characteristics of the offers they rejected. At the same time, a major limitation arose where only slightly more than one-third of our original sample of 436 was able to sufficiently characterize these other offers. This problem is best mitigated by contacting customers as soon as possible after they have chosen a supplier. Given that customers make their choices at different times, however, significantly shortening the lag between choice surveys and customers' decisions is difficult in practice.

The findings of this analysis also should be viewed in light of an important caveat. The attributes of the suppliers were obtained, as explained above, by surveying the customer. Customers might have a tendency to justify their choice, ex post, by attributing less desirable characteristics to the suppliers that they did not choose versus the supplier that they chose. For example, a customer that chose a particular supplier might say during the interview that he/she was comfortable with that supplier and not comfortable with the other suppliers, whereas in fact the customer might have been equally comfortable with all the suppliers at the time of the choice. The customer's response could be simply an attempt to understand and give meaning to his/her own choice, or perhaps to justify it to the interviewer. The estimated model necessarily incorporates both the impact of this factor on customers' choices as well as the impact of customers' choices on their reporting of suppliers' characteristics.

APPLICATION OF DOUBLE-BOUNDED WITH FOLLOW-UP STATED PREFERENCE METHOD TO CALIFORNIA COMMERCIAL CUSTOMERS

This section examines the likelihood that California commercial customers would be willing to switch electric suppliers. Customers of California's investor-owned utilities obtained the ability to choose their electricity supplier in November 1997, although power from new suppliers did not actually flow until March 31, 1998. The survey underlying the approach described in this section was conducted in late spring of 1997, before small and medium commercial customers were eligible to choose a supplier. Surveyed customers were asked various questions about whether they would switch to a new supplier if the new supplier offered lower prices and different levels of service. The amount of price discount varied, as did the service characteristics in order to determine the share of customers who would switch at each discount level and the relative importance of each service attribute. The method used for analyzing the subsequent data is an extension of the dichotomous choice procedure, which was primarily developed to uncover the value of natural resources. Rather than simply trying to estimate a customer's willingness to pay, this study also accounts for the nonprice attributes of electric service that can affect the customer's choice of electric supplier. The approach employed is referred to as a "double-bounded plus follow-up" method and is described further on the next page.

There are several reasons to ask customers whether they would switch suppliers, as opposed to conducting the type of conjoint study that is described in the next section. First, questions about switching are relatively easy to phrase and easy for the respondent to understand. For example, the customer might be asked "Suppose an electricity supplier offered you electricity at a 10 percent lower price than you currently pay. Would you switch to that supplier?" Or: "Suppose an electricity supplier offered you electricity at an 8 percent lower price but did not have any energy efficiency programs. Would you switch to this supplier?" These types of questions can be asked over the phone during a short interview. In contrast, the conjoint-type questions that are discussed on page 50 often are so complex that the analyst is required to mail material to the respondent prior to the phone interview.

Second, the questions can be geared such that the respondent's answer to one question affects the content of subsequent questions. For example, if the customer had said that they would not switch at a 5 percent discount, the interviewer can follow up with a question asking whether the customer would switch if the discount were 8 percent. Similarly, a customer who says they would switch for a 10 percent discount could be asked if they still would switch if the supplier did not use renewable generation sources. Essentially, the answers to earlier questions can be used in later questions to "hone-in on" the customer's decision criteria. The analysis of the responses needs to account for the fact that later questions depend on responses to previous questions; we describe an appropriate analysis procedure in the following section.

The third reason for phrasing questions in terms of switching is that, in many areas, competition will take a form such that customers remain with their current supplier unless they make a positive decision to switch. Questions regarding switching therefore accurately reflect the customers' decision when competition arises.

Double-Bounded Plus Follow-up Survey Approach

A telephone survey was conducted of approximately 600 small- and medium-sized commercial customers in California. The survey instrument first asked questions about whether the customer would switch at various price discounts, holding other service attributes constant. Then, the customer was asked follow-up questions that were designed to determine the importance of various service attributes relative to price. The details of each of the surveys are described below.

In order to implement the double-bounded approach, a series of price discount questions were asked to provide upper and lower bounds on the price discount required for the customer to switch suppliers, assuming a constant quality of electric service. The questioning approach is referred to as "double-bounded" because the respondent's discount rate is bounded on two sides. The question sequence is depicted graphically in Figure 4–5 and described here:

- The customer was first asked if they would switch suppliers with a 0 percent price discount.
  - If the customer answered "yes," they were asked no further percent discount questions, since no discount was required.
  - If the customer answered "no," they were asked if they would switch suppliers at some specific (randomly assigned) discount rate (y) off the total electric price.
    - If the customer answered "yes," a follow-up question was asked concerning a lower discount rate ($y-k$).
    - If the customer answered "no," a follow-up question was asked concerning a higher discount rate ($y+k$).

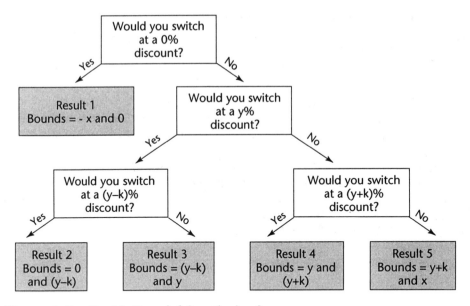

Figure 4–5 *Double-Bounded Questioning Sequence*

For this study, y was randomly chosen to be 10 percent or 15 percent and k was set at 5 percent. Thus, customers were offered possible discounts of 5 percent, 10 percent, 15 percent, 20 percent.

The second part of the survey considered nonprice attributes of electric service. For the questions in this part, the price discount information from the first portion of the survey was retained. The customer was then asked whether they would still switch if they were offered a discount equal to the lowest rate that they said they would switch at, but if the new energy provider did not provide a particular service that was currently provided. Five types of nonprice service degradation were considered:

- increased number of momentary power outages per year;
- more difficulty reaching the supplier with questions about service;
- non-local company (headquarters out of state);
- no energy efficiency programs provided; and
- generation of electricity using fossil fuels versus renewable power.

Model Approach and Generic Results

The data generated by this survey were analyzed using the maximum likelihood techniques described by one of the coauthors in a 1998 article in *The Energy Journal*.[2] The maximum likelihood estimates were obtained using a two-step approach.

1. First, a model was defined that estimated the overall level of price discount required for 50 percent of the surveyed cus-

tomers to switch (the critical level of discount), assuming no change in the quality of electric service. This model included parameters for customer characteristics such as size of electric bill, participation in energy efficiency programs, customer satisfaction, and telephone carrier switching behavior.

2. The base level price discount and its associated standard error from the first model were then used in the second-stage model to estimate the value of various nonprice attributes of electric service.

The final joint product of these two maximum likelihood procedures provides the key components to identify the critical level of price discounts under a variety of situations. Specifically, the overall expected critical level of discount is the starting point. The coefficients associated with specific customer characteristics can then be added or subtracted to arrive at a "customer-type level of critical discount" for a given type of customer. Then the values of additional nonprice service factors are added in to account for the presence of specific utility characteristics.

A number of customer characteristics were examined during the first-stage modeling process. It was found that many customer attributes were the most statistically significant indicators of customers' propensity to switch. Examples of the types of characteristics that were tested are: customer size, participation in utility programs, switching behavior for other services, (such as gas and long-distance telephone), satisfaction with the current electric utility, and whether the site was part of a chain.

The second stage of the estimation process examined two basic questions:

1. Would the customer still switch suppliers if certain utility service components were perceived as being worse with the new supplier?

2. How large of a price discount would be required to induce customers to switch when the "undesirable" utility service characteristics were present in the competitor?

The same type of maximum likelihood algorithm used for Stage 1 was employed in Stage 2. Normal distributions are assumed for the base threshold discount, while the additional discount required for a customer to accept the undesirable service attributes is assumed to follow a log-normal distribution.

The results of the Stage 2 model indicated that customers seemed willing to pay "premiums" for the value-added-services that utilities are able to provide. The coefficients for all of the discount premiums associated with the nonprice attributes were statistically significant at greater than the 99 percent confidence level. In addition, very large differences were found in the relative importance of the alternative attributes. These differences are shown generically in Figure 4–6.

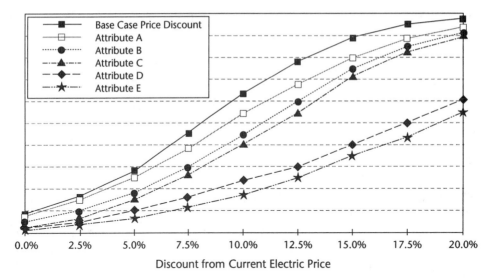

Figure 4–6 *Generic Representation of the Relative Effects of Nonprice Service Attributes on Predicted Switching Rates from Double-Bounded Plus Follow-ups Method*

The most important aspect of Figure 4–6 is that predicted switching rates dropped markedly when customers considered the importance of key non-price attributes. If customers were asked only whether they would switch to an alternative supplier for a given discount rate, assuming that the other supplier is identical in all other respects, switching rates were reported that exceeded those observed when choices were actually made available. The incorporation of the nonprice attributes provides a more realistic basis for estimating switching during the early stages of a market's opening when it is more likely that customers will be apprehensive about whether alternative suppliers will meet their needs as well as incumbent providers.

APPLICATION OF CONJOINT STATED PREFERENCE METHOD TO CALIFORNIA COMMERCIAL CUSTOMERS

This section describes our application of conjoint-type questions[3] for a subsample of the small- and medium-sized commercial customers sample described in the previous section. Each customer was mailed a package that contained six cards. Each card described a hypothetical supplier, giving the price, contract terms, and other attributes. The customer was asked to choose the supplier that it preferred most. The responses then were used to infer statistically the importance of each supplier attribute in the customers' decisions. This type of analysis is very common in new product market research and other fields. In the current context, it has some advantage relative to the switching questions described in the previous section. First, it can represent more accurately the type of choice that customers might actually face in a competitive market, where customers generally can choose among many suppliers. In contrast, switching questions present the customer with just one alternative supplier and ask whether the customer would choose to switch to that provider. Second, greater variation in attributes can be introduced into the analysis, since many suppliers are described and the attributes of the customer's current supplier do not need to be included. In contrast, switching questions ask about only one alternative supplier in each question and that supplier is always compared against the current provider. The disadvantage of these conjoint-type questions is that they are more complex than switching questions. Usually it is necessary to mail the respondent a package that contains the descriptions of the suppliers; switching questions, by contrast, usually can be asked over the phone without mailing any materials to the respondent. Finally, in some competitive markets, customers remain with their incumbent utility unless they make a positive decision to switch. In these cases, switching questions might capture the situation more accurately than conjoint-type questions that ask the customer to choose among a set of suppliers without reference to their current supplier.

Survey Data

A mail survey was sent to each of the customers who had participated in the phone survey described in the previous section on the double-bounded method. As stated earlier, the survey included a packet of six cards that described the attributes of six hypothetical suppliers. The customer was asked to indicate which of the six suppliers he or she would choose. A total of 215 customers completed and returned the mail survey.

Several attributes of suppliers were included in the descriptions. In particular, each supplier was described with respect to the following: whether the supplier's price was fixed or variable; the price, if fixed; the range of price, if variable; whether the supplier provided conventional or enhanced billing and customer service; the level of energy services that the supplier provided (none, standard, or full); whether the supplier used environmentally friendly generation; and whether the supplier was the current utility, a new in-state supplier, or an out-of-state supplier. Table 4–4 gives the details on the terms and levels for each attribute. Given the number of possible levels for each attribute, 216 distinct combinations of price and service options were possible. A package was developed for each customer by randomly choosing six combinations from the 216. Combinations that were clearly superior to all others, such as a lower price and better service, were eliminated when choosing the six for each customer.

TABLE 4-4		Attributes Utilized in Creating Conjoint Packages
Dimension	**Option**	**Description**
Price	1	Fixed. Price is guaranteed at 2.5 cents/kWh.
	2	Spot Market Hourly. Price varies from hour to hour in relation to spot market. Price ranges from 1.3 cents/kWh to 3.3 cents/kWh.
	3	Fixed. Price is guaranteed at 3.0 cents/kWh.
	4	Spot Market Hourly. Price varies from hour to hour in relation to spot market. Price ranges from 1.8 cents/kWh to 3.8 cents/kWh.
	5	Fixed. Price is guaranteed at 3.5 cents/kWh.
	6	Spot Market Hourly. Price varies from hour to hour in relation to spot market. Price ranges from 2.3 cents/kWh to 4.3 cents/kWh.
Billing Services	1	Premium—Customer service staff are available 24 hours every day, or account information may be accessed via the Internet. All accounts are itemized and combined into a single bill. Any day of the month can be selected for bill receipt.
	2	Standard—Customer service staff are available during normal business hours. Separate bills are received for each account. The date of the bill is determined by the electric company.
Energy Services	1	Full—telephone consultations, energy audits and other general reports and recommendations are provided as part of service. Engineering design services, maintenance, financing and other services are provided at competitive prices.
	2	Standard—telephone consultation, energy audits and other general reports and recommendations are provided as part of service.
	3	None—no special energy services are provided.
Power Type	1	Green—Power is supplied by hydroelectric and renewable resources such as biomass, wind, and solar.
	2	Conventional—Power is supplied by conventional fossil fuel and nuclear resources.
Supplier Type	1	Current Electric Utility Company
	2	A large national electric utility company with headquarters outside of California.
	3	A new nonregulated commercial for-profit electric company.

MODELING APPROACH AND COMPARISON OF FINDINGS WITH DOUBLE-BOUNDED PLUS FOLLOW-UP METHOD

The customer's responses were examined with a logit model, as described in the equation on page 41, with the summation in the denominator being over the six suppliers that were given in the customer's packet. We then compared the results of this analysis to the results obtained from the switching questions described on page 48. Two important findings were obtained.

Both methods attempted to predict the market share of the incumbent utility under a variety of scenarios. As expected, the incumbent's share was consistently higher in the analysis based on switching questions than for the conjoint-type analysis. This finding emphasized the importance of applying whichever method is more appropriate for the form that competition will take in a given area. The switching questions are more appropriate when customers will stay with the incumbent utility unless they choose to switch. The conjoint-type questions are more appropriate when customers will be choosing among several suppliers without the default being to remain with the incumbent. (Of course, this concept needs to be considered in the context of other issues, such as the extra variation that conjoint-type data allow and the relative ease of switching questions.) Second, and most important, the two methods gave very similar estimates for the relative importance of supplier attributes. That is, the switching questions and the conjoint-type questions obtained similar estimated values for customers' willingness to pay for environmentally friendly power, energy services, a price guarantee, and the incumbent. This finding implies that either method can be used for analysis of the value of service attributes.

SUMMARY

The purpose of this chapter has been to demonstrate that there are a variety of analytical methods available to estimate the importance of price and non-price electricity product attributes on customers' choices of their suppliers. Each method has strengths and weaknesses in any given application and may or may not be appropriate, depending on the research questions being addressed. The pilot programs in Pennsylvania offered the best opportunity to date to test revealed preference methods for analyzing customers' choices of electricity providers. Our experience shows that revealed preference modeling can be used to untangle an otherwise convoluted web of factors to explain the success of particular supplier approaches, when underpinned by comprehensive collection of mass market and customer-specific data. The application of stated preference methods to the California market before it opened is an example of how choice models can be used for forecasting and baseline purposes. With many markets still isolated from retail competition, stated preference methods will continue to be important research tools. In addition, stated preference methods will retain their value as means of testing and evaluating new product concepts. Finally, in all of the revealed and stated preference studies that we have conducted, it has been found consistently that the importance of non-price factors can be quantified relative to price. Why does this matter? Because in the quest for success in highly competitive commodity markets, understanding the value of attributes for which retail customers are willing to pay premiums will be critical to both surviving and prospering in the brave new world of customer choice. Specifically, quantifying the value of desirable nonprice attributes provides the information needed to perform financial analyses of whether the incremental market share attributable to such attributes justifies any added expense associated with their provision.

NOTES

1. Retail Wheeling Multi-Client Studies, Phases I, II, and III. Prepared by XENERGY Inc., 1996–1998.

2. Y. Cai, I. Deilami, and K. Train, "Customer Retention in a Competitive Power Market: Analysis of a 'Double-bounded Plus Follow-ups' Questionnaire," *The Energy Journal*, Vol. 19, No. 2 (1998).

3. Conjoint analysis is a method of obtaining the relative worth or value of each level of several attributes from rank-ordered preferences of attribute combinations. The questionnaire was styled after conjoint analysis.

CHAPTER 5

How to Hear the Voice of the Customer

Lisa Wood
Sheryl Cates
Sue Winemiller

Two approaches are described in this chapter that help identify what customers really want in a new product or service. We start by describing a qualitative approach—Voice of the Customer (VOC)—to uncover and identify customer needs. Then we describe a quantitative approach—choice-based conjoint analysis—for determining customer preferences for product features that satisfy those needs. Finally, we discuss adjusting the choice share estimates resulting from choice-based conjoint analysis to account for loyalty to the incumbent, inertia, and likelihood of switching. These techniques can be used with customers in general or with specific market segments. If customer needs are likely to differ by segment, then both the VOC analysis and the preference analysis should be conducted at the market segment level. Segmentation approaches are not discussed in this chapter.

LISTENING TO THE VOICE OF THE CUSTOMER

Well-established research suggests that communication among marketing, manufacturing, engineering, and research and development departments leads to greater new product success.[1] However, because developing new products and services is a difficult task, the degree of communication in the new product development (NPD) process varies widely across firms. Over the past decade, companies have begun to incorporate VOC into their NPD process to identify customer needs. In addition to uncovering needs, incorporating VOC can improve communication among the various participants in the NPD process.

VOC is a qualitative technique that uses in-depth, one-on-one interviews with customers to identify their needs; to categorize those needs as tertiary, secondary, or primary; and to prioritize needs based on estimated importances.[2] These interviews typically last about 45 minutes and are couched in a specific experience, such as the last time a customer called his utility to inquire about a bill payment. VOC offers several key advantages over traditional qualitative research methods. With this technique, the interviewer offers scenarios to prompt discussion but the respondent is the primary conversation driver. Interviewers allow respondents to introduce their own ideas in their own words, listen for needs statements, and prompt respondents to describe the benefits they expect to realize if their needs are met.

VOC interviews typically result in needs related to service, image, price, and quality. After all of the needs are expressed, specific interviewees are invited back to a group setting to sort the needs. During this process, needs are sorted into three bins—very detailed or tertiary (150 to 300 needs), secondary (20 to 40 needs), and very general or primary (5 to 10 needs) (see Figure 5–1). We use the tertiary needs to determine the primary needs. This preserves the voice of the customer throughout the hierarchy. This hierarchy of needs then can be used for process improvement, product development, or marketing/positioning. When used for product development, the secondary needs become the attributes for developing new products and services. (The next section discusses how to quantify preferences for those attributes.)

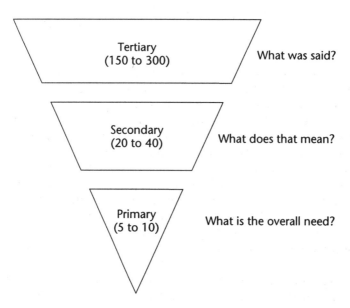

Figure 5–1   *Hierarchy of Customer Needs*

Is VOC more effective than traditional methods? Griffin and Hauser report that, while a single focus group uncovers more needs than a single one-hour, one-on-one interview, two one-on-one interviews are about as effective as one focus group, and four interviews are about as effective as two focus groups. Hence, one-on-one interviews are far more efficient. Their research also indicates that 90 percent to 95 percent of the customers' needs can be uncovered with 20 to 30 interviews.[3]

Over the past decade, as companies have involved customers in their NPD process in varying degrees, they have discovered that customers do not always know what they want, what they need, or what they will want in the future. As a result, companies have begun to ask themselves whether they are talking to the right customers and whether they are asking the right questions. Some companies have found that talking to a small group of "right" customers is far more effective than talking to customers in general about new products and services. In addition, if a product is very new or different, it may be difficult to obtain customer feedback. For very new products, researchers have suggested immersing the customer in a future environment, referred to as "future conditioning," and then asking about the product. Future conditioning presents customers with a purchase and usage environment as well as with a set of competitive products that they might face in the future and asks them to make choices or to evaluate products in that setting.[4] Multimedia or virtual reality might be used to communicate this future environment more effectively to customers.[5]

As the U.S. electric utility industry faces restructuring and customers have choices among electricity suppliers, incorporating VOC into the design and development of products and services is critical for maintaining existing customers and for attracting new ones. Successfully listening to customers to identify their needs involves asking the right customers the right questions.

Asking the right customers is important because customers have limited or no experience with electric utility deregulation and it is difficult for them to imagine things they have not experienced. By April 1998, the electric utility industry in California, Rhode Island, and Massachusetts was opened to competition and customer choice. Several other states across the country have implemented pilot choice programs since then. Although many electric utilities are accustomed to gathering information from their existing customers, because some customers already have been exposed to deregulation and competition, talking to customers who have experienced choice will be far more effective for VOC and identifying needs. Hence, to address the first issue, utilities can talk to the "right" customers; in this case, the right customers are those who have experienced competition and choice.

Asking the right questions is important because customers are accustomed to one electricity supplier and it is difficult for them to imagine how this supplier might improve because they do not have a comparison supplier. It is useful, therefore, to present customers with a reference point from another industry. For example, to discuss the concept of customer service, we might ask respondents to think about the type of customer service offered by a retail company such as Land's End, and then ask them to discuss utility customer service in comparison. This helps customers focus on the new product or service concept in a meaningful way and helps the interviewer uncover needs.

Listening to the right customers and asking the right questions to identify needs and then incorporating those needs into the development of new products and services is a fuzzy and difficult task. When the product is very new or different, the task is even more difficult. When VOC is executed properly, customer needs, wants, and expectations for specific products or services are identified and translated into concrete, discreet attributes or features. This is the cornerstone of developing new products and services.

DETERMINING CUSTOMER PREFERENCES

Understanding customer preferences for new product and service offerings and then developing appropriate offerings is a crucial step toward meeting customer needs. This section discusses a conjoint analysis approach for quantifying customer preferences for features of new electricity-related products and services. This approach typically begins by identifying a set of product or service features associated with customer needs. As discussed in the previous section, VOC is one method for identifying customer needs; using this approach, secondary needs are translated into attributes or features for quantitative assessment. Other methods include focus groups with customers, one-on-one interviews, and internal discussions.

Once we have defined a set of specific features or attributes for a potential new product or service based on customers' needs, how do we determine customer preferences? Choice-based conjoint analysis (choice) is one quantitative technique that can be used to quantify preferences for new products or services. We recommend a choice approach for the electric utility industry where competitive effects now matter.

Using this technique, new products or services are defined by a set of features or attributes, and a trade-off approach is used to assess the relative importance of alternative product or service features. This technique is often used in the early stages of product and service design. Instead of asking customers how much they like or dislike a particular product or the importance of a particular feature, we indirectly elicit customer preferences by forcing them to make trade-offs among product features.

It is well known that direct methods often produce results that say, "everything is important;" this is not particularly useful for designing new products and services. Trade-off methods overcome this limitation by forcing respondents to make trade-offs among product attributes so that some attributes are relatively more important than others. In general, we recommend trade-off methods for determining customer preferences rather than direct methods such as questions that elicit an importance or degree of like/dislike. Some trade-off methods measure preference for particular product or service features using a rating or ranking approach. A choice approach, which simply asks respondents to make a choice, provides more accurate estimates of competitive effects because choice tasks more accurately reflect real-life decision making.

To illustrate the use of the choice approach, look at an example that used this technique to determine preferences for a set of electricity supply options among residential customers. This example is based on an EPRI-funded study, "Predicting Customer Choices Among Electricity Pricing Options."[6] This study focused on several market segments but the discussion here is limited to the residential customer segment.

For residential customers, preliminary qualitative market research identified three features or attributes that matter to customers in choosing an electricity supply option—the rate plan, the length of the contract, and the electricity supplier and customer service provider. For each of these features, we defined a range of levels. For example, we incorporated three levels for contract length—none, one year, and five years. Table 5–1 presents the complete set of features and levels. Each respondent received detailed descriptions of the rates and definitions of the contract and the electricity supplier/customer service provider options. The detailed descriptions are presented in the fact card in Figure 5–6 (see pages 65 and 66).

Choice-based conjoint analysis elicits preferences by asking respondents to choose from among a set of hypothetical product or service offerings. In this case, we asked respondents to select a single preferred electricity supply option from four alternatives.[7] Figure 5–2 shows a sample choice question for residential customers. Each respondent answered a set of choice questions and, for each question, the levels were varied, based on an experimental design.

TABLE 5–1   Features and Levels for Residential Customers
• Rate plan —Fixed rate of 7¢/kWh —Fixed rate of 9¢/kWh —Rates vary by season —Rates vary by time-of-day (TOD)   • Contract length —No contract —1-year contract —5-year contract
• Electricity supplier and customer service provider —Local utility supplies electricity and provides customer service —Well-known company supplies electricity and provides customer service —Well-known company supplies electricity and local utility provides customer service —Unfamiliar company supplies electricity and provides customer service —Unfamiliar company supplies electricity and local utility provides customer service

Discrete choice analysis typically is used to quantify the relationship between choices that respondents make in the choice questions and the features that comprise each option or alternative. This type of analysis provides the following types of results for each customer segment:

- **The utility weights for each level of each feature.** Figure 5–3 shows, for example, that residential customers prefer no contract rather than a five-year contract and prefer that their local utility supply electricity and customer service rather than an unfamiliar company.

- **The importance of each feature or attribute on a scale ranging from 1 to 100.** Figure 5–4 shows that the importance of rate plan and electricity supplier/customer service provider is very similar for residential customers. This indicates that customers are not choosing their electricity supplier based solely on price.

- **The premium customers are willing to pay for a specific feature.** Table 5–2 shows that customers are willing to pay a premium to

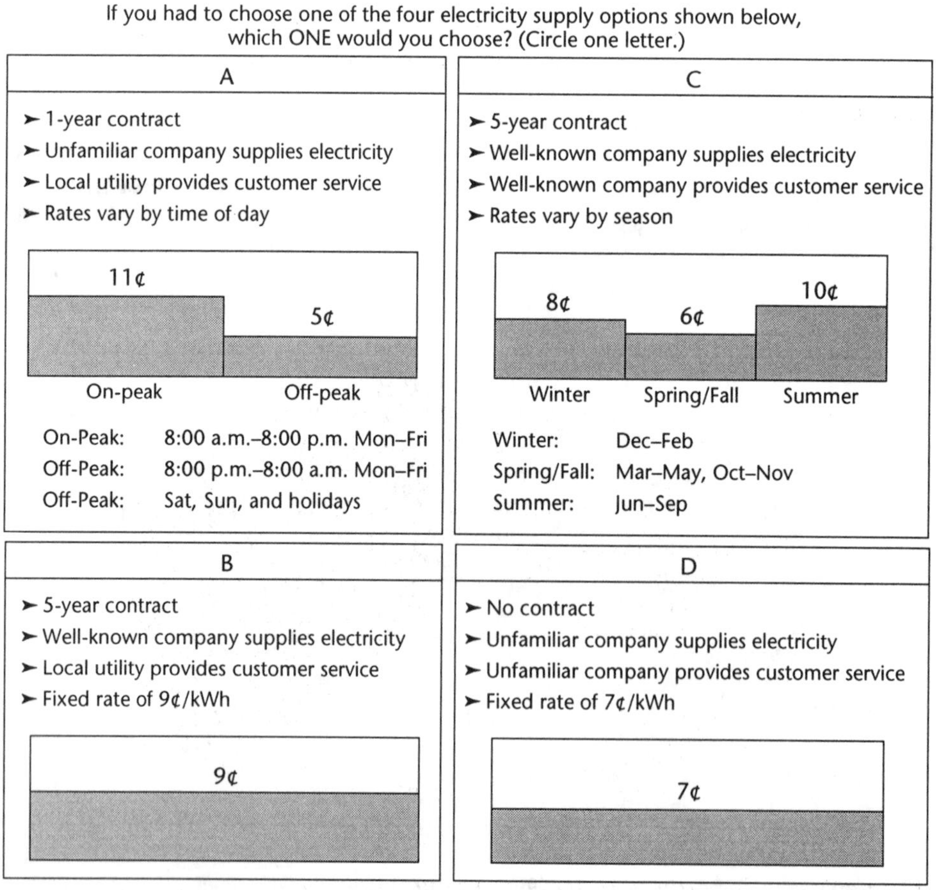

Figure 5–2 *Sample Choice Question from the Residential Survey*

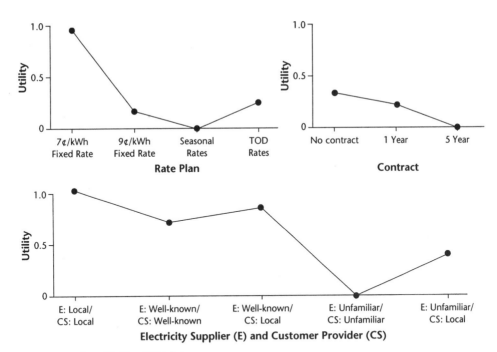

Figure 5–3   *Utility Weights*

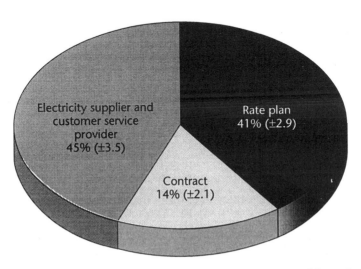

Figure 5–4   *Importance Weights for Features (95 percent confidence intervals in parentheses)*

TABLE 5–2   Willingness to Pay (WTP) Estimates

	WTP (¢/kWh)	95% Confidence Interval (¢/kWh)
Contract		
No contract versus 1-year contract	0.15	–0.01 - 0.31
No contract versus 5-year contract	0.78	0.65 - 0.91
1-year contract versus 5-year contract	0.63	0.51 - 0.76
Electricity Supplier and Customer Service Provider		
Local utility versus well-known company (supplier and customer service)	0.74	0.49 - 0.98
Local utility versus well-known company (supplier only)	0.58	0.35 - 0.82
Local utility versus unfamiliar company (supplier and customer service)	2.47	2.19 - 2.75
Local utility versus unfamiliar company (supplier only)	1.52	1.25 - 1.80

avoid an unfamiliar supplier. In this case, they are willing to pay about 2.5 cents more per kilowatt-hour to have their electricity and customer service provided by their local utility rather than by an unfamiliar supplier. This is an indication of customer value for a specific feature and should not be interpreted as a price.

• **The capability to simulate how the choice share for a product or service offering(s) changes as the price or other features of the product or competing product change.** For example, examining the choice share for Options 1 and 2 in Figure 5–5 shows that the choice share for the local utility increases from 26 percent (Scenario A) to 55 percent (Scenario B) as the local utility switches from seasonal rates to a fixed or flat rate and continues to offer time-of-day rates.

PREFERENCES VERSUS ACTUAL CHOICE

In the scenarios in Figure 5–5, given the set of competitive offerings, a large percentage of customers are predicted to switch from their local utility and choose a well-known company. Do we really expect this to happen in a competitive environment? The choice results provide quantitative estimates of relative preferences and the importance to customers of alternative product features. These estimates assume full awareness of all offerings and no barriers or transaction costs associated with making a choice. In many ways, these can be viewed as long-run estimates of choice and are useful for examining relative preferences. However, these estimates alone are not likely to reflect actual behavior in the marketplace in the near future.

To predict actual short-run behavior, we must adjust the choice share estimates to account for factors such as inertia, loyalty to the incumbent, and

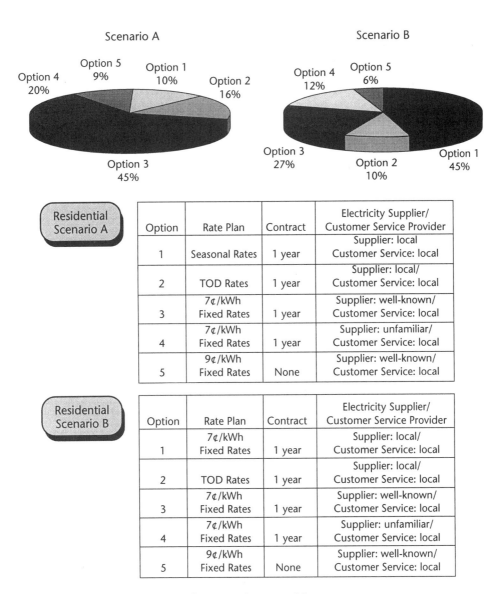

Figure 5–5 *Choice Shares for Scenarios A and B*

the likelihood of switching electricity supplier. We typically accomplish this by asking respondents a battery of loyalty, inertia, and switching questions and then adjusting the choice share estimates.

January 1998 marked the beginning of a four-year transition period to full competition in California. Although customers could choose a new electricity supplier in January, the actual switching of service did not take place until March 31, 1998. What happened? As of July 1998, very few residential customers (about 0.7 percent) had actually switched. Those who did so switched to a familiar company, including affiliates of their local utility, and some switched to purchase green power.

Based on EPRI-sponsored research,[8] many switchers in California said they switched to save money and think (perhaps incorrectly) that they will save more than the mandated 10 percent rate reduction. Those who did not switch said they wanted to wait and see, that the savings were not sufficient, or that they were happy with their current supplier. These early results indicate a high degree of incumbent loyalty, a high degree of inertia, and a low probability of switching.

CONCLUSION

Listening to customers and determining their preferences for new products and services is a difficult task. We recommend using VOC to identify needs and product attributes and choice-based conjoint analysis to quantify preferences for attributes or features of new products and services. However, to move from preference shares to realistic estimates of choice shares in a market undergoing deregulation, we must go one step further and include factors such as loyalty to the incumbent, inertia, and the likelihood of switching. The exact method for making this adjustment is not well defined. As demonstrated in California, very few residential customers actually switched when faced with choice.

Rate Plan (in ¢/kWh)	Your total electric bill includes a monthly service charge plus an energy usage charge. The energy usage charge is based on the amount of electricity you use (in terms of kilowatt-hours) and the price of electricity or rate in cents per kilowatt-hour (¢/kWh).
	Under competition, you will need to choose a rate plan. The four rate plans described below differ in terms of whether or not the rate varies, and if so, how the rate varies. For this survey, assume the rates shown below are being offered.

✔ *Fixed rate of 7¢/kWh*

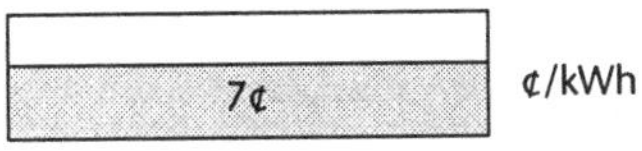

| 7¢ | ¢/kWh |

With this rate plan, **you pay 7¢/kWh.** The rate does not vary by season or time of day.

✔ *Fixed rate of 9¢/kWh*

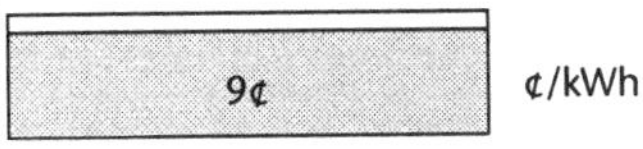

| 9¢ | ¢/kWh |

With this rate plan, **you pay 9¢/kWh.** The rate does not vary by season or time of day.

✔ *Rates vary by season*

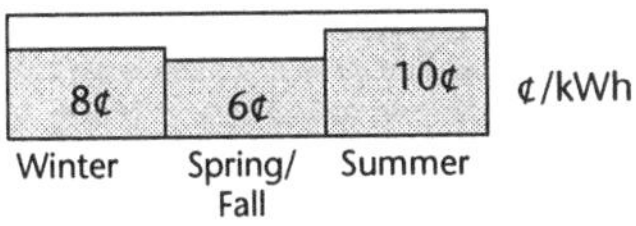

| 8¢ | 6¢ | 10¢ | ¢/kWh |
| Winter | Spring/ Fall | Summer | |

Winter: Dec–Feb
Spring/Fall: Mar–May, Oct–Nov
Summer: Jun–Sep

With this rate plan, rates vary by season. **Your annual electric bill could increase or decrease depending on the months you use the most electricity.** For example, if you use the most electricity during the summer, your bill could increase compared to a fixed rate. However, if you use the most electricity during the spring or fall, your bill could decrease compared to a fixed rate.

The rates and months for each season are shown on the left.

✔ *Rates vary by time of day*

| 11¢ | 5¢ | ¢/kWh |
| On-Peak | Off-Peak | |

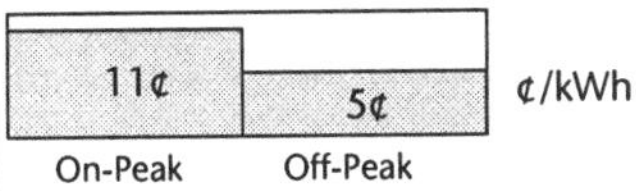

On-Peak: 8:00 a.m.–8:00 p.m. Mon–Fri
Off-Peak: 8:00 p.m.–8:00 a.m. Mon–Fri
Off-Peak: Sat, Sun, and holidays

With this rate plan, rates vary by time of day, meaning that you will be charged a higher rate during the hours when there is the highest demand for electricity (on-peak period) and a lower rate at other times (off-peak period). Under this rate plan, you would have to be careful about using electric appliances that consume a lot of energy—such as your dishwasher and washing machine—during the hours when electricity is most expensive.
Your annual electric bill could increase or decrease depending on the hours you use the most electricity. For example, if you use the most electricity during the day from 8:00 a.m. to 8:00 p.m., your bill could increase compared to a fixed rate. However, if you use the most electricity at night, you bill could decrease compared to a fixed rate.

The rates and times for the on- off-peak periods are shown on the left.

continues

Figure 5–6   *Residential Fact Card*

continued

Electricity Supplier and Customer Service Provider	You will need to choose which company will supply your electricity and which company will provide basic customer services—that is, the one company you call if you have service or billing problems. **Regardless of the supplier you choose, you local electric utility will still continue to deliver electricity over its wires to your home and will be responsible for reliable service.** If the electricity supplier from which you choose to buy power cannot produce the electricity that you have purchased, your local electric utility will purchase power for you at market prices so that your service is not interrupted. You may be charged a penalty for this service. Five supplier/provider options are described below.
✓ *Local utility supplies electricity and provides customer service*	You can continue to have your electricity supplied and customer service provided by your local electric utility.
✓ *Well-known company supplies electricity and provides customer service*	You can choose a well-known company *other than your local electric utility* to supply your electricity and provide customer service.
✓ *Well-known company supplies electricity and local utility provides customer service*	You can choose a well-known company *other than your local electric utility* to supply your electricity; however, your local electric utility will continue to provide customer service.
✓ *Unfamiliar company supplies electricity and provides customer service*	You can choose a company that you are not so familiar with to supply your electricity and provide customer service.
✓ *Unfamiliar company supplies electricity and local utility provides customer service*	You can choose a company that you are not so familiar with to supply your electricity; however, your local electric utility will continue to provide customer service.
Contract	You will need to choose the conditions under which you contract with an electricity supplier, that is, how long you must remain with that supplier and rate plan. Three options are described below.
✓ *No contract*	You can change electricity supplier at any time by paying a switching or exit fee of $25. The supplier must provide a 3-month notification for rate changes.
✓ *1-year contract*	You will have to remain with the same electricity supplier for a period of 1 year. Rates are guaranteed for the length of the contract. (Contract is canceled if you move to a new location.)
✓ *5-year contract*	You will have to remain with the same electricity supplier for a period of 5 years. Rates are guaranteed for the length of the contract. (Contract is canceled if you move to a new location.)

Figure 5–6   *Residential Fact Card*

NOTES

1. See, R.G. Cooper, "New Product Strategies: What Distinguishes the Top Performers?" *Journal of Product Innovation Management* 2:151–164 (1984); R.G. Cooper, "How New Product Strategies Impact on Performance," *Journal of Product Innovation Management* 2:5–18 (1984); A.J. Griffin, and J. R. Hauser, "Integrating R&D and Marketing: A Review and Analysis of the Literature," *Journal of Product Innovation Management* 13(3):191–215 (1996); A.J. Griffin and A.L. Page, "The PDMA Success Measurement Project: Recommended Measures for Product Development Success and Failure," *Journal of Product Innovation Management* 13(4):478–496 (1996).

2. A.J. Griffin and J.R. Hauser, "The Voice of the Customer," *Marketing Science* 12(1):127 (1993).

3. Ibid.

4. G.L. Urban, B.D. Weinberg, and J.R. Hauser, "Premarket Forecasting of Really-New Products," *Journal of Marketing* 60(1):47-60 (1996).

5. See, for example, H.H. Adamjee and J.H. Scaife, "The Face of the Customer: The Use of Multimedia in Quality Function Deployment." Masters Thesis, Sloan School of Management, 1993; and L. Wood, D. Hering, M. Bala, S. Cates, and T. Romig, "Using Virtual-Reality Based Conjoint for Capturing the Voice of the Customer," *Quirk's Marketing Research Review* (May 1998): 34–39.

6. Electric Power Research Institute, "Predicting Customer Choices Among Electricity Pricing Options," EPRI TR-108864-V2.

7. For examples of conjoint studies for other products and services in the electric utility industry see, Electric Power Research Institute, "Designing New Utility Programs and Services: Case Studies on the Use of Conjoint Analysis," [1996] TR-105778.

8. Electric Power Research Institute, Internal Document, "What Drives Customer Choice in Competitive Power Markets?" [October 1998] TR-111806.

SECTION II

WHAT IS DRIVING
CUSTOMER CHOICE?

CHAPTER 6

Convergence of Utility Services: Technological Challenges and Opportunities

Karl E. Stahlkopf

A fundamental transition is taking place in the way public utilities provide essential services to their customers. In the United States, deregulation has already swept through the telecommunications industry and is now transforming the gas and electric power industries. As a result, many utilities are restructuring—generally moving from vertical integration, based on economies of scale, toward functional integration, based on economies of efficiency. The convergence of gas and electric service represents one important aspect of this trend, but further integration of utility functions such as telephone, cable television, and Internet services are coming.

The need to reduce costs has been the key driver of the present transition, which has been furthered by a combination of market-pull and technology-push. The utilities that thrive in this era of increased competition will be those that can adopt new technologies quickly and use them for strategic advantage. This chapter focuses on how technology can contribute to the convergence of gas and electricity—and eventually to the integration of other utility services.

This article originally appeared in *The Electricity Journal* (Elsevier Science, Inc.) in July 1997. It is reprinted here with their permission.

In order to anticipate the future, however, we first need to understand the combination of economic and technological forces that have produced the current wave of deregulation and convergence.

UTILITY DEREGULATION: TECHNOLOGY PUSH/MARKET PULL

In each of the utility industries, deregulation has been driven by a combination of forces, involving a technology push and a market pull. Let's consider each in turn:

In telecommunications, for example, the initial technology push was commercial development of microwave technology in the 1970s, which provided a wireless alternative to the existing, cable-based infrastructure for long-distance calling. At the same time, corporate computer users were providing a market pull by demanding new data network services. These forces created the first wave of telecom deregulation, which culminated in the court-ordered breakup of the Bell System in 1984. Subsequent development of satellite communications and fiber-optic networks have combined with exponentially rising demand to accelerate the trend toward deregulation.

In the gas industry, deregulation began in response to severe supply disruptions that occurred in the 1970s. The Natural Gas Policy Act (NGPA) of 1978 removed price restrictions that had contributed to the disruptions, which led to a burst of new exploration and technology development. In particular, computer-enhanced seismic imaging provided three-dimensional pictures of subterranean gas and oil fields, allowing gas producers to locate and drill more precisely at lower cost. Advanced sensors and new drilling techniques also enabled producers to guide horizontal drilling toward more precise, distant targets. The result was a stunning surge in available gas resources: Estimates of remaining gas reserves in the U.S. have grown 700 percent over the last decade and are still climbing.

Deregulation of the electric utility industry began in the 1970s when steeply rising electricity prices during the 1970s led to passage of the Public Utility Regulatory Policies Act of 1978 (PURPA), which required utilities to purchase electricity from cogenerators and independent power producers. Since then, technological development steadily has lowered the cost and increased the efficiency of combustion turbines—eroding the economic advantage of large coal and nuclear plants, and exacerbating regional electricity price disparities. Development of high-voltage electronic controllers and sophisticated control software for power systems also have enabled less expensive power to be transferred to more distant customers. Large industrial customers in regions with higher electricity prices have demanded open access to utility transmission systems in order to take advantage of lower cost power from more distant suppliers. Such access was provided at the wholesale level by the National Energy Policy Act of 1992 (NEPA).

DEREGULATION: THE CURRENT SITUATION

The same powerful combination of technology push and market pull is now driving deregulation of utility services in the U.S. down to the retail level, so that individual consumers are gaining unprecedented levels of choice among potential providers.

Most telephone customers in the U.S. can now choose among several competing long-distance companies, and soon will be able to pick an alternative local-service provider. In addition, from the perspective of inter-industry convergence, one of the most important regulatory changes was passage of the Telecommunications Act of 1996, which removed long-standing restrictions that prevented electric utilities from offering telephone service.

The gas industry in the U.S. is currently focused on complying with provisions of Order 636, issued by the Federal Energy Regulatory Commission (FERC) in 1992. This order furthers deregulation by requiring pipeline companies to "unbundle" their services and provide open access that is equal for all gas suppliers. It also establishes a secondary market for reserving pipeline capacity.

In 1996, FERC issued similar orders for unbundling electric transmission services. Order No. 888 provides open access to transmission networks by requiring the utility owners to offer non-discriminatory tariffs to third parties. This order not only affects individual utilities but also power pools, which are encouraged to set up an independent system operator (ISO). At the same time, the order allows utilities to recover stranded costs resulting from this unprecedented transition. Order No. 889 requires establishment of an Open Access Same-time Information System (OASIS) to provide all market participants with critical information about tariffs and available transmission capacity in real time. This order also sets up a Code of Conduct, which requires separation of utility power marketing and transmission system operation functions.

Meanwhile, at the state level—where rates between neighboring utilities sometimes differ by as much as a factor of two—deregulation is reaching the level of utility distribution systems. A variety of retail wheeling proposals are now under consideration by state utility commissions.

CONVERGENCE, STEP 1: INTRA-INDUSTRY MERGERS

As a result of simultaneous deregulation of various utility industries, a massive restructuring is taking place, generally involving some convergence of services—the trend toward functional integration based on economies of efficiency, mentioned above. The first step of this convergence has focused primarily on mergers of utilities in the same industry. Between 1985 and 1995, some 43 utility mergers were consummated[1] and the pace is accelerating.

In December 1996, FERC revised its guidelines on utility mergers to place more emphasis on ultimate benefit to customers and to reduce the time required for approvals. When these new guidelines were proposed, FERC was considering nine merger applications with a total value of about $25 billion. Most of these mergers involved electric utilities seeking to strengthen their regional competitiveness and cut costs in the face of proposed state-level deregulation of retail markets.

CONVERGENCE, STEP 2: GAS & ELECTRIC UTILITY MERGERS

Further cost savings and competitive advantage can come if a combined utility offers customers multiple energy services. Over the last couple of years, several gas and electric companies have proposed mergers that would provide this opportunity.

Duke Power Company and PanEnergy, for example, have proposed a merger that would unite an electric utility that serves 1.8 million customers with a gas transport company that owns 37,000 miles of pipeline and handles more than 15 percent of the natural gas consumed in the U.S. The companies expect that their merger will increase their income $225 million by the year 2000 through marketing of integrated energy products and services beyond what either utility could do alone.[2] Significantly, this merger will also unite the companies' respective marketing units, which are run as joint ventures with Louis Dreyfus and Mobil.

CONVERGENCE, STEP 3: COMMUNICATIONS AND OTHER SERVICES

Eventually, we will see a new type of utility emerge from the current market turmoil, as a logical extension of the market-pull, technology-push forces discussed earlier. This utility will be able to provide customers with fully integrated utility services at lower prices, in addition to offering the convenience of having a single bill and a single number to call to report service problems. Already we are seeing the early signs of convergence involving communications and other services.

About 85 electric utilities have established or are planning to offer telecommunications services, using their own private communications networks. Electric and gas companies own a total of 600,000 miles of high-capacity, fiber-optic cable, now employed mainly for their own control purposes, but with excess capacity that could easily be used to provide telephone, data, or video services to retail customers.[3] Utilities involved in this trend range in size from such giants as American Electric Power and Duke Power to the small Otter Tail Power Company of Fergus Falls, Minnesota.

Conversely, AT&T has developed a retail energy portfolio that it is planning to introduce as part of a diversification program. The company is expected to be a powerful competitor in the coming utility services convergence because of its marketing expertise and its brand-name recognition.[4]

Security services are also being added. Western Resources of Topeka, Kansas, for example, purchased Westinghouse Security in a friendly transaction. John Hayes, Jr., chairman and CEO of Western Resources, says his company's goal to become "an effective marketer of security, comfort, and economy to the public."[5]

INTERNATIONAL ASPECT OF CONVERGENCE

Utility convergence also has an important international aspect. On February 15, 1997, 68 nations agreed to wide-ranging changes in the global telecommunications industry. The agreement will remove many protections from local telephone monopolies and open national telecom markets to international competition. In the long term, this agreement will have a profound impact on convergence of other utility industries, because information technology provides a critical link for offering more widely integrated services. Already, many energy utilities are seeking international alliances and mergers. As global telecom markets become more open, additional opportunities will emerge for seeking synergies across the boundaries of traditional utility services.

In addition, some of the projected impacts of the new agreement on customers worldwide provide an insight into the potential economic and humanitarian benefits that could be achieved through broader utility services convergence. Today, the average cost of an international phone call is about a dollar a minute; but with open markets, that cost is expected to fall by 80 percent. Such a decline in price will make telephone service more widely available to the one-*half*(*!*) of the world's people who have never made a phone call. The convergence of energy and information technologies will likely have benefits of similar magnitude for people around the world who now are trapped in lives of physical hardship and economic malaise.

TECHNOLOGY FOR STRATEGIC ADVANTAGE

Just as technologies already available have helped provide the "push" needed to create a convergence among utility industries, future technological development can provide the strategic advantage needed by individual utilities to thrive in the newly competitive marketplace. Utilities can use emerging technologies in several ways to further their competitiveness along the whole continuum of business strategies.

The most immediately important strategy in an era of increased competition is cost reduction. With the coming of competition, low-cost power and gas are being transferred over longer distances. Innovative technology can help facilitate such transactions, while providing utilities with additional ways to lower costs, particularly in the areas of operations and maintenance. Reliability-centered maintenance, for example, can provide savings of about 25 percent, compared to previous utility maintenance practices.

Utilities must also find new ways to increase customer satisfaction. As consumers gain access to a broader choice of electricity and gas suppliers, ensuring customer retention requires moving toward a more service-driven approach. New technology can help utilities provide integrated services seamlessly to customers. The Utility Communications Architecture (UCA), for example, provides a common interface for electric, gas, and water services, as well as Internet compatibility. Developed by EPRI and based on internationally recognized standards, UCA is endorsed by the American Gas Association and the American Water Works Association. The establishment of such standard communications protocols allows a utility to "plug and play" equipment from different vendors over the same data network.

Another key business strategy is enhancement of asset utilization. Advanced technology can help gas and electric utilities take fullest advantage of their combined physical assets as they try to increase efficiency, reduce costs, and provide integrated services. One of the most far-reaching technological trends affecting the convergence of gas and electricity is a shift away from large, central-station power plants (1,000-plus megawatts) towards dispersed generators (typically hundreds of kilowatts to tens of megawatts), which can be built closer to customer loads. Small, gas-based technologies such as fuel cells and small combustion turbines offer a better match between capacity additions and load growth, reduced energy losses during delivery, potentially lower emissions, and less need for additions to transmission and distribution systems. [In 1996], for example, the first 2-megawatts molten carbonate fuel cell using natural gas was installed at a utility substation in California, and combustion turbines with capacities as small as 24 kW are now entering the market.

As deregulation creates new market opportunities and challenges for both electricity and gas, a variety of analytical technologies can also identify better ways to manage a company's participation in these more complex markets. One example of such a tool is POWERCOACH, a software package that supports utility decision making related to bulk power transactions—and, potentially, wholesale gas transactions. POWERCOACH uses an expert system to analyze transaction decisions on the basis of multiple objectives, such as minimizing costs while fulfilling net margin requirements. Typical output includes risk profiles associated with specific transactions, together with a comparison of alternative strategies and the sensitivity of a decision to specific uncertainties. Other analytical tools are being developed to help utilities optimize their own retail service designs, which could include integrated services.

UTILITY OF THE FUTURE

Finally, looking toward the future, let's consider how deregulation and technological developments are likely to foster even greater convergence of utility services and change forever the electric power industry as we know it.

Make no mistake: convergence has begun. Consolidation among utilities in the same industry is well under way, and innovative experiments involving combined electricity and gas services are emerging in many areas of the country. Legal barriers to convergence of energy and telecom services are falling and international agreements are allowing utilities to look increasingly toward a global market.

In the near term, I believe we will see a patchwork of service arrangements emerge, as both traditional utilities and their competitors offer customers a variety of energy, communications, and entertainment packages. Sorting through these offerings is just the sort of task that an increasingly dynamic marketplace can handle efficiently, and I'll make no attempt to second-guess the outcome for specific cases.

Over the long term, however, I believe that we can safely predict that numerous economic and technological forces will drive utility convergence beyond anything yet seen. Electricity and gas services, of course, make a particularly attractive combination for the majority of customers who are interested primarily in comfort and convenience. Advanced real-time metering and single-source billing will contribute to this convergence. Adding communications services to this one-stop-shopping utility package then becomes a natural—and I believe, inevitable—next step. Already, electricity and telecom companies have facilities that extend to virtually every household in the country. What remains is to determine how best to build on this existing infrastructure to offer value-added services, using technologies already becoming available in both industries.

Such convergence will create unprecedented challenges and opportunities—both for traditional utilities and their employees. Beyond the need to develop new business strategies and to take advantage of advanced technologies, a new perspective will also be necessary. For most of our professional lives, people such as I have considered ourselves a member of the *electric* utility industry. The time is rapidly approaching, I believe, when we will say, instead, "I belong to the *utility* industry."

NOTES

1. "Study Finds Most Mergers of Energy Utilities Have Not Enhanced Profits," *Electric Utility Week*, 16 December 1996, 11.

2. "Duke Will Buy PanEnergy for $7.7 Bn," *Electricity Daily*, 26 November 1996, 1. [*Editor's note*: The formal merger occurred on June 18, 1997.]

3. *Wall Street Journal*, 27 January 1997, B1.

4. "Electric Utilities Better Brace for Turbulent Restructuring," *Energy Daily*, 13 December 1997, 2.

5. "Western Resources Buys One Security Firm, Launches Hostile Bid for Another," *Electric Utility Week*, 23 December 1996, 1.

CHAPTER 7

Preparing For Gas/Electric Convergence: Mergers Or Alliances?

*Robert J. Michaels**

It is common knowledge that the markets for gas and electricity are converging. Beyond arbitrage, cross-market transactions will bring new tools for dealing with risk and new energy packages to compete for sales to increasingly sophisticated users. Market participants already are exploiting the choice between purchasing power directly and purchasing it as gas molecules that will be burned in generators.[1]

With these choices have come mergers between electric and gas utilities.[2] Some of these combinations carried high acquisition premiums that reflect very optimistic earnings expectations for the new company. Enron Corporation, for example, paid Portland General Electric's shareholders a 48 percent premium over the current value of their shares. According to analyst Ed Tirello of NatWest, the typical hostile acquisition carries a 30 percent premium. Enron paid the premium without any known competing bidders for its target.[3] The average corporate control transaction creates shareholder value, but only for those who hold stock in the acquired firm.[4] Gas/electric mergers may bring more competition to the production, marketing, and transmission of both commodities. If the two industries are converging, firms with one foot in each may have strategic choices that are unavailable to those who remain specialists.

---

*The views expressed in this article are not necessarily those of the author's affiliations or clients.

There is also a case, however, for remaining specialized and not crossing market boundaries. The coevolution of gas and electricity need not imply that individual firms will gain by holding extensive assets in both. The increasing perfection of the two markets might just as well imply that interindustry mergers are unsound strategies, undertaken in opposition to important long-term economic trends. The more efficiently markets function, the smaller are the benefits from heterogeneous mergers. Modern information technology has made access to transactions, market intelligence, and risk management easier than ever before for entities of all sizes and scopes. It makes the most sense to put gas and electricity under a common roof when markets are costly to use and offer few choices. In common parlance, such markets are "thin." If thinness limits external transactions, a firm that controls both gas and electricity may be able to provide services that cannot be assembled by those who can only rely on markets.

The costs of using gas and electric markets are not only those of a bidweek telephone call or a marketer's fee. If markets are thin, more than one price may prevail for the same commodity at any instant. (There also may be numerous possible contractual provisions for delivery of the commodity, but in an efficient market the commodity itself will trade at the same price regardless of who is buying it or selling it, or for what purpose.) Thin markets are more costly to use than thick ones. Buyers and sellers who trade in a thin market may bear high transaction costs. For example, if a buyer can contact only a small sample of all sellers because communication is costly, the random sample it selects may not include any seller who offers a bargain price. This unfortunate buyer foregoes profits or incurs losses that are avoidable if the market is thick enough to have a uniform price, and it might well prefer to purchase through a reasonably priced specialist who can sample the entire market. A seller in a thin market likewise can sample only a limited number of buyers, and may take a corresponding loss because it does not find a purchaser willing to pay a high price. Thicker markets increase the potential for mutually gainful trades by lowering the cost of bringing buyers and sellers together. Thick markets also may extend to allow better risk management, for example, if a futures market grows up alongside the spot market. If buyers have more perfect markets at their disposal, it becomes easier for them to transact on their own. To succeed, the firm produced by a gas/electric merger must be able to transact more advantageously in these changed markets than either buyers or specialist sellers.

DIVERSIFICATION OR SPECIALIZATION?

What Is Convergence?

The concept of convergence, seemingly a term from the economics of industrial organization, receives virtually no attention in that literature. At minimum, an economic theory of the phenomenon would identify the characteristics of those industries likely to converge or diverge, predict how individual firms will respond to the forces of convergence (for example, when is merger rational?), and examine its market power consequences.[5] A useful definition of convergence should distinguish its consequences for the market from its effects on individual sellers.

At the macro level, convergence is marked by the coming of efficient markets for both gas and electricity, where competitive forces bring uniform prices and quickly eliminate any possibilities for profitable arbitrage. Open access and expanded interconnections have unified formerly disjoint regional gas markets into a single national market. Prices at the wellhead, citygate, and futures markets have become largely arbitrage-proof; that is, the law of one price holds, given the costs of transportation and transacting.[6] Electricity has started down the same path, but end-users are still excluded as buyers in most states, transmission losses limit feasible trading areas, and products and contracts are relatively unstandardized. Open access to power transmission is less developed than open access to gas pipelines. As electrical markets develop, correlations of short-term energy prices among regions are becoming stronger.[7] Links between electricity and gas prices are still diffuse, but there are increasing attempts to arbitrage by trading of gas, power, and financial instruments.[8] The emerging linkage between short-term power and gas prices will develop further as today's thin power markets become thicker.

At the micro level, convergence enhances a firm's ability to manage transactions in the two goods. Firms that handle both goods may be able to construct product packages, financial instruments, and marketing strategies superior to those that specialized firms can offer. If combination gas/electric firms are superior competitors in the converged markets, a merger can quickly give a single-product business dual-industry capability, while it avoids the risk and delay of building that capability on its own. To examine strategy more closely, consider next the broader economic trends within which convergence operates.

The Decline of Scale and Scope

Markets are places where buyers and sellers compare their valuations of goods and then make the exchanges they find most advantageous.[9] Goods and money change hands, but only after buyers and sellers communicate information to one another. Markets will allocate resources more efficiently the lower the cost of arranging transactions and acquiring information about potential trades. When these costs are high, administrative decisions may produce more efficient outcomes than the market.[10] For example, if transacting is costly, centralized management of an entire chain of production may be more efficient than a system in which each link has an independent owner that buys and sells goods-in-process from owners of other links. For yesterday's electric utility, ownership of both generation and transmission was efficient because markets for generated energy were thin, operating technologies did not allow extensive and timely bidding for resources, and the reliability cost of power from an unfamiliar source might be high. Unitary management yielded coordination economies that probably exceeded the benefits of using the small and constrained markets that existed in the past. As nonutility generation has become competitive and purchased power has ceased to endanger reliability, utilities are deintegrating (sometimes at the request of regulators) and markets are replacing some internal decisions. Changes in the industry's environment have rendered decentralized markets superior to centralized management.

If markets are thin, a firm that manages both electric and gas resources might be able to offer deals that a single-product seller cannot. Assume, for example, that users view two goods as substitutes. Market sourcing of the first is costly because the minimum feasible market purchases far exceed the typical buyer's needs, while the second is in unreliable supply, perhaps because of weather risks. A seller who maintains substantial inventories of the unreliable good alongside bulk-breaking facilities for the other may be able to sell small customers more economical packages than they can arrange for themselves. For example, a package might include provisions that are (in effect) call options or insurance policies that pay off in the bulky good if the risky good becomes unavailable. A marketer who handles both products may have competitive strategies at its disposal that are unavailable to a marketer who handles only one of them. As the trading infrastructure grows, that marketer can be either a stand-alone entity or an affiliate of a vertically integrated utility.

Now assume that markets have become thicker. New technologies allow purchase of the first goods in small lots, while the opening of import markets renders supplies of the second more reliable. Now any transactions the two-product marketer could do in the past can be done as efficiently by two single-product sellers, or possibly by buyers themselves. Note that the two-product seller need not incur a loss if it can sell its facilities for handling one of the products. The two-product seller may not be as competent in either good as a seller who specializes in it. As markets thicken, the scope for profitable sale of ancillary services by a goods' original provider may also shrink. For example, if futures markets for a good do not exist, a large seller that can self-insure its inventory risk will have an advantage in offering price stability to buyers. If futures markets are available, a buyer can individualize its risk-bearing rather than taking the consequences of whatever method the seller chooses. The manufacturer of a newly invented good might offer financing because conventional lenders cannot easily assess its quality as collateral. As the good becomes familiar, third parties might better take on the finance function because unlike the manufacturer they specialize in evaluating applications for credit and tailoring loan terms. The easier it is to access markets, the smaller the scope for heterogeneous transactions administered within a single firm.

Information is an important cost of any economic activity, and arguably the most important economic development of the past 50 years has been a revolutionary drop in its cost.[11] Telecommunications and computers have rendered all types of information easier to disseminate, receive, and process, 24 hours a day around the planet. Cheaper information reduces the cost of obtaining market intelligence, learning about alternative deals, evaluating the facts one has found, and consummating transactions. As a consequence, the amount of market activity undertaken by the typical business firm has increased markedly. In 1940, the value of commodity and service inputs purchased by American business averaged 20 percent of the selling price of finished goods. Correlated with the greater ease of using markets, by 1990 that figure was 56 percent of the selling price.[12] With the increase in external supply has come a decline in the range of functions that the typical firm performs in-house.[13] This dramatic but little-noticed organizational change is the result of competitive forces that favor firms that best take advantage of cheaper information and in consequence use markets more intensively.

A gas/electric merger swims against these global currents. The product of a gas/electric merger will be a firm that intends to seek its fortune by expanding the scope of in-house activity, created at a time when the costs of using markets for both goods are in drastic decline. Gas/electric mergers also go against global trends in the size of businesses. In every developed country for which we have statistics, the average size of a business firm has fallen. In the U.S., since the mid-1970s, an increasing fraction of every major type of good or service is being produced in smaller business units.[14] The percentage of industry output produced in small firms has increased by more in manufacturing than it has in services and other industries.[15] Even among the largest firms, shrinkage has taken its toll. The median number of employees in a Fortune 500 firm fell from 16,018 in 1973 to 10,136 in 1993, a 37 percent drop.[16] Mergers to achieve a "critical mass" can be rationalized only with evidence on economies of scale in marketing, and that evidence does not yet exist in power or gas.[17]

COMPETING IN CONVERGENT MARKETS

Deregulation and Mergers: The Supply Side

Merger activity has increased during regulatory transitions in other industries. Firms suddenly faced with deregulation are of shapes and sizes that reflect past regulation and the barriers to competition that institution provided. Instead of market forces, politics determined the boundaries of utility territories, the customers they had to serve, and the range of services to be offered. A quick change may be necessary if such a firm is to become a viable competitor. Airline mergers followed deregulation so companies could rationalize the uneconomic route patterns that had been awarded by past politics and protected by limitations on competitive entry. Financial institutions are combining to meet the new imperatives of global finance, new competition in retail markets, and the disappearance of such protections as state anti-branching laws. In 1997, U.S. bank mergers were valued at a record

$95.09 billion, up 100 percent from 1996.[18] Some other major mergers are combinations of commercial banks with financial institutions that range from insurers (Citicorp/Travelers Group) to families of mutual funds (Mellon Bank/Dreyfus). Electric utility mergers are reshaping service territories that were delineated in accordance with past regulatory mandates and constrained by technologies that until recently required self-sufficiency in generation. Since their inception, electric utilities have merged to exploit scale economies. The current flurry of mergers resumes this trend, which inexplicably slowed in the 1970s and 1980s.[19] Mergers were common long before retail competition threatened, and prior to recent policy changes by the Federal Energy Regulatory Commission (FERC), utilities merging with neighboring systems had to show that agency evidence that the merger would produce a lower-cost entity.[20]

Mergers between electric systems are costly to arrange and finance. They carry other costs that may be larger than the direct expense, such as lost time and risk. The typical merger must be approved by FERC, by state commissions, and by the Securities and Exchange Commission if it involves a holding company. Few mergers have gone through the entire process in fewer than two years, and some have been terminated by unfavorable decisions from a single commission. Executives particularly may be affected as the merger consumes time and resources that they might better devote to operational and competitive matters. With market choices growing, the feasible realm of interutility contracts short of merger expands. A merger is an all-or-nothing matter in which sources of post-merger efficiency as well as sources of inefficiency go into a single package. If the sources of efficiency can be united by contract, both organizations may be better off without the merger than with it. Additionally, the post-merger problem of determining which part of which company goes where vanishes, as does the scope of a possible clash between the cultures of the two organizations.

Mergers can be viewed as investments in which a firm attempts to acquire resources (including personnel and market access) by purchasing a firm that already has them rather than developing them in-house.[21] Transition costs apply to all of the acquired firm's resources, both nonhuman and human. The value of an investment that must be made at a fixed date is easily calculated (in principle) as the present value of its expected net revenues, discounted by a factor that reflects the returns on the next-best investment. An opportunity whose financial commitment can be postponed until some uncertainty is resolved (such as future regulatory policy, market conditions, and so forth) probably will be more valuable than one that must be made at a certain date or not at all.[22] This increment, the "option value" of flexibility, may be important in markets changing at the pace of today's markets. (Counter to this is an industry-specific difficulty stemming from the Public Utility Holding Company Act. If an electric utility's electric merger partners are limited to those contiguous to its territory, not moving quickly may leave it with fewer and less desirable choices.)

An electric utility that merges with a gas entity (whether supplier, marketer, pipeline, or distributor) will face most of the same costs as the electric utility that merges with one of its brethren, including both management distraction and loss of option value. Market power issues may threaten a heterogeneous merger less than they do a merger between two electric systems, but the clear economies of operating two power systems as one will not be present in an electric/gas merger.[23] The oft-cited source of gain in these mergers is gas/electric convergence, which is said to offer opportunities for service offerings and customer penetration that cannot be achieved by standalone entities. Oddly, there is no record (or reason to anticipate) that existing combination electric/ gas utilities are better positioned to do what the products of recent and proposed mergers expect to do.[24]

According to the logic of convergence, an electric/gas merger offers the merged entity more than just an additional product to sell. It offers both the chance to repackage its products more attractively and the chance to reach new customers in new ways. Repackaging will take place using what one executive has called the "common currency" of BTUs.[25] Transactions will include both variations on the spark spread and certain physical transfers of power and gas. As before, however, the coming of open access to all segments of both industry's networks raises the question of why the necessary facilities must be controlled by the same company. Implementation of open access will be imperfect and time-consuming, but the institution will affect ever-more competitive commodity markets for both power and gas. Electric systems across the nation will probably surrender operational control of their transmission to ISOs, depriving them of the chance to exercise the monopoly power of ownership.[26] Open access has largely nullified yesterday's market advantages of gas pipelines, and in an increasing number of states it is now eroding those of gas distributors. Open access means that any user of a facility can duplicate any transaction that the facility's owner can make. The coming of improved electronic information systems in both industries (and restrictions on information transfer between transmission or pipeline owners and their marketing affiliates) further makes it unlikely that information generated in the converged firm will give it a lasting competitive advantage.

The Demand Side

A commonly cited counterpart to BTU supply integration in converged utilities is the potentially enhanced ability to bundle gas and electric services and sell them in personalized packages, creating a metaphorical "supermarket" that offers "one-stop shopping."[27] This principle only creates competitive advantage if consumers value the bundle at a premium over the cost of assembling it, and if the merged firm can achieve a cost advantage or construct barriers to the entry of competing assemblers. This seems to be the point of sellers who propose services beyond conventional electricity and gas, such as home security and telecom. There is no obvious reason why an electric utility, or a merged gas/electric operation, should have any advantage in selling a package that goes even further beyond the bounds of utility experience.

As markets grow, those competing assemblers include the user itself. Even if the converged seller can offer a wider range of personalized packages (green resources, sophisticated metering, and demand-side management, for example) sellers of generic packages probably can still compete, and all will sell to both residential and nonresidential customers. If markets, however, have become cheaper for assemblers of packages to use, they are also becoming cheaper for end-users who wish to do the job themselves. A convergence merger may be of particular value to those who prefer having the packaging done for them by their supplier. The elements of packaging go beyond physical commodities, and include the value of a brand name's reputation and the quality of individualized service that some will demand.

If shopping entails a fixed cost for buyers, such as driving to the mall regardless of the size of purchase, putting two commodities under one roof may make their seller more attractive. That seller can profit by capturing some of the buyers' gains from convenience, but its powers are limited by potential competitors who intend to let buyers keep more of their savings.[28] Likewise, if economies in marketing, such as from volume warehousing, arise when a single seller handles two goods, it can keep those gains only by deterring the entry of imitative competitors.[29] When a supermarket stocks everything from soup to nuts, there are economies in marketing and lower fixed shopping costs for buyers. Competition nevertheless makes them a low-margin business, with a range of competitors that includes other supermarkets, specialty butchers and bakers, restaurants, and diet plans.

Another variant on the supermarket theme views convergence mergers as ways of buying customers for strategic advantage. Because competition will give alternatives to all customers, however, customers will only stay with the company if they are offered competitive prices and packages. The benefits of the August 1997 Houston Industries/NorAm Energy merger, for example, were said to include access by HI to the 2.1 million NorAm retail gas customers outside of HI's electrical service territory. A Houston spokesperson referred to them as "good marketing targets," for "non-gas energy products and services," but with open access everyone will have roughly the same access to those customers as the merged company.[30] With competition, profitable customers become either marginal customers or lost ones.

Pressure by large users to shop for their own power supplies has driven much of electrical deregulation, and they will become mobile and market-wise. Utilities that mass market will be seeking those customers whose load characteristics make them the most unattractive and whose bills can no longer be cross-subsidized by others, whether those others are consumers of electricity or gas. Services for small customers are probably those with the highest transaction costs as well. Converged or not, a utility that branches into competitive lines of business can mark up the price of a bundle of services only to the extent that it adds to their convenience value.

The Strategic Alliance Alternative

Deregulatory changes in electricity and gas are taking place against a world-wide backdrop of innovations in information technology and finance. As the latter developments give rise to new markets and reshape old ones, they offer existing utilities new arenas for experimentation. Utilities will have little choice but to experiment, because their nonutility competitors certainly will be doing so. Since all experiments are risky, a utility might best diversify over multiple activities, limiting its exposure as the situation warrants. If a utility sees its future in non-traditional activities, only in rare cases (perhaps mass marketing to low-margin customers) will it want to acquire the full bundle of resources of another utility, some of which will be unneeded or redundant. (Divestiture is also a costly and risky process.) Often it will be better to grow the resources in-house or to construct a strategic alliance or joint venture with an entity that owns them. Alliances (and to a lesser extent joint ventures that entail formation of a new enterprise) are usually based on less complex contracts than are mergers. Alliances usually will entail simpler financial arrangements than will mergers and not carry as heavy a burden of regulatory approvals. Regulatory rejection of a merger (or acceptance with stringent conditioning) will leave the utility with substantial bills and lost time that management was unable to utilize in other competitive endeavors. Unlike a merger agreement, a contract that enables an alliance almost surely will include explicit termination and reopener provisions that allow the parties to end unproductive experiments at predictable costs.[31]

In today's power industry, alliances short of a merger offer an immense range of possible partners, types of commitment, and degrees of commitment. The geographic and legal barriers to effective utility mergers need not be obstacles to successful alliances in nonutility activities, and can extend a utility's local sphere of influence in multiple directions. As one of numerous examples, Columbia Electric Corp., an arm of a gas pipeline company based near Washington D.C., is forming a joint venture with Westcoast Energy, of Vancouver, British Columbia, to develop merchant power plants in the northeast, whose output will be contracted for in part by utilities there. Columbia also owns plants in conjunction with other eastern utilities, and Westcoast has projects in Mexico and Brazil. Engage Energy, the product of a joint venture between Westcoast and Coastal Corp. of Houston, will enter a strategic alliance with HEC, a subsidiary of Northeast Utilities, to provide customized energy services for customers throughout North America.[32]

Unlike mergers, where two parties are usually the limit, the number in an alliance and the distribution of responsibility among them is limited only by the ingenuity of contract writers. Several unregulated business units of a single company can cooperate with outsiders to provide specialized services. All four of PG&E's unregulated business units, for example, will ally themselves with Ultramar Diamond Shamrock to manage that company's $2 billion annual power and gas purchases, build a cogeneration plant, and provide energy efficiency services.[33]

Public power, too, has begun to see the value of alliances. Santa Clara, California's municipal utility has formed alliances with Engage Energy and with Illinova Energy Partners, an unregulated unit of an Illinois utility holding company, for business planning, risk management, and wholesale power acquisition.[34] As electricity becomes "commodified," it also will become a financial industry, with specialized needs for intermediation and risk-bearing competency that may best be met by alliances with banks and other financial institutions. One example is Baltimore Gas & Electric Co., which has a joint venture with investment bankers Goldman, Sachs, and Co. to acquire plants being divested by utilities, with Baltimore Gas & Electric's Constellation Power Source unit marketing their power.[35]

The range of activities that can be undertaken plausibly by strategic allies or joint venturers is wider than the range that utility mergers can facilitate. Conventional plans to market gas and electricity in conjunction remain feasible, as do sharing the risks of plant construction. PacifiCorp and Washington Water Power, neighboring electric systems in the competitive northwest market, are forming separate alliances with neighboring gas distributors Northwest Natural Gas and Cascade Natural Gas.[36] Alliances between electricity and gas producers may help alleviate transmission bottlenecks by coordinating gas flows with power production.[37] Other alliances are experimenting with new products on the periphery of traditional utility markets, such as the sale of micro-cogenerators and waste heat from power plants.[38] As states begin to allow competitive metering and billing, alliances are moving into that area as well.[39] Still more original packages include those of EnergyPact, a joint venture of PacifiCorp and ABB Inc. to offer "plant enhancement services," including repowering, fuel procurement, and risk management to owners of power plants who do not want to become merger partners. Another unit of PacifiCorp is available to market the plants' energy production.[40] In another twist, Western Resources is attempting to sharpen its focus on competitive electricity by transferring its gas operations to ONEOK, a neighboring distributor. Western will acquire a noncontrolling equity stake in ONEOK while retaining ownership of its gas assets.[41]

Alliances also probably are superior to mergers for coping with differences between utilities whose cultures are still those of a regulated industry and those with more competitive attitudes (which includes some utilities).[42] A merger between a progressive and a more traditional utility provides no guarantee that culture will be unidirectionally transmitted from the former to the latter. In a strategic alliance, there likely will be fewer clashes between personnel from different lines of business than between those with the same jobs in two utilities. The very smallness of alliance activity relative to utility size may increase the likelihood that employees in it can be insulated from the less competitive climate that may prevail elsewhere in the organization.[43]

The principles behind a successful alliance are like those of any other successful competitor. Regardless of legalisms, the partners, in effect, are acquiring specialized resources from each other to create an entity that is more competitive than either can create otherwise. Recent research has found that alliances on average increase the stock prices of the partners, whether they are in the same or different industries. Value increases are greater when a same-industry alliance pools technical knowledge than when it is formed for marketing purposes, possibly because the former better maintains the focus of the partners. Interestingly, firms that enter into alliances tend to be superior performers relative to their industries before they join forces, while those entering joint ventures are not.[44]

Alliances are by nature less dependent on long-term investment commitments that might otherwise warrant the formation of a stand-alone corporate entity. This being so, it is unlikely that the allies can keep competitors at bay by scale alone. Even intangible capital such as brand names may be of less importance, given the more specialized services that the typical alliance will offer. Because alliances seldom will be commodity businesses, they more likely will create markets or innovate in existing ones than take current market conditions as facts of life.[45] Novelty does not necessarily imply that an alliance be made with all possible speed to gain a "first-mover" advantage. In economic theory, the first move matters if it restricts the choices of later movers in ways that redound to the advantage of the first mover. In the new markets for electricity services, however, a first mover probably will be unable to guess the future accurately enough to make meaningful strategic plans directed at competitors who do not yet exist, and later movers may learn from the mistakes of the earlier ones. In 1995, UtiliCorp (joined in 1997 by PECO Energy) founded EnergyOne, a first attempt to institute one-stop retail shopping, achieve economies of scope in marketing, and introduce a nationwide franchised brand. EnergyOne ceased its efforts in mid-1998, citing the slow development of markets as the cause of its inability to attract franchisees.[46] Its attempt to obtain a first-mover advantage in retrospect produced little more than valuable information for those who will come later.

CONCLUSIONS

Utility mergers invariably promise reductions in operating costs, but the record of recent electric combinations is mixed. In some cases, not enough time may have passed to accomplish the full savings, and in others market changes unrelated to the merger may have affected them. Whatever the likelihood of such savings in a gas/electric merger, there are also possible diseconomies. There may be loss of focus and dilution of managerial control because important parts of the new entity are now beyond either of the original firms' core competencies. Many of the first power marketers began in gas, for example, but in the early days none of them touted economies that stemmed from joint marketing or coordination of gas and electric operations. Businesses that are units of diversified firms have lower market values than otherwise similar stand-alone operations, and the more diversified any firm is, the lower is its overall return.[47] Electric utility diversifications of the mid-1980s hardly lived up to their advance billing, one sample producing an average return of zero.[48] Those diversifications, like the current crop of gas/electric mergers, sometimes were announced as attempts to capture synergies. The fact that both gas and power are energy sources does not imply that an entity will be more competitive if it invests in facilities to produce or market both of them.

Real-world competition is a search for the high returns that reward innovation. Competition is worthwhile only if it promises supernormal profits for successful risky ventures. In open markets, competition erodes these returns and gives rise to further attempts at innovation and imitation. To sustain high returns, an innovator must be able to suppress competition. In important ways, the firms that are born in electric/gas mergers will be easy to compete against. In contrast to monopolies that result from patents or trademarks, electricity transmission, gas pipelines, and most gas distributors are open access facilities whose owners must allow competitors to use them on nondiscriminatory terms. Gas/electric mergers are not clear ways to win in competitive markets, and they may not even be good ways to monopolize.

If the competition that will matter is for an innovative, short-term "monopoly," an alliance or joint venture makes more economic sense. It is easier to form an alliance by contract that uses only those resources that are needed for a well-defined activity than to consummate a costly and risky merger that leaves the successor company with a heterogeneous mass of resources, some valuable for competition and some not. Convergence mergers may embody no more than the hope that bigness will protect utilities as it has in the past.[49] With open access at wholesale in both commodities and retail direct access on the way, expanding into a new line of business is unlikely to protect a utility that cannot compete. A firm that operates in two competitive markets has no clear advantages over a firm that only needs to operate in one. Having experienced the poor performance of past conglomeration mergers and acquisition, it is hard to see the point of convergence energy mergers that move a company away from its core competence and expose it to risks in another, unfamiliar, market. At the announcement of its merger with Travelers Group, Citicorp Chief Executive Officer John Reed was questioned about the frequent failure of convergence mergers in finance to meet expectations. He responded that one "could argue that the fact that companies keep trying to do it indicates there's something there."[50] Will history show that convergence mergers in energy had a similarly shaky basis in reality?

NOTES

1. Liane Kucher, "Igniting Interest in the Spark Spread," *Gas Daily's* NG 4 (August 1996), 14.

2. Enron Corporation and Portland General Corporation, Dockets No. EC-96-36-000 *et al*, 78 FERC ¶ 61,179 (1997); NorAm Energy Services and Houston Industries, Docket No. EL97-25-000, 79 FERC ¶ 61,108 (1997); Duke Power Company and PanEnergy Corporation, Docket No. EC97-13-000, 79 FERC ¶ 61,236 (1997); Enova Corporation and Pacific Enterprises, Docket No. EL97-15-000, 79 FERC ¶ 61,107 (1997); PG&E Corporation and Valero Energy Corporation, Docket No. EC97-22-000, 80 FERC ¶ 61,041 (1997); Long Island Lighting Company and Brooklyn Union Gas Company, Docket No. EC-97-19-000, 80 FERC ¶ 61,035 (1997). The possible convergence between electric utilities and telecommunications is not discussed in this chapter.

3. See, "The New Face of Competition: Enron Buys an Electric Utility," *Energy Daily*, 23 July 1996.

4. Michael Jensen, "The Modern Industrial Revolution, Exit, and the Failure of Internal Control Systems," *Journal of Finance*, July 1993.

5. One author uses the language of convergence while investigating corporate coherence, which he defines as the nonrandom distribution of activities in multiproduct firms:

    Technological paths are shaped not only by changes in a firm's history, but also by changes in the public knowledge base. These changes, often driven by developments in basic science, may be such that the technological foundations of businesses shift dramatically. This may cause the technological knowledge base of industries to converge or diverge. Thus digital electronics is presently causing a convergence in the public knowledge base supporting computer and telecommunications firms. This has implications for corporate coherence because products which may be aimed at rather different customer groups might suddenly develop common technological and production roots.

    David J. Teece et al, "Understanding Corporate Coherence," *Journal of Economic Behavior and Organization* 24 (1994), 21.

6. Robert J. Michaels and Arthur S. De Vany, "Market-Based Rates for Interstate Gas Pipelines: The Relevant Market and the Real Market," *Energy Law Journal* 16 (No. 2, 1995), 299–346.

7. Samuel A. Van Vactor, "Establishing an Electric Energy Trading Market in the West." Presented at Executive Enterprises Western Electric Market Conference. San Francisco. October 17, 1995; Robert McCullough, "Trading on the Index: Spot Markets and Price Spreads in the Western Interconnection," *Public Utilities Fortnightly*, 1 October 1996, 32–38.

8. Michael Hsu, "Spark Spread Options Are Hot!" *The Electricity Journal* 11 (March 1998), 28–39.

9. Armen A. Alchian and William R. Allen, *University Economics*, 3rd ed. (Belmont, CA: Wadsworth Publishing Co., 1972), 56.

10. For more on these issues, see Oliver Williamson, *Markets and Hierarchies* (New York: The Free Press, 1974).

11. Charles Jonscher, "An Economic Study of the Information Technology Revolution," in Thomas J. Allen and Michael Scott Morton, eds., *Information Technology and the Corporation of the 1990s: Research Studies* (Oxford University Press, 1994), 5–41.

12. John McMillan, "Reorganizing Vertical Supply Relationships," in Horst Siebert, ed., *Trends in Business Organization* (Mohr, 1995), 203.

13. Information technology potentially allows the firm to coordinate a wider range of activities internally. The data show, however, that these effects are overwhelmed by the benefits of

turning to the market. See Erik Brynjolfsson et al., "Does Information Technology Lead to Smaller Firms?" *Management Science* 40 (December 1994), 1628–1644.

14. Zoltan Acs and Daniel Gerlowski, *Managerial Economics and Organization* (New Jersey: Prentice-Hall Business Publishing, 1996), 363–366.

15. "The Rise and Rise of America's Small Firms," *The Economist*, 21 January 1989, 73–74.

16. McMillan, "Reorganizing Vertical Supply Relationships," 203.

17. Such economies are not necessarily related to those in transmission or distribution. One reason for uncertainty is that these economies probably depend on the mix of mass-marketed products and customized ones that comes to prevail under competition. Note that "critical mass" originally referred to the amount of uranium one must accumulate before it self-destructs and takes the neighborhood with it.

18. Ivan Cintron, "Mergers, High Volume Boosted Bank's Muni Holdings in 1997," *The Bond Buyer*, 20 May 1998, 7. Banks accounted for approximately 10 percent of merger volume in that year.

19. Richard J. Gilbert and Edward P. Kahn, "Competition and Institutional Change in U.S. Electric Power Regulation," in Richard J. Gilbert and Edward P. Kahn, eds., *International Comparisons of Electricity Regulation* (Cambridge University Press, 1996), 195.

20. "Inquiry Concerning the Commission's Merger Policy Under the Federal Power Act: Policy Statement," Order No. 592, III FERC Stats. & Regs. ¶ 31,044 (1996). The Public Utility Holding Company Act further restricts utilities to mergers with neighbors whose systems can be electrically integrated with their own. Proposals to repeal the act have surfaced in every recent Congress, and as competition grows in electricity the pressure for repeal grows with it.

21. George Bittlingmayer, "Merger as a Form of Investment," *Kyklos* 49 (No. 2, 1996), 127–153.

22. Avinash K. Dixit and Robert S. Pindyck, *Investment Under Uncertainty* (Princeton, NJ: Princeton University Press, 1994).

23. Compare statements by Michael Morris, president of Northeast Utilities ("Gas service workers are not electric line persons, never will be, never have been") and Larry Brummett, chairman and chief executive officer of gas distributor ONEOK ("Few combination utilities have been able to achieve excellence in both electric and gas operations"). "Convergence Is Here, More's Coming, But Is It the Right Path? Some Ask," *Gas Utility Report*, 10 October 1997, 1.

24. Some of today's mergers, however, are combinations of qualitatively different entities. They include, for example, diversified electric utilities and pipeline systems [Duke/PanEnergy] or diversified energy firms [Enron/Portland General], and others are expansions into non-overlapping territories [Enova/Pacific Enterprises]. PG&E is reconsidering its recent acquisition of Valero Energy Co., a Texas pipeline that is not performing as expected financially or strategically. "PG&E Reviews Texas Strategy, May Shed Assets," *Wall Street Journal*, 26 May 1998, A4.

25. "Duke Energy Sees BTU as 'Common Currency' in Converged Marketplace," *Inside FERC*, 24 November 1997, 1. The company's president then cited an assortment of hypothetical transactions, none of which obviously requires ownership of the pipes or wires needed for it to take place.

26. See Paul Kemezis, "Why Enron Paid a Premium for Portland General," *Electrical World* 210 (September 1996), 57: "Enron will soon acquire PGE's 25% ownership of the line capacity into the California-Oregon Border (COB) interconnect, giving it more control of that strategic transmission hub than any other non-federal utility . . . Because the grid is so near the point where electricity futures are priced and exchanged for physicals . . . Enron's credibility for backing paper trades with real kilowatt-hours will be enhanced." Under open access, Enron's ownership of the lines will not allow it to act monopolistically, and if it wants to use those lines to back up its credibility with physical deliveries, so can anyone else with credibility concerns. In any case, the lines in question may soon be under the control of INDeGO, the Northwest's on-again off-again independent system operator.

27. "As in the retail merchandise field, the customer will ascend to the throne in the energy industry with convenience centers, providing one-stop shopping for multiple utility-like services, responding to the new consumer power." Howard A. Christensen, "Looking Ahead to a New Era," *Hart's Energy Markets* 3 (February 1998), 14.

28. One consultant refers to this expectation in the proposed merger between Citicorp and Travelers Group as "product polygamy." See, "For Consumers, Going to Citicorp Has an Appeal, but Some Are Wary," *Wall Street Journal*, 7 April 1998, C17.

29. Despite frequent assertions that such economies exist, evidence is sparse. There is no econometric consensus on even the algebraic sign of cost differences between straight utilities and combination gas-electric systems.

30. See, "HI Touts Gain in Market Skills, Customer Access as Benefits of NorAm Acquisition," *Electric Utility Week*, 19 August 1996, 5.

31. "Eastern Enterprises divorces NEES; Drops Out of Joint Marketing Venture," *Electric Utility Week*, 8 December 1997, 1. Eastern found that the initial losses it was taking in the venture, along with future necessary investments, were depriving it of other competitive options.

32. See, "Columbia Unit, Westcoast form Link for 3 Northeast Projects, 1,000 MW," *Global Power Report*, 6 February 1998, 1; "Engage Energy and HEC Form Strategic Alliance," *Foster Natural Gas Report*, 4 December 1997, 27.

33. See, "PG&E Alliance with Ultramar Seen Saving the Oil Company $355-Million," *Power Markets Week*, 16 March 1998, 6.

34. See, "Santa Clara Selects Engage, Illinova; New Muni Name is Silicon Valley Power," *Power Markets Week*, 22 December 1997, 5.

35. "BGE, Goldman Sachs to Buy Capacity, Market it through Constellation Power," *Power Markets Week*, 16 March 1998, 4.

36. See, "PacifiCorp, NW Natural Gas Set Strategic Marketing Pact," *Electricity Daily*, 21 July 1997; "Cascade Natural Gas and Washington Water Power form Strategic Alliance," *Foster's Natural Gas Report*, 19 March 1998, 31. For a statement by PacifiCorp and NWNG on the benefits of their alliance, see "PacifiCorp, NWNG Seek Merger's Gains Without Regulatory Hassle," *Restructuring Today*, 18 July 1997, 2.

37. Pauline Whitcomb, "Fishing in the BTU River," *Oil & Gas Investor*, May 1997, 22–35.

38. "Comed Parent, Allied Signal Link to Produce, Market Small Plants," *Electric Utility Week*, 9 June 1997, 15; "Trigen Looks to Send Pepco Plant Steam to D.C. for Heating/Cooling," *Energy Services and Telecom Report*, 4 December 1997, 9.

39. "Strategic Alliances Link Enron, ABB, TransData, Motorola, Mtel, and ARDIS to Offer a Revolutionary and affordable Wireless Electricity Metering System," *PR Newswire*, 15 December 1997.

40. "PacifiCorp / ABB Venture to Improve Clients' Plants, Market their Power," *Power Markets Week*, 12 May 1997, 4.

41. "Western Resources Closes Alliance Deals with ONEOK and Protection One," *Electric Utility Week*, 8 December 1997, 11.

42. The stereotypes remain. "Like it or not, the utilities are going to have to visit venues much less familiar to them than regulators' offices and PGA golf courses," Whitcomb, "Fishing in the BTU River," 24.

43. Executives hired by utilities from other industries are subject to high attrition, which some executives attribute to cultural differences. See "Utility Newcomers Hired From Other Industries Defect at High Rate," *Wall Street Journal*, 2 April 1998, A1.

44. Su Han Chan et al., "Do Strategic Alliances Create Value?" *Journal of Financial Economics* 24 (1997), 199–221.

45. Compare "Creation of Strategic Organizations Called Key to Marketing in New Era," *Gas Utility Report*, 15 April 1994, 10.

46. "Hard Times for One-Stop Franchises: EnergyOne Gives Up, Enable Struggling," *Power Markets Week*, 4 May 1998, 1.

47. Sanjai Bhagat, et al., "Hostile Takeovers in the 1980s: The Return to Corporate Specialization," *Brookings Papers on Economic Activity Microeconomics 1990*, 1-84; Amar Bhide, "Reversing Corporate Diversification," *Journal of Applied Corporate Finance* 5 (Summer 1990), 415–439.

48. Charles Studness, "Earnings from Utility Diversification Ventures," *Public Utilities Fortnightly*, 1 June 1992, 28–30.

49. "[Merger] is the favored response of Wall Street bankers and lawyers to structural change in any industry for any reason." Vinod K. Dar, "The Future of the U.S. Electric Utility Industry," *The Electricity Journal* 8 (July 1995), 25.

50. "One-Stop Shopping Is the Reason for Deal," *Wall Street Journal*, 7 April 1998, C16.

CHAPTER 8

Lights Out for Regulated Utility Monopolies

Michael R. Peevey

Summer has just begun, and already industrial production facilities in the U.S. are shutting down for lack of electricity. As it gets hotter and demand for electricity grows, blackouts are expected to spread through much of the Midwest and Eastern U.S.

Utilities respond by urging customers to turn off air conditioning, by cutting off power to industrial users with interruptible supply contracts, and by sweating the possibility of power plant breakdowns. As [*The Wall Street Journal*] reported June 26, [1998] some utility employees are walking up office stairways because their employer has resorted to the absurd symbolic gesture of shutting down elevators to reduce energy use.

There's no reason for the U.S. economy to tolerate such costly and inconvenient power shortages. A simple cure is available—deregulate the electric industry. By providing price signals to both suppliers and users of electricity, competitive markets would eliminate power shortages almost overnight.

Utility deregulation has been proceeding state-by-state, and the results of early competitive initiatives are clear. California—the first state to fully open its electricity markets to competition—expects no power supply problems this summer, and its future will be brightly lit. More than 7,000 megawatts of new electric generation capacity is currently working its way through California's environmental regulatory process that oversees power plant siting. That's enough additional generating capacity to meet the needs of the

This article originally appeared in *The Wall Street Journal* on July 14, 1998. It is reprinted here with their permission.

city of Los Angeles. Take note: it's not utilities (or more precisely, their captive customers) putting up the money to build these power plants, but rather risk-taking independent power producers who expect to realize a good return on their investments by selling electricity into a competitive marketplace.

Similarly, in New England, where deregulation is moving relatively swiftly as a cure for the region's chronic high energy costs and power shortages, prospects of competitive opportunity have stimulated the launching of an additional 5,000 megawatts of new power plant projects—all nonutility projects.

At the national level, both the Administration and several Members of Congress have proposed legislation that would encourage all states to open their markets to competition by 2002. Realistically, however, this issue hasn't been high enough on Congress' list of priorities to qualify as "likely to become law" in the current session of Congress. Perhaps the current power emergency will help lawmakers to realize that there's more at stake than the price of energy.

Historically, planning for system generation capacity, design, construction, and operation of power plants has been the province of vertically integrated monopoly electric utilities. Three-quarters of the nation's electricity today is distributed by investor-owned utilities; the remaining one-fourth comes from utilities owned by government entities ranging from small municipalities to the federal government. In either case, the utilities' power production and procurement practices and their retail pricing systems function almost wholly without regard of market forces.

Under the old monopoly system, neither electricity providers nor users have effective means to respond to the system's capacity shortages. Historically, utility planners created complex 10-year demand projections and made their case to state regulators for the need to build new power plants. The regulatory arena became a debating club for advancing different theoretical models about how much new generation capacity was necessary. Meanwhile, retail energy prices were set based on the cost of providing the service established by the regulator's generation capacity decisions, rather than through a balancing of actual supply and demand. This approach more closely resembles the old Soviet Union five-year central plan system than anything in the U.S. economy. Results of such centralized planning predictably include both higher costs and lower reliability than can be achieved through competitive markets.

The existing (and rigid) regulated rate structure provides little means to send price signals; energy consumers therefore have little incentive to cut back on use when generation resources are taxed. Exhortation and symbolic elevator shutdowns are a poor substitute for price signals. In a fully competitive marketplace, consumers—especially business users, who use about two-thirds of the nation's electricity—will get very clear messages from the marketplace when energy supplies become tight.

This is possible because non-utility energy service providers can provide "smart meters" for each customer that record and report energy use on a real-time basis. Energy service providers can combine detailed energy price information with "real-time pricing" that reflects the balance of energy demand with available power plants and their production costs. Armed with both real-time information and real-time pricing, businesses will respond by installing energy-efficient equipment and lighting, adjusting the hours of operation of their facilities, or investing in their own small-scale electric generating equipment. You can be sure that they will respond to minimize their energy costs and optimize (for their needs) the trade-offs between reliability and cost.

Utilities and utility planners historically have been wed to large central-station power plants, whether fueled by coal, oil, natural gas, or uranium. When even a few large power plants break down—as we currently see—large amounts of capacity disappear from the system. New technologies and competition can combine to eliminate such problems. New plants being built by independent power producers rely on gas turbine generators closely akin to aircraft jet engines that can economically add capacity in much smaller increments than old steam plant technology.

Competition will usher in many such smaller power plants. Supply risks will be lower, because generation resources will be more dispersed; if one plant is forced out of service it will have less impact on the system. Studies show that large power plants (over 600 megawatts capacity) are five times more likely to suffer forced outages than small plants (roughly 100 megawatts). Moreover, nonutility power plants have a huge economic incentive to keep running during peak demand periods. Put simply, if they don't make electricity, they don't get paid. It's thus not surprising that nonutility power plants in the U.S. are available to produce energy more than 95 percent of the time, in contrast with just 80 percent for utility-owned baseload plants (those that are intended to run all the time), and only 72 percent for utility-owned nuclear power plants.

Non-utility energy service competitors also are introducing new generation technologies to the marketplace that provide even more encouraging news. These "distributed generation" technologies (microturbines, fuel cells, and even solar cells) can create electricity more cheaply and reliably than the existing combination of large central power plants and the utility grid system. Within a few years, distributed generation technologies could represent a significant portion of the nation's incremental (newly installed) electric generating capacity. Distributed generation users will bask in air conditioned comfort, keep their factories and stores open, and pay less for their electricity than other, less far-sighted enterprises. However, *all* utility customers will benefit from distributed generation, because it will reduce the load and stress on the utility distribution system and large central-station power plants.

Without competition, businesses in power-scarce states may not see the benefits of these new technologies, because regulated monopoly utilities have no economic impetus to introduce distributed generation to their customers. Indeed, utilities have strong incentives to fight this form of progress, as we are now witnessing in California. But the forces of competition and change are irresistible, as Mikhail Gorbachev learned a decade ago and utilities are learning now.

As California already is demonstrating, a competitive market will stimulate construction of new power plants, bringing both lower energy costs and greater system reliability. Competition also has already stimulated the development of new information systems that will allow customers to respond effectively to the varying prices of a truly competitive commodity market for electricity.

The message is clear for businesses and citizens in states where power is costly or in short supply. Tell your state legislature to move—fast!—to get rid of the old regulated monopoly utility system.

SECTION III

WHAT OPPORTUNITIES ARE CREATED BY CUSTOMER CHOICE?

CHAPTER 9

Electric Restructuring and Consumer Interests: Lessons from Other Industries

Robert Crandall
Jerry Ellig

As major industry players rush to accommodate electricity restructuring, some also rush forward with speculations about how regulatory reform will affect consumers. Advocates of restructuring promise lower rates and better service; critics warn that retail competition will be a boon only for large, sophisticated customers.

Fortunately, the United States has a decade or more of experience with regulatory reform in a variety of industries that share some similarities with electricity. Natural gas, telecommunications, airlines, railroads, and trucking are all industries in which competing producers use a network of wires, pipe, roads, or rails to reach their customers. Economists have analyzed the effects of deregulation and restructuring in these industries in great detail, aided by a wealth of data collected by regulators and trade associations.

An overwhelming consensus emerges from these scholarly studies: real-world competition, though not necessarily perfect, is far better for consumers than economic regulation. Our reading of the record suggests that the results in electricity should be no different.

This article originally appeared in the January/February 1998 issue of *The Electricity Journal* (Elsevier Science). It is reprinted here with their permission.

EXAMINING ANALOGOUS INDUSTRIES

From an economist's perspective, these five network industries share a number of similarities with electricity.

Like the electric industry, all five industries have a production, transmission, and distribution stage. Gas wells, telephone equipment, trains, airplanes, and trucks are the production stage, analogous to power plants. Interstate gas pipelines, long-distance phone lines, railroad trunk lines, airways and air traffic control, and interstate highways offer long-distance transportation similar to high-voltage transmission lines. And local gas pipes, local telephone lines, rail sidings, airports, and local streets are economically similar to electric distribution lines.

Because these industries have several stages, discussions of "deregulation," "regulatory reform," "customer choice," "restructuring," and similar topics can get quite confusing. In economic terminology, "deregulation" means the partial or complete elimination of governmental restrictions on prices and entry. The other terms may describe deregulation of some parts of the industry, but they also may characterize changes in regulation of one stage of the industry thought necessary to facilitate competition in other parts. Thus, "customer choice" means the elimination of price and entry regulation in electricity generation, coupled with "open access" regulation that would turn the electric wires into highways that would allow customers to deal directly with producers of the electric power.

The production side of the other network industries was deregulated in the late 1970s and early 1980s. Most natural gas wellhead prices were deregulated between 1978 and 1984. Competition in telephone equipment came in the late 1970s. Airline route and rate regulations were phased out beginning in 1978, and surface freight companies were partially or fully deregulated in 1980.

Large segments of these industries are also subject to some form of open access regulation. Interstate natural gas pipelines became open access transporters during the late 1980s. When AT&T was broken up in 1984, local phone companies were required to let competing long-distance companies use their lines to reach customers. The federal government has long had authority to impose open access on a railroad in specific cases. Airlines and trucking companies, meanwhile, both use publicly owned infrastructure that is generally open to all competitors.

None of these industries is identical to the electric industry—but neither are they identical to one another. If common patterns emerge in all of these industries as a result of regulatory reform, we can be reasonably sure that similar results will occur in electricity.

SIGNIFICANT CUSTOMER BENEFITS

In all five industries, regulatory reform produced significant customer benefits that grew over time. Consider prices. As Table 9–1 shows, inflation-adjusted prices fell within two years after regulatory reform—often by 10 percent or more.[1] Within 10 years, prices were at least 25 percent lower, and sometimes 50 percent lower.

Regulatory reform did not cause all of these price reductions, but it is worth noting that most predictions of price reductions from electric restructuring fall comfortably within this range. In an August 1997 report, the U.S. Energy Information Administration estimated that in the next two to three years, competition would lower the average retail price of electricity by between 6 percent and 22 percent; by 2010, the price would be roughly 11 percent to 28 percent lower than it is today. The now-famous 1996 study by Citizens for a Sound Economy Foundation, meanwhile, projected that retail competition would reduce the price of electricity by 13 percent in the short run and 42 percent in the long run.[2]

Statistical studies of transportation industries that control for other factors affecting prices consistently have shown that regulatory reform produced more than $50 billion annually in price reductions and other consumer benefits.[3] A significant portion of these benefits came in the form of improved quality of service. In the airline industry, for example, one study found that increased flight frequency accounted for more than half the value of consumer benefits.

TABLE 9–1	**Summary of Trends Following Regulatory Change**			
	% Real price reduction after . . .			**Annual value of consumer benefits from deregulation**
Industry	**2 years**	**5 years**	**10 years**	
Gas	10–38% (1984–86)	23–45% (1984–89)	27–57% (1984–94)	N.A.*
Long-Distance Telecom	5–16% (1984–86)	23–41% (1984–89)	40–47% (1984–94)	$5 billion
Airlines	13% (1977–79)	12% (1977–82)	29% (1977–87)	$19.4 billion
Trucking	N.A.**	3–17% (1980–85)	28–58%*** (1977–87)	$19.6 billion
Railroads	4% (1980–82)	20% (1980–85)	44% (1980–90)	$9.10 billion

Note: All figures are real, in $1995.
*For natural gas, no controlled studies quantify the separate effect of deregulation on gas prices.
**For trucking, no studies have documented the effects for the first couple of years.
***No trucking figure is available for 1980–90; figure quoted is for 1977–87. Because regulation made it difficult to cut trucking rates, the bulk of these rate reductions occurred after 1980.
Source: Robert Crandall and Jerry Ellig, *Economic Deregulation and Customer Choice: Lessons for the Electric Industry* (Fairfax, VA: Center for Market Processes, 1997).

Surface freight deregulation also generated billions of dollars in shipper savings due to improved reliability of rail and truck transportation.

WIDESPREAD CUSTOMER BENEFITS

No one can reasonably claim that regulatory reform benefited every single customer in the United States. That standard is impossible for any public policy to meet. Nevertheless, regulatory reform did benefit all major customer groups.

Natural gas provides a good case in point. The average wellhead price fell by $2.32 per million cubic feet (mmcf) in the 10 years following 1985, when most wellhead prices were deregulated. (See Figure 9–1.) Prices paid by every customer class fell by even more: $2.93/mmcf for residential customers, $3.13/mmcf for commercial customers, $3.53/mmcf for industrial customers, and $3.41/mmcf for electric utilities. The additional price reductions came largely out of the margins earned by interstate pipelines for transporting gas.

Despite such figures, even executives in the gas and electric industry sometimes speak as if large industrial customers got most of the benefits of wellhead deregulation. This misperception survives because the savings often are expressed as percentages. Since residential gas rates are higher than industrial and electric utility rates, residential customers received only a 32 percent savings, compared to 57 percent for industrial customers and 63 percent for electric utilities.

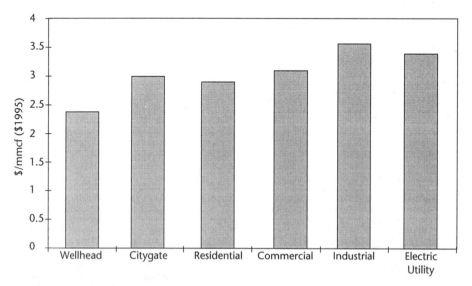

Figure 9–1 *Reduction in Natural Gas Prices, 1984–95*
Source: U.S. DOE, Energy Information Administration

This pattern stems from the fact that the local gas utilities' share of residential and commercial gas bills is much greater than their share of industrial and electric utilities' gas bills. Distribution, metering, and billing expenses per mmcf are higher for residential and commercial customers, and thus the total cost per mmcf is higher. As a result, even an equal price reduction for all customers would amount to a smaller percentage of residential and commercial bills.

Airlines provide another example of where the conventional wisdom is just plain wrong. Few people dispute that deregulation lowered air fares on average, but most simply know that airlines are gouging passengers unlucky enough to travel to "fortress hubs" or on routes where one airline carries most of the traffic.

Less well known is that even "captive" markets enjoy lower fares than under regulation. For example, in "hub" cities dominated by one carrier, real fares were 19 percent lower in 1995 than they were in 1979. On routes dominated by a single carrier, fares were 27 percent lower. Fares for small, medium, and large cities were all lower in 1995 than in 1979.[4] It is true that fares for some cities and routes dominated by one carrier are generally higher than for other cities and routes, but these fares are still lower than they would have been under continued regulation.

LOW-COST CUSTOMERS STILL SAVED MONEY

Another major fear about electric regulatory reform is that lower prices for high-cost regions will come at the expense of higher prices for low-cost regions. After all, common sense suggests that electricity producers in the Pacific Northwest and Kentucky will ship their power to California and New York if they can get a better price for it there.

The experience of other industries, however, demonstrates that high-cost customers' relief does not come at low-cost customers' expense. Rather, all customers gain as increased efficiency and productivity make it possible to reduce rates for all.

Railroads provide several prominent examples. Prior to 1980, the Interstate Commerce Commission's "value of service" pricing dictated that bulk commodities like coal and farm products paid relatively low rail rates compared to "high-value" products like automobiles and other manufactured goods. Deregulation, it seemed, would lead to lower rail rates for manufactured products and higher rates for bulk commodities.

In reality, all rail rates fell. Within two years after deregulation, average inflation-adjusted rail rates had fallen by 4 percent, coal rates had fallen by 1 percent, and rates for farm products had fallen by 18 percent. Within 10 years, coal and farm rates had fallen by 38 and 50 percent, respectively. Low-cost and high-cost shippers alike got lower rates, because railroad productivity more than doubled in the ten years following deregulation, after a decade of stagnation.[5]

Trucking provides a similar example. "Less-than-truckload" shipments are more expensive to haul than truckload shipments, but the cost of both fell following trucking deregulation. Between 1977 and 1993, the inflation-adjusted operating cost per mile for less-than-truckload shipments fell by 35 percent, while the cost for truckload shipments fell by 75 percent.[6]

WHAT'S SEEN AND WHAT'S NOT SEEN

A great deal of the electricity debate focuses on how an existing pool of costs will be divided up among arbitrarily defined groups of customers—industrial, residential, captive, low-cost, high-cost, and so forth. From a consumer perspective, this debate is highly misleading, for two reasons.

First, even if some big customers receive the lion's share of deregulation's benefits, that does not mean the rest are left with a pittance. If Ford, McDonald's, or Safeway receives lower utility bills, they do not get to keep all of those savings as profits. Their competitors also will save money as a result of deregulation, and competition will force them to pass some or all of the savings on to their customers. For this reason, consumer advocates who focus only on residential electric rates overlook many of the more significant consumer benefits deregulation could produce.

Second, experience shows that regulatory reform does not simply redistribute costs from one group of customers to another. Instead, competition unleashes waves of entrepreneurial ingenuity that we simply do not find in highly regulated industries. Exposed to both increased competition and greater risk, gas pipelines increased their operating efficiency, railroads doubled their productivity, and truckers found ways to move more freight with less resources. Airlines deployed hub-and-spoke route networks—and then Southwest Airlines challenged the dominant hubbing airlines with a distinctly different route strategy and corporate culture. Such innovations were difficult if not impossible to predict in advance.

For these reasons, the most important regulatory reform lesson is also the oldest lesson of economics. When considering alternative policies, it is crucial to consider not just the obvious things that are seen, but also the secondary effects that are not seen.

NOTES

1. Unless otherwise noted, all figures in this article for industries other than electricity have been converted into 1995 dollars to adjust for inflation.

2. U.S. Energy Information Administration, *Electricity Prices in a Competitive Environment: Marginal Cost Pricing of Generation Services and Financial Status of Electric Utilities* (Washington, D.C.: U. S. Department of Energy, August 1997); Michael T. Maloney, Robert E. McCormick, and Raymond Sauer, *Customer Choice, Consumer Value: An Analysis of Retail Competition in America's Electric Industry* (Washington, D.C.: Citizens for a Sound Economy Foundation, 1996).

3. Robert Crandall and Jerry Ellig, *Economic Deregulation and Customer Choice: Lessons for the Electric Industry* (Fairfax, VA: Center for Market Processes, 1997).

4. Airline deregulation legislation was enacted in 1978, and the Civil Aeronautics Board took some steps toward deregulation as early as 1976. Unfortunately, the Department of Transportation's annual fare data for hubs and particular routes begins in 1979. Since fares were higher in 1978, our percentages understate the effect of deregulation on fares. All figures are calculated from data drawn from DOT Data Bank 4.

5. Rail rate data were supplied by the Association of American Railroads. Our productivity measure is the Bureau of Labor Statistics' rail productivity index, which rose from 54.6 in 1980 to 118.5 in 1990.

6. Authors' calculations. For original data sources, see Crandall and Ellig, Appendix.

CHAPTER 10

Lesson from the Natural Gas Business

H. Dean Jones II

For the past 12 years, I've been selling a deregulated product to companies that are predominantly still regulated. In the process, I've learned a great deal. The primary lesson learned: competition works. It takes time and can be painful to move from deregulation to competition, but the transition is worth the trouble. Contrary to popular belief, most utilities have made the transition well; in fact most of them have thrived. This chapter will explain the steps they've taken to success.

IT WON'T WORK IN OUR INDUSTRY

Prior to 1984, "strategic planning" at an interstate natural gas pipeline focused primarily on where to hold the next customer meeting. Not long after I accepted a position as manager of strategic planning for a major interstate pipeline company, the Federal Energy Regulatory Commission issued a Notice of Proposed Rulemaking (NOPR) that would alter forever the pipeline industry, eventually culminating in Order 636. Strategic planning now took on an entirely new meaning for interstate pipelines. I consider myself fortunate to have had the job I did at such a unique point in industry history.

I can still recall those first planning sessions with the interstate pipeline officers as we began planning how to respond to the commission's proposed deregulation plan. Initially, about a third of the officers felt it was a waste of time to even talk about it; they believed it was impossible to deregulate the gas industry. Another third were sufficiently interested to discuss it but felt it was a bad thing for the industry that should be strongly opposed in our comments to the FERC. The remaining third felt it was inevitable that the

FERC would act, therefore it was in our interest to try to become part of the process of shaping the rules.

Fortunately for our company, the president and chief executive officer favored accepting deregulation and moving forward to try to make it work. He was able to convince most of the other officers that this was the correct strategic path. Ultimately, several who disagreed with him left the company.

My earliest recollection of how and when deregulated pricing began goes back to the "special marketing programs" approved on some pipelines. Pipelines that were drowning in "take-or-pay" petitioned the FERC to allow them to transport natural gas to certain large end users on their systems. This access to interstate transportation actually put the first gas marketers into business. The Maryland People's Council (MPC) eventually protested this limited gas transportation as being discriminatory, which led to a much broader opening of the interstate systems. The MPC court ruling actually led to the NOPR and ultimately Order 436, the initial basis for open access transportation.

The rulemaking process at FERC for Order 436 was a most interesting battle. Interstate pipeline companies that opposed competition tried to make it difficult to use their systems for transportation. Large imbalance penalties and credit requirements that Fort Knox couldn't satisfy were proposed as standard by the pipelines. By contrast, some pipelines felt a strategic advantage as transporters and were therefore much more reasonable in their certificate negotiations. The FERC was a great place for pipelines to fight marketing companies because it takes lots of time and expensive D.C. lawyers. Marketing companies such as Entrade, Hadson, and Clearinghouse were tenacious in those early proceedings and won many battles for the upstart marketers. Fighting big companies with unlimited funds turned out to be difficult. The marketers soon devised an even more effective strategy: take the customers!

My first real marketing job—director of sales for an interstate pipeline—started in 1986. The first six months on the job were a disaster. The job of protecting our market share in this time of deregulation proved to be impossible. Within six months we lost 30 percent of our share in a market we'd served for nearly 40 years. We were trying to sell gas against deregulated competitors who waited for us to post prices in the public tariffs then discounted that price by 20 percent to 30 percent. We were losing badly. After six months of getting our teeth kicked in, we decided "if you can't beat them, join them." It was time to become an affiliated marketer.

After much debate on whether to "control" affiliate transactions, the FERC issued Order 497. The stated purpose of this order was to make sure that deregulated marketing affiliates of interstate pipelines did not benefit by unfair competitive advantages when doing business on the sister pipelines system. At the time the order was issued, I went to Washington to testify against it on behalf of affiliated marketers. There are still strong opinions on both sides of the 497 debate. Some believe that without such an order competition might have been reduced. On the other hand, in many ways Order 497 actually handicaps affiliated marketers. This is a debate that is likely to be with us for some time.

WE DON'T WANT CHOICE

In the mid-1980s, when the FERC issued the NOPR that preceded Order 436, the energy world as we had known it was about to come to an end. One of my company's first actions after the NOPR announcement was to arrange a meeting to inform customers about the proposal. I can still remember the unexpected reaction our customers had to "the right to choose." One of our long-time customers interrupted the meeting by saying he didn't understand why he was being offered the choices and didn't want them. Several other customers expressed similar feelings.

Keep in mind that these were utility customers. They were buying and selling gas at no profit, and "choice" did nothing for their bottom line in the short term. The industrial customers served by those same utilities, however, had a very different reaction to acquiring access to the interstate gas systems and soon became the target of every gas marketing company in the country. The retail customers of today now gaining access to deregulated products are reacting in ways similar to the industrial customers of 1986. They expect a decrease in their commodity price as a result of deregulation and an improvement in service and the number of products offered by their new suppliers. Meeting their expectations may be difficult sometimes but ultimately rewarding for those companies who find new ways of doing business with these customers.

As natural gas deregulation began to unfold, large industrial customers and utilities that wanted to "experiment" with marketers became the target of newly incorporated "competitive" gas marketers. Significant discounts were offered to certain "marquee" customers to establish credibility for these up-start energy companies. It worked. Companies with no balance sheets, no experience, and no reserves became the suppliers of choice for many large industrials and gas distribution companies. Commitments made by interstate pipelines to producers and by gas utilities to interstate pipelines were quickly undermined. It was time to take "choice" to the masses and capitalize on the trophy customers who were now in the portfolio. We see this same phenomenon today. Large retail customers are announcing "deregulated" deals significantly less than utility rates and everyone wonders how the supplier could make money at those rates. They aren't. It's called advertising. Loss leaders! Welcome to the world of competition.

When deregulation first occurred, a gas buyer at a regulated utility told me that he was not going to buy any "spot" gas. He was concerned that it might have some "bad stuff" in it. Of course, most of the spot gas being sold by marketers was exactly the same gas interstate pipelines would have bought had they been able to serve the market in the same flexible way as the deregulated marketers. This is a very important point. Pipelines and regulated utilities were required to sell gas according to standard terms and conditions and at the same price to all customers. The only difference in the rate charged to a local distribution company (LDC) in Georgia and one in New York was a transportation differential, not a gas cost differential, reflective of the spot market. It seemed common sense that customers with different risk profiles and different sensitivities to price could be better served by suppliers who could address those particular needs with various terms and conditions and supply alternatives. Even Henry Ford quickly learned he could sell more cars if he didn't paint them all the same color. Pipelines and utility suppliers could not compete with marketers who had flexibility on how they priced and packaged their products. Pipelines and utilities couldn't even change their product pricing, much less differentiate it, without a rate filing that could sometimes take months for approval. As time went on, the industry came to accept the presence of marketers, and every utility in the country felt obligated to experiment with these new companies.

At that point something critical happened. Competition worked. The good suppliers continued to draw business and the bad suppliers (as well as the handcuffed ones) were quickly driven out of the business. Within a two-year period, utilities went from buying no spot gas to buying almost 100 percent of their requirements from marketers. Industrial customers left their utility suppliers like rats abandoning a sinking ship. For their part, marketers were making healthy margins by pricing their products just under the rate industrials could get from their utility or just under the filed rate of a pipeline. Times were good. Times would change.

CUSTOMERS LOVE CHOICE

In 1986, one of my first deregulated sales was to a regulated gas-fueled power generating station. I never will forget that deal. Deregulation was still new and the gas buyer at the power plant didn't know many suppliers. He and I had known each other during regulated days. He called to see if I could teach him how to use the unregulated process. I made a deal to sell gas to his plant for one month at five cents less than his current regulated supplier. His plant had a very large load, about 100,000 decatherms a day. At the time we made the deal, we hadn't bought any gas for the month he wanted, but we were reasonably sure we could get gas for him at a price that would provide a small profit. When we actually bought the gas I was shocked to find we were able to buy his full requirements at 62 cents below our price to him. This meant that our newly formed company with about 10 employees would make some $62,000 per day on this one deal. That's what you call "low hanging fruit." The customer was happy with his nickel price savings. The producer was happy because he landed a market for gas previously shut in by his pipeline customer. And needless to say, we were very happy with a transaction delivering that kind of money. That deal lasted one month. The next month the plant buyer wanted a "market" price rather than one based on his utility alternative. We were again his supplier but our 62 cent margin fell to about 30 cents. That still wasn't bad.

Within a matter of months, with four or five suppliers calling on him, the plant buyer was conducting a competitive bidding process. Some months we'd get the business; others we didn't. But even in this competitive bidding process we sometimes realized margins of 10 to 20 cents. We were making money, buying advertising, and entertaining customers like there was no tomorrow. The cost of entry into gas marketing was minimal, and soon there were literally thousands of gas marketers selling to every utility, power plant, and industrial customer in the country with access to transportation capacity. Wholesale margins continued to erode. Companies growing great guns with margins of 20 to 30 cents began having trouble making money when margins fell to 5 to 10 cents. Once again competition was working its magic, and the young industry was learning an important but costly lesson.

I would dare to say that in 1986 not one deregulated gas marketer knew what its cost structure was or why it was important. With margins in excess of 10 cents, about 4 percent, who cared? Nearly all of the business processes were handled manually, primarily because there were no systems for this new industry and people costs were small relative to the other major operating costs of gas and transportation. In less than three years, however, things had changed dramatically. Consultants were telling companies to watch their cost structures. Software systems were on the market to help automate business processes. Margins were eroding and small companies were finding it harder to compete. Companies began tracking expenses as religiously as they tracked gas volumes. Volume was becoming king in order to lower one's cost-per-unit sold. Numbers such as a "B" a day and then "2 B's" a day became the talk of the industry. As with many other industries, people soon figured out that the fastest way to grow and add market share was to buy it. Companies with large volumes and poor cost structures became targets for those who had learned to conduct business efficiently. The urge to merge was ripe.

I can't remember the first major merger that occurred but I can surely remember the first one that convinced me that "big" was going to be "really big." In the early 1990s, Enron Corporation bought Access Energy for a tidy sum, and, as they say, the race was on. Pipelines began merging and combining their market affiliates, as did large independents and producer marketers. Associated/Panhandle, Transco/Texas Gas, and Tenneco/Entrade are just a few examples of how companies combined to gain both scale and scope.

Today in the deregulated natural gas business there are mega-competitors with national scope and volumetric scales of "10B's" per day or more. Sophisticated systems to rouse, nominate, and bill gas are a must with the extremely small margins that exist in this highly competitive market. The cost to enter today's market is quite high relative to 1986 costs. There are still a few niche players but the top 15 marketers in natural gas serve nearly 80 percent of the market share. It's not enough today to be a highly efficient natural gas marketer. System costs and other infrastructure costs must be spread over other products such as financial offerings and other commodities. Pipelines and some LDCs have exited the merchant business and nearly every unit of natural gas sold is handled at least once by the competitive market. In roughly 10 years, the natural gas experiment has proven highly successful and has begun evolving to the next level of competition (i.e., new products). I believe today's retail deregulation of electricity will be founded more on new products than on commodity savings.

THE PAPER WORLD

In 1989, an accountant who worked for me was attending night classes to obtain his MBA. One morning after his previous evening's class he came to me, said he'd been learning about financial swaps and wondered if I thought they'd ever have any application in the gas business. In my best "physical gas delivery voice" I assured him that those things may work on Wall Street but no one in his right mind would want one in the gas business. Was I ever wrong! I sell more swaps today than I do molecules of gas. Physical natural gas products of all kinds have emerged from the early days of "best efforts" contracts. Firm baseload, setup swing, requirements contracts, and the like are a few of the physical products offered today and priced uniquely because of their inherent differences. Financial products too numerous to mention are bought and sold every day in far greater numbers than the physical volumes themselves. Customers with price risk now can manage that risk with products offered by deregulated marketers that would never have been offered by regulated marketers given their conservative risk profiles. At the New York Mercantile Exchange, triggers, swaps, collars, ceilings, floors and options of infinite variety are bought and sold every day to customers who either are managing a known risk or choosing to speculate. New products are invented every day by 20-year-old whiz kids who don't know a smart pig from a dumb one. New products create market differentiation, which can create new margin opportunities. Margins in a mature commodity industry are thin at best and extremely thin in highly traded products. The new deregulated markets for both gas and power will be more like a mature market than a new one.

The sizzle these days for deregulated marketers is to be highly efficient in lowering transaction cost and to provide custom-tailored energy solutions that allow marketers to earn margins in excess of what the individual commodities and financial products would bring if sold separately.

If there's one thing I've learned from the deregulated market in the last 10 years, it's that customer service is something all customers want and in most cases are willing to pay for. Solving customer problems is what competition and deregulation are all about. The successful mega-marketers today pay a great deal more attention to customer service now than they did when the markets were first opened to them, when they thought everyone would compete on price alone. Take that to the retail level and the successful marketers of tomorrow will have call centers, web pages, regional offices, and so on— all designed to provide customers more immediate access to customer service centers. I believe natural gas and the other energy commodities will follow the lead of the telecommunications industry, which has evolved into a highly competitive customer service business that lives or dies on its ability to offer simple products to the mass market. We're not there yet in energy, and the transition will be made even more difficult in our industry given the highly variable cost of the products we sell. But we will get there. It's only a matter of time.

Competition in the wholesale natural gas industry has been a huge success. We still have a long way to go to get retail access to all customers but progress at the wholesale level has been phenomenal. It built the groundwork for today's deregulation to the retail gas and power customers. Customers and suppliers with price risk can find many ways to manage that risk. The physical delivery system is highly efficient and has worked flawlessly in cold winters as well as warm ones. New products are being offered every day. Producers, pipelines and local distribution companies are all focused on what they do best and seem not to have been harmed by deregulation. The people who said it would never work in the gas business were wrong. Most of them are no longer in the industry. In most cases the industry passed them by while they were so busy saying it couldn't happen. There are still nay sayers among us in the electric industry. If they persist in the premise "not in my industry," they may find themselves among the dinosaurs. The retail markets will open, and creative, customer service-driven companies will be highly successful. I know this is true because I've seen it happen for the last 12 years.

SECTION **IV**

WHAT STRATEGIES SHOULD ENERGY SERVICE PROVIDERS PURSUE?

CHAPTER 11

Creating Competitive Advantage by Strategic Listening

Ahmad Faruqui

Electric utilities are preoccupied today with creating competitive advantage by lowering costs through restructuring their organizations. This short-term focus on cost control has created an inward focused corporate culture. Drawing upon case studies from within and outside the utility industry, this chapter highlights the importance of creating an outward focused corporate culture that generates long-term competitive advantage by strategic listening.

WHAT IS STRATEGIC LISTENING?

Strategic listening is a management process that involves listening to all players in the utility's value chain—from customers to competitors to suppliers—with a sole purpose: improving business strategy and operations. This process conveys information in real time to senior management, and results in decisions and actions that confer continuing competitive advantage on the utility. It is designed to shape and influence strategic decisions about fundamental business issues:

- What business are we in?
- What is our core competency?

This originally appeared in the May 1997 issue of *The Electricity Journal* (Elsevier Science Inc.) It is reprinted here with their permission.

- Who is our competition?
- What products and services should we provide?

Strategic listening differs from conventional listening that is not carried out as part of an actionable, strategic process. Conventional listening does not influence the corporation's business performance, since it produces information that is too diffuse to be of interest to senior management. And it often produces a glut of useless information, as revealed in a recent survey of 1,300 senior managers which found that 43 percent of the respondents had become ill because of information overload.[1] According to psychologist David Lewis, who analyzed the survey results, "information overload can lead to a paralysis of analysis, making it far harder to find the right solutions or make the best decisions." By itself, information cannot confer competitive advantage on anyone. To paraphrase T. S. Eliot:

Where is the information that is lost in data?

Where is the knowledge that is lost in information?

Where is the wisdom that is lost in knowledge?

WHY STRATEGIC LISTENING?

The planning paradigm in the electric utility industry has conclusively shifted from regulated business planning to competitive market planning. The paradigm of the 1980s was centered on the concept of integrated resource planning (IRP). The utility submitted an IRP filing every few years that identified a portfolio of demand and supply-side resource options designed to meet the energy needs of its "ratepayers" at the lowest cost. It minimized "revenue requirements" to meet a "load forecast." The words 'customers,' 'markets,' and 'prices' did not enter the utility's lexicon. Utility customers had no choice of supplier. They bought as much electricity as they needed, at a cost-based rate. The utility was obligated to serve them, in return for having a monopoly franchise. As shown in Figure 11–1, the regulated planning process could be characterized as a straight line from the utility to its regulator to its ratepayer.

In the first half of the 1990s, the planning paradigm shifted to market planning. The planning process expanded from a straight line to a triangle, with the utility and its competition at the base of the triangle, and the customer at the apex. The regulator was still present, but had a diminished role.

[In] the second half of the 1990s, this process is shifting to competitive market planning. Utilities now seek to understand not only their customer, but also their customer's customer. In some cases, they are also trying to understand their competition's customer. And they are recognizing the important role of trade allies such as equipment manufacturers, architecture and engineering firms, wholesalers, retailers, and contractors in leveraging their limited resources to gain competitive advantage.

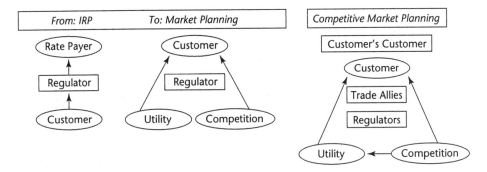

Figure 11–1   *The Evolving Planning Paradigm*

Utilities have now realized that customers have a wide range of choices:

- Buy from their current utility.
- Self-generate their power using cogeneration or distributed generation.
- Buy from a power broker or marketer.
- Switch from electricity to natural gas or other fuels such as oil and coal.
- Relocate to other service areas in the United States.
- Move off shore.

COMPETITIVE STRATEGIES

Faced with customer choice a utility that wants to succeed has to radically rethink its competitive strategies. It can no longer win by satisfying the regulatory commission that it has filed a comprehensive IRP. It has to win by succeeding in the real marketplace.

Like firms in all competitive industries, utilities have three strategic options:

- **Become a low-cost power provider.** This involves re-engineering, downsizing, and renegotiating power contracts. This option cannot make a winning competitor out of a high-cost provider.
- **Become a niche player.** This involves becoming the supplier par excellence to a small, undiscovered, but profitable market segment.
- **Expand their product horizons.** This involves introducing new products and services to existing customers and entering new markets. Many people think that after a monopoly is exposed to competition its fortunes will deteriorate, because its market share will decline from 100 percent to some smaller number. They often state that AT&T's market share in the long-distance telephone market went from 100 percent in the regulated world to 60

percent in today's market. However, it is less often noted that by expanding product horizons AT&T was able to expand earnings while losing market share in the old markets.

For many utilities that cannot become the low-cost provider of electricity, the only practical option is to expand their product horizons. They can expand into downstream services, such as end-use energy services. Some may want to go even further, and enter the consulting business. They would advise customers on how to improve the productivity of their businesses, and how to comply with increasingly stringent environmental regulations.

This process can change the relationship between the utility and its customers, from that of a vendor competing to provide services to the customer, to that of a strategic partner in the customer's business.

WHAT IS NOT STRATEGIC LISTENING?

As Ludwig Wittgenstein said: To know something, you need to know what it is not. To understand strategic listening, it is useful to understand two types of conventional listening: anemic listening and unfocused listening.

Anemic Listening

This creates information through assumptions, not through active listening. It assumes that the utility knows what the customers want, and it assumes that the utility has the core competency to create products and services to meet these customer needs. As a result, smart competitors engaged in true strategic listening skim away the utility's most profitable customers by offering them more attractive products and services. The utility is left with unprofitable customers and has to raise its prices to these customers to stay afloat. This, in turn, creates more unhappy customers and invites more competitive entry.

Anemic listening happens insidiously. The utility engages in listening, but this listening is passive and not active. It collects "megabytes" of information. But this information is collected slowly and is often focused on irrelevant issues. It is reactive in nature, rather than proactive. For example, information is derived only from customer complaints. Moreover, even this information is filtered before it reaches the decisionmaker. In some ways, it is analogous to the information that was collected in countries with totalitarian regimes. Year after year, the polls showed that the regime was liked by no less than 98.6 percent of the populace. Then, a revolution occurred and swept away the "very highly liked" regime into the dustbin of history.

Anemic listening tempts the utility to substitute its assumptions for market evidence. It is a dangerous trap for utilities. A classic example of anemic listening is provided by the first crop of power quality programs. These programs were based on responding to the needs of those customers who called to complain about power quality. As a consequence, utility companies missed the opportunity to market premium quality power as a new value-added service. They ignored customers who were suffering but had not called their utility, because the utility had other priorities. Such customers were lost to competitive firms that came to see them, even though they were their "non-customers." Through these visits, the competitive firms developed an understanding of their problems because they engaged in strategic listening. Ultimately, they offered customers a solution to their problems. Unfortunately, this solution invariably included switching power companies.

Unfocused Listening

Unfocused or casual listening involves listening without a strategic process. Here, the utility is unable to separate the "wheat from the chaff." Strategic information about customers and competitors is mixed in with the useless chatter (and sometimes dangerous gossip) of the market place. The signal-to-noise ratio of the information is very low, so the utility is unable to act on the information. Listening becomes an end in itself rather than a means to an end.

Pursuing Strategic Listening

The pursuit of strategic listening does not happen by accident. It requires recognition by senior management that candid, focused, and timely information about customers, competitors and trade allies is an indispensable part of strategy development and execution. Only in this way will market feedback about the utility's new product offerings, obtained through strategic listening, be given sufficient weight by product designers to modify designs that are not working well and to leave unchanged designs that are working well. Thus, business strategy will adapt to changing customer needs in real time, rather than having to wait for the annual planning cycle.

The success of many Japanese companies is credited to the speed with which they respond to changing customer needs. Rather than relying on formal market research studies, they rely on informal market feedback that comes in daily through delivery channels. For example, when Canon introduces new cameras, the features in these cameras are often based on questions about nonexistent product features that potential buyers ask store salespeople.

This example highlights several attributes of strategic listening:

- It gathers both qualitative and quantitative information from the market place.
- It collects information from all components of the value chain: customers, trade allies and competitors.

- It ensures that product designers act on the information.
- It avoids tunnel vision by talking to people in other industries.

UTILITY EXAMPLES

This section presents two examples of strategic listening in the electric utility industry.

Utility A

Utility A is a high-cost producer of electricity. A large chunk of its electricity sales are made to industrial customers. Concerned that several customers might switch to lower-cost suppliers, it began a proactive program of strategic listening. Before the program, it saw the customer as a buyer of electricity. What the customer did with the electricity was the customer's business. It was not for the utility to understand the customer's business.

Through the program, Utility A discovered that the customers' view of the world was very different. They manufactured and sold products that added value to their customers' business. Their main focus was making a profit for their shareholders, which meant that they needed to achieve high standards of productivity and product quality in their manufacturing process. This set standards for the efficiency of the various processes that were in use at their manufacturing facilities. Energy, and in particular electricity, entered this process efficiency calculation at some point, but it was not in the "driver's seat." The cost of energy was no more than 5 percent of the production costs of most customers.

To obtain additional insights, Utility A boldly asked these customers to share their views about itself. Examples of the responses are given in Inset 1.

INSET 1:

What Some Customers Said

- You are very fixed in what you do. You aren't very flexible.
- You are not very innovative, and prefer to give us "take it or leave it" propositions.
- You have no concept of customer service, because you are a monopoly supplier.
- You are not focused on our needs. You don't take the time to understand our business. You are only interested in reducing your costs.
- You are very expensive. You are not competitive.
- You are always trying to discourage us from self-generation, even though it may be in our interest.
- We look to you for help in managing our energy bills by giving us rate discounts.
- You don't come to see us very often, unless you're trying to sell us something.
- You take a long time to make your decisions.

These customer responses caused Utility A to rethink its entire relationship with industrial customers. After conducting additional in-depth interviews with these customers, it came to the conclusion that except for the very biggest industrial customers that had large engineering departments, most customers valued many attributes aside from the cost of electricity. These attributes included:

- **Direct attributes of electricity.** Besides cost, these included voltage level, power reliability and power quality.
- **Indirect attributes: energy services.** These included assistance in choosing equipment efficiency, size and O&M practices; working with vendors; and general technical assistance.
- **Indirect attributes: process attributes.** These included a variety of technical characteristics, including: product quality, through-put rate, scaling losses, controllability and precision, versatility, thermodynamic potential, equipment size, pollution emission characteristics, noise, esthetics, and worker safety.
- **Indirect attributes: business attributes.** These included plant productivity, business competitiveness and market share, risk management and profitability.

Utility A decided to redo its customer segmentation scheme once it became aware of the multiple attributes that its customers valued in an electric utility. It created a new set of segments constructed around the different weights that different types of customers placed on each of these attributes. An example of this approach is shown in Figure 11–2.

Through additional research, Utility A found that segmentation needs to involve both "hard" and "soft" factors. Two firms in the same industry of the same size may yet have very different needs for new products being sold by the electric utility, because of different management styles, different financial profiles, and so on.

Segments	Segment 1	Segment 2	Segment 3
Attributes			
Cost of Electricity	■	■	○
Voltage Level	○	●	■
Power Reliability	●	■	○
Profitability	○	●	■

● Very Important	■ Somewhat Important	○ Not Important

Figure 11–2   *Example of a Weighted Customer Segmentation Scheme*

Thus it developed an integrated segmentation approach, as shown in Table 11–1. In this table, hard factors such as demographics, economics and technology are shown in the first two columns, and soft factors such as buyer-seller relationships, stage of purchase, and the buyer's risk-taking preferences are shown in the last three columns. Segments based on these factors allowed Utility A to market its products and services more effectively to its customers.

Utility B

Utility B knew that many of its customers were at risk to switch to competitive suppliers. It also knew that some of these customers were profitable, while others were not, but it did not know which customer fell into which category. Through strategic listening, it found that customers with the following attributes were at risk:

- Spent a lot on electricity.
- Spent a lot of time talking to other suppliers and to trade allies about other electricity suppliers.
- Had multiple plants in several different utility service areas.
- Had ability to self-generate or [use] distributed generation.
- Could switch from electricity to other fuels.

It then ranked customers based on their probability of leaving the system. Because it was not possible to rank all of its customers, it individually ranked the top 250 customers, grouping the remaining customers into segments, and then ranked these segments. Next, it estimated the profitability of each of these customers and market segments.

It eliminated from further consideration those customers and customer segments that were at risk but also unprofitable. For the remaining customers and customer segments, it identified a variety of customer retention strategies. These strategies were identified by reviewing the experience of other utilities, reviewing notes of meetings with customers, and internal brainstorming sessions. They included developing a wide range of new products

TABLE 11-1	Integration Segment Approach			
"Demographic"	Operating Environment	Purchasing Approach	Situational Factors	Personal Characteristics
Product(s) (SCI) Code	Technology of Production	Elements of Buying Center	Stage of Purchase	Motivation
Size of Plant	Economies	Buyer-Seller Relationships	Buying Situation	Risk Taking Preferences
Location of Plant	Capabilities		Application/ End Use	Empathy with Seller
	Growth Outlook	Purchase Policies		

and services and pricing strategies. It used focus groups to identify strategies that had high profit potential, and then the top five strategies were offered to its customers. This process is in Figure 11–3.

OTHER INDUSTRY EXAMPLES

This section presents five examples of strategic listening in other industries.

Compaq Computer Corporation

In 1994, Compaq faced stiff competition in the market for personal computers (PCs).[2] Hewlett-Packard (HP) had entered the market. New Pentium machines were being introduced by Dell Computer and Gateway 2000. Its own sales force was insisting that unless Compaq introduced Pentium machines, it would lose market share on a large scale. On the other hand, it still had a large inventory of machines based on the 486 microprocessor that would be cannibalized if it introduced a Pentium line prematurely.

To deal with this strategic issue, Compaq undertook a large scale strategic listening project. It conducted focus groups with customers to gauge interest in and price sensitivity to new features, and polled resellers on inventory levels and available cash to gauge the timing of new-model introductions. It tracked suppliers' production levels and, with these figures, developed alternative production plans in the event market forecasts turned out to be wrong. Collecting this data was not easy, since Compaq's marketers had never estimated demand elasticity for PCs.

Based on this information, Compaq developed a spreadsheet model that simulated the behavior of customers, suppliers (such as Intel) and competitors (such as HP and Dell). The model used market behavior rules derived from game theory. It was essentially a giant spreadsheet that could simulate conditions such as component price changes, fluctuating demand for a given feature or price, and the impact of rival models. Most important, this tool could let managers consider the risk of certain actions before taking them.

Results from this exercise that merged strategic listening with strategic thinking indicated that Compaq should delay the launch of its Pentium line by a

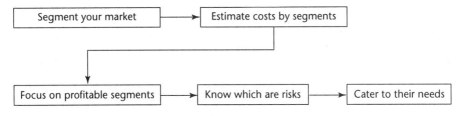

Figure 11–3   *The Process Used by Utility B*

few months, until the inventories of the existing line based on the 486 microprocessor had been lowered to acceptable levels and prices of the Pentium chip came down. This decision added $50 million to its quarterly profitability. CEO Eckhard Pfeiffer commented: "It's about time we're doing these things."

Ford Motor Company

While its Ford Taurus was still the best selling car in America, Ford was coming under increasing threat from competitors such as Honda and Toyota.[3] Also, the incumbent model had matured and a replacement was needed.

Ford decided to keep the Taurus name but created an entirely new car based on a design that used advanced engineering techniques and materials to produce a "tear drop" look with high-tech features, with a vehicle that remained competitive at a price of $20,000.

Strategic listening helped Ford create the new design. The unique feature of Ford's strategic listening was that it involved listening to the competitors' customers, in addition to listening to its own customers. A whole series of focus groups was conducted with owners of the Honda Accord. They were asked about the features that had led them to purchase the Accord, and about the importance and weight of these features. Ford found out that customers in some cases bought the Accord thinking it had air bags, when it did not. Nevertheless, these features were highly valued by customers.

VF Corporation[4]

One out of every three pairs of jeans sold in the United States is a VF Corporation brand such as Lee, Rustler, or Wrangler. Because of rapidly changing consumer tastes, it became difficult to know ahead of time which jeans were in demand. VF wanted to avoid having unsold inventories, and also avoid frustrating customers by having items out of stock. According to CEO Mackey McDonald, "We realized several years ago that the apparel industry is changing from a manufacturing industry to an industry that studies consumers and their needs. Rather than trying to predict demand, we have to learn to respond, which means we need to improve our ability to collect and use information." Thus, VF decided to change the entire nature of its planning process, from being one that was designed to meet a given demand, to one where the need to predict demand was minimal: In short, it incorporated strategic listening into its production process.

It now uses a planning process driven by real-time sales information that tells it which jeans are selling well and which are not. This process is so flexible that it doesn't need more than a day's notice to switch from one style to another.

This high-tech, computer-driven market response system has reduced inventory, widened customer choice and cut restocking time from weeks to days. Even as other companies struggle with 50 to 70 percent in-stock averages, VF has increased the odds to 90 percent that its customers will find the style, color, and size they want. In its application of a market response system based on the principles of strategic listening, VF is a pioneer. Its approach has now been adopted nationwide.[5]

Nintendo Company Ltd.

This Kyoto-based entertainment company is well known for its line of futuristic electronic games. Today it commands some 40 percent of the video game business—not bad for a company that began selling playing cards in 1889. By the 1980s, Nintendo had defined itself as being in the business of delivering interactive entertainment.[6] However, this decade saw the rapid decline of the coin-operated arcade market. Nintendo repositioned itself quickly into the home video game market. It chose customer service as the linchpin for maintaining market share. The centerpiece of this strategy was an extensive hot-line network aimed at creating an ongoing source of consumer research. By 1995, over 40 million consumers had picked up the phone and called Nintendo with ideas.

According to Howard Lincoln, chairman of Nintendo of America, "The reason we know what will work when we create a new game or system is because we just talked to the customer." This is strategic listening in action. This approach also opened up new channels of communication with the customer. In 1988, it launched *Nintendo Power* magazine. Aimed at boys 8 to 15 years old, this magazine claims a subscription-based circulation of one million. Strategic listening is not only helping Nintendo with strategic decisions, but also with day-to-day operational decisions.

R. S. Means Company

This company is a leading provider of cost information to the construction industry. It recently discovered the power of two-way communication carried out on the World Wide Web.[7] The company has sold enough of its products through the Website to recoup its investment within the first seven months. Seventy-five percent of those who have bought products through the Web are new customers. Says Greg Wurtz, Webmaster for R. S. Means, these customers are the "cream of the crop," because they are younger, more affluent, and more techno-savvy than the average customer. The Web is now integrated into R. S. Means's marketing plan. As such, the company's Web address is on its 800-number greeting, in its catalogs, on its direct mail pieces, and in its trade-magazine advertisements.

This example points out the advantages of having a two-way communication medium which opens up entirely new possibilities for businesses. First, because it can advertise its products and take orders for them. And second, because the Web is interactive, consumers can get information about many other products, while marketers can collect information about consumers.

Rich Anderman of Mercedes Benz of North America says that the "Web is completely redefining how we communicate or build relationships with our customers." In a similar vein, Mary Lou Floyd of AT&T adds: "You must throw out all traditional thinking and start from scratch."

WHAT IS WORTH LISTENING TO?

As shown above, winning companies are continuously engaged in strategic listening. Specifically, their listening processes are tuned in to three major issues:

- How do customers use our product?
- What factors influence customer use of our product?
- How satisfied are our customers with our product?

To explore the first issue, utilities should ask their customers the following questions:

- What are your ultimate end uses of electricity?
- What are your hours of operation?
- What is your operating schedule?
- Can you change your operating schedule? At what cost?
- What are your competitive options?

To explore the second issue, they should ask the following questions:

- When you buy energy-using equipment, what features do you look for?
- How much weight do you place on each feature, and how much are you willing to pay for it?
- How many alternative products do you consider before making your purchase?
- Who is involved in the final purchase decision? How long does it take to make the purchase?
- What type of after-sales support do you need?

Finally, to explore the third issue, utilities should get information on customer satisfaction, a key determinant of customer loyalty. Specifically, they need to understand customer satisfaction in the areas of price, quality and service. Armed with this information, utilities can decide how best to position themselves in the rapidly changing marketplace. For example, should they battle for market share in existing markets by lowering costs or by offering new products and services? Or should they be creating new markets?

METHODS OF STRATEGIC LISTENING

Because of time and budget constraints, it is best to use a continuation of primary and secondary methods. More important, the specific method is the attitude and timeliness with which listening is carried out. As mentioned earlier, it is critical that the results of the listening exercise are passed on to the appropriate decisionmakers in real time.

Primary methods include customer survey, focus group, in-person interviews, and electronic, two-way interactive communication systems. Secondary methods include a review of the trade and professional literature to identify what customers, competitors, and trade allies are saying, a review of conferences and workshops, and networking with industry experts. These reviews can often generate penetrating insights into the future of customers and markets.

According to Peter Drucker, "the most important structural trends are those that many executives have never heard of."[8] Drucker emphasizes the need to look deeply at structural trends, because they hold the key to understanding the future. "Uncertainty in the economy, society and politics has become so great as to render futile, if not counterproductive, the kind of planning most companies still practice: forecasting based on probabilities." Drucker argues that instead of asking, "What is most likely to happen?" companies should ask, "What has already happened that will create the future?" The answer to this question will define the potential opportunities for a given company or industry. To convert this potential into reality requires matching opportunities with the company's core competencies. "This will create a plan of action, enabling a business to turn the unexpected into advantage. Uncertainty ceases to be a threat and becomes an opportunity."

CONCLUSIONS

Strategic listening is a concept shared by winning companies in a wide range of industries. It goes beyond passive listening, and involves two key components: (1) asking the right questions of the right people at the right time and (2) acting on this information in real time. Used effectively, it can help utilities retain their profitable customers.

NOTES

1. Charles Leroux, *Chicago Tribune* (quoted in *San Ramon Valley Times*, Oct. 20, 1966).

2. Gary McWilliams, "At Compaq, a desktop crystal ball," *Business Week*, March 20, 1995.

3. Cover story, *Business Week*, July 24, 1995.

4. Liz Seymour, "Custom Tailored for Service: VF Corp.," *Hemispheres*, March 1996.

5. Marshall Fisher et al., "Making Supply Meet Demand in an Uncertain World," *Harvard Business Review*, May-June, 1994.

6. W. W. Williams, "Game for Growth: Nintendo Co.," *Hemispheres*, April 1996.

7. Evan L. Schwartz, "The Lure of the Link," *Momentum*, Vol. 2, No. 1, 1996.

8. Peter Drucker, "Planning for Uncertainty," *Managing In A Time Of Great Change* (Truman Talley Books, 1995).

Chapter 12

Is Anyone Listening?

Joseph T. Ewing

Over the last few years, the energy industry has undergone massive changes. The natural gas industry has seen advances toward a competitive market. Federal Energy Regulatory Commission Order No. 636 significantly changed pipeline capacity. Many local distribution companies opened their market areas to retail competition. Electricity has been open to competition on the wholesale market. Some states have started down the path to retail competition. Mergers of the traditional providers have occurred at a rapid rate. New entrants have emerged and are starting to make a major impact on the industry. Regulatory organizations and regulators are trying to determine what their role will be in this transition. End-use customers are looking at these changes with a critical eye.

One of the most important interpersonal skills that a person can possess is the ability to listen to others. It is essential in all aspects of our life and particularly important in a competitive business. Relationships are built between suppliers and customers on this basic principle. Suppliers must ask themselves, "What relationship do I want with a customer? Will I be listening to respond, which may lead to a short-term event? Will I be listening to understand?" If it is the latter, the potential for a long-term customer/supplier relationship is possible.

'Who is the customer?' is a subject of great debate in the energy industry. In a regulated monopoly, the answer is easy: it is the regulators. Power is an exclusive right of the regulator. It does not make business sense for the monopoly to do anything other than meet the needs of the regulator. If end users (industrial, commercial, or residential) want to influence the system, they must resort to an intervention process. However, the customer seldom will be able to match the resources of the regulated monopoly in this process. The customer should be able to influence a business based on the dollars paid for products and services.

The nature of the energy industry has been adversarial. Interventions have been the main method of interface between the various participants. In a regulated environment, this process makes sense. It is the nature of the beast. Many of those in leadership positions have been raised on this concept. Key skills have been rules interpretation and legal-type activities, where it was important to build a case and then argue its points in the appropriate proceeding. Competition will change this situation considerably. Face-to-face negotiation with a customer requires a different set of skills. These new skill sets will be listening-based.

In a competitive energy market, supply participants will be required to have two "customer service" strategies. The first will be for the regulator; the other will be for the end user. Listening to the needs of the customer is essential if the second customer service objective is to be reached.

Evolving from a focus on the regulator to the end user will be difficult. Creative evolution is needed for a change of this significance. The needs and rewards systems of regulators are very different from what both customers and suppliers want. The rewards system of the monopoly and its employees has been focused on the regulator. In the 'new world' the focus will need to be on the end-use customer. Money "talks" and the end user will have a choice that it has not had before. If the current participants cannot meet the customer's needs, then others will emerge to create the match.

In the new world, the customer faces many decisions, such as: Should I/we build an organization positioned to take advantage of these changes? Is it now appropriate to outsource the responsibility for energy sourcing and management? The changes also provide electricity suppliers a once-in-a-lifetime opportunity to separate themselves from others in the industry. Customers are looking for alternatives and suppliers that add value. A competitive market will provide that opportunity. The current state of customer service will not be adequate to meet the future's needs. A creative evolution is needed. The winners of the evolution will be those who are able to adapt to this environment and who make the fundamental business behavior changes that will be required to prosper.

MEETING THE NEEDS OF THE CUSTOMER

Success in a competitive business is meeting the needs of the customer. How will a company ever approach that objective if it does not have the ability to meet the needs of its customer? As discussed earlier, the customer and its needs are very different in the new, emerging market. Seldom has a regulator been required to meet a profit or market share commitment. Also, with a franchise territory and a guaranteed rate of return, most of those working in a monopoly have had limited exposure to the market. Being able to take customer's needs and translating them to a competitive business opportunity will be the key for survival. Many in the business world have refined these skills to a high level of expertise. If the participants in the electricity industry are unable to make this transition, others will learn the industry. These will be the new players.

Whenever I'm asked, "What is your management style?" I smile and respond with the question, "For which person?" One can build a framework of style but it is impossible to have one style that will fit all people and all situations. The same will hold true in a competitive energy market. It does not make sense for someone to try to operate in an area that does not match with your skills. If the match does not exist then it will be a challenge to produce a profit.

I doubt that many of the end-use customers will be willing to sacrifice their needs to match the skills of a particular vendor. They will continue to search and look until they find a company that will meet their needs. Clearly, if a profit potential is available, someone will fill the gap in the market. The one exception to this concept may be the valuing of a strong business relationship. If a strong business relationship exists, it has been built on meeting each other's needs. How does one meet the needs of the other? There is an easy answer. Knowing how they want to be treated and what they want. How does one determine these great mysteries, other than through listening? Ask the questions. You may be surprised to have the end user tell you exactly what they want.

Competition means more than choosing the unregulated affiliate of a utility. Recent studies show that as soon as choice is available, very few will stay with their incumbent supplier. Why are so many anxious to change? A great deal of history exists between the customer and supplier. Reliability should not have been a problem. Pricing is a problem but competition will lower prices. What is driving this other than a belief that needs may not be met?

A competitive energy market will offer unlimited choices. Buyers will be able to select the services and products that meet their needs and add value. As the market matures, we can expect to see creativity flourish. This will occur with products and services, as well as with vendors with different skills. Vendor skills will segment the market. Customer service will be a cornerstone for many. Customer service skills can take many forms. A common thread of any excellent vendor is the ability to know their customer. Various techniques will be used to gather customer knowledge. Listening to the customer will be the common thread of this research.

The supplier of the future may be a surprise to many. I doubt that we have seen the elite as of yet. Think about those companies whom we consider to be the best today. Do any of the businesses in the energy field come to mind? Probably very few, if any. So where does that leave us? Whom do we trust? Now look at that list of companies and ask, "What prevents the companies on it from becoming an energy supplier?" Maybe a partner has energy expertise. Will they leverage their service skills and hire the technical expertise to compete? Stay tuned. After the suppliers of the future emerge, then we all will be able to look back at our list and check our guesses.

LISTENING TO WHAT CUSTOMERS WANT

A saying on a motivational poster reads, "Rule #1: If you don't take care of your customer, someone else will." That may become the theme for the competitive energy industry. As a buyer of energy for 50 sites in North America, I am always surprised at the decline in customer service after the first year or two of the relationship. Rotating suppliers and customers is the norm. This does not make sense. Once a supplier has a customer, why would it give them up? This seems like a huge risk in a competitive market. In a regulated market one does not need to worry about this problem. Will the current providers be able to transition or will they do more of the same and wonder what happened?

The priority of any supplier should be to understand what is important to the customer. I measure the value of any discussion with a vendor on the amount of time I spend talking versus listening. Typically the first discussion is a barrage of what someone is able to do for me. That is great if the "pitch" meets my needs, but seldom is the message on target, and we both have wasted a great deal of time. It makes a lot more sense for the supplier to ask the potential customer a few simple questions, such as:

1. What are you trying to accomplish?
2. What does your reward system require you to do?
3. What are you not interested in or not willing to do?
4. Do you plan to do everything internally or do you want to outsource part of the work?
5. What was it about previous suppliers that lead you to switch?
6. What are the characteristics of your long-term suppliers?
7. What are the products and services you have not been able to find that meet your needs?
8. What is your strategy?
9. What are the areas where you will need our expertise?
10. What are the criteria you use to select suppliers?

This format makes a supplier/customer discussion more productive. I feel more valued than if I receive the "canned" presentation and it gives me a chance to share my vision and strategies. After this dialogue both parties should be able to clearly assess the prospect for a business relationship. If the ratio of talking to listening is not close to 50/50, there is a problem. If I am listening and have a limited opportunity to describe my needs, why continue with the process? It tells me that I'm dealing with an order taker and not a value-added service provider. If the reverse is true and I do all of the talking, I doubt I will turn into a valued customer. How would I ever know if the supplier is capable of meeting our needs? Balance in discussions and negotiations is an essential ingredient of the relationship.

Energy organizations are not created equal. As a result, energy buying customers will have very different needs from their vendors. We are comfortable outsourcing some activities but for other areas we would not consider doing so. There is a growing belief that most industrial buyers are interested in outsourcing energy purchasing as much as possible. This is not the case for us. There clearly are areas that make sense to be handled by others, but other areas will stay in our control. How will a vendor ever know that division unless they ask questions? I don't want a supplier to guess what my needs are. A very high level of trust must be built with someone who will be responsible for delivering our needs. The first criteria for selection of the outsource vendor will be the need for a highly developed skill of listening and asking questions. If they fail that test, what would lead me to believe that they will add value to my needs?

Supplier selection for our company is based on the following six criteria:

1. Reliability
2. Customer Service
3. Related Services
4. Price
5. Financial Stability
6. Vision and Alignment

Reliability is the key component of all of our sourcing. It is understood that everyone will deliver what is promised. Cheap electricity that does not arrive has no value. On the other hand, delivery under the obligation to serve is too expensive. Therefore, we head to the market to look for a fair value for the level of reliability requested. Reliability levels will vary by site and may vary from one period to the next. It is essential that both parties to an energy supply agreement have a clear understanding of what is needed and wanted. This should be one of the areas that is frequently checked for clarity. This is not an area for guessing or assumptions.

Customer service brings us back to the question, "Is anyone listening?" I often have felt that what I am buying is customer service and using electrons as an efficient way to exchange money. We are not talking about tee shirts, hats, and golf outings. The focus is questions and listening; listening to understand the need. Listening to respond will not meet the need. My definition of customer service can be defined as follows: Delivering what I want, when I want it, and how I want it. Vendors should understand my internal processes. If I am forced to use the vendor's process, I will be looking for another vendor to match to my process. High-maintenance suppliers are beyond reasonable capability of any end user, especially when others are available to meet my needs at a lower transaction cost.

Related services are the answers to all of the questions asked by the buyer. It is time for a role reversal and the test of the buyer's abilities to listen. Hopefully, the energy buyer is always looking for opportunities to create additional value for their company. A characteristic of that behavior is a relentless search to learn. What better way does a buyer have than to ask questions of those who have a view of the different ways to execute strategies? If the buyer is going to ask the questions, then the buyer should be prepared to listen and learn. This is the time to lead, by modeling the art of listening. Developing a strong relationship between the buyer and seller will pay dividends for both. Retaining customers and suppliers is less time-consuming than finding new ones.

Price may be the only criteria for many, but not for all. Everyone is looking for the lowest price but it's important to recognize that there are no free rides. If the end-use customer is asking for a specific set of services, then a fee may be associated with that need. It always concerns me when an additional service is offered at no charge. If that service requires resources, how can it be provided at no charge? If I have been paying for the service but not getting it does that mean that I am entitled to a refund? Another possibility is that I am paying for services that I am still not using. A potential solution to this dilemma is for the customer to ask for completely unbundled prices. Pick the services wanted and pay for what is received. Buying a bundled service is fine if you understand how it is priced.

Financial stability is an essential criteria for both the buyer and seller. This area does not require much listening skill. The facts and data should speak for themselves. The electricity futures market has emerged and we can expect to see much of the pricing moving to this area. The volatility of the natural gas market has given us all many anxious moments, and I'm sure there are more to come. Don't participate in the market if credit is an issue.

Vision and alignment are an acid test for listening. It does not make sense to attempt to develop a business relationship when two organizations or peoples do not share the same vision. This takes us back to the price question. Is it really worth the last few dollars of savings to select a supplier that may consume great amounts of time trying to convince you to do something you are not capable of doing? Did the proposal include processes that you do not feel deliver the best value to you?

THE CUSTOMER-SUPPLIER RELATIONSHIP

Early in my buying career I had requested a vendor to triple the normal inventory level. In exchange, the time from production to payment of the inventory would be extended from 30 to 90 days. After a few discussions, the plant personnel finally agreed to the change. Production schedules were developed, and the inventory plan was ready to be executed. Suddenly numerous production issues developed and the objectives were not reached. As had happened many times when the needs were not being met at the plant, we ignored their concerns and demanded compliance. Sure enough, the inventory started to build but we had a deteriorating relationship with the plant contacts. After several weeks, we discovered a valuable piece of data. The inventory increase we requested eliminated the operational bonus for key plant personnel. Here is a classic example of not listening. If we had made the effort to listen to both the verbal and nonverbal messages, we could have developed a solution jointly. If this type of situation happens during negotiations for the new business, the outcome may be very different. The group not listening will not have the same option we had. How do you go to management and demand compliance? Probably what happens is you are removed from the list of potential vendors or customers. When this happens it is important that it is called to the attention of the other person. After all, listening is part of the portfolio of a business agreement. It is very difficult to listen if communication is lacking.

Customers are looking for best value and competitive advantage. One of the best ways to accomplish this is to provide the option for custom-designed products. A Request for Proposals (RFP) is the mechanism many buyers use to begin the process. A significant amount of time and effort go into the development of an RFP. The process usually involves the opportunity for potential bidders to ask questions of the buyer. It also is an opportunity to make suggestions to the buyer about additions to the process. After the response to the RFP has been submitted, the buyer should review it and have a discussion with the bidder if there are any questions. When a buyer goes through this process it should be obvious that communication is going to be an important part of the business relationship. Custom-designing a product requires a considerable amount of interaction.

A response to an RFP is a strong indicator of a vendor's approach to listening. The indicators of the response are as follows:

1. Were all of the questions answered?
2. Was the requested format followed?
3. Did the response arrive by the time requested?

Answering all the questions is a clear indication that the RFP was reviewed and effort was made to meet the needs of the buyer. The lack of responses to questions indicates the opposite. Following the correct format is a great aid to the buyer. Once the response arrives, the buyer wants to complete the analysis as soon as possible. If the format is not followed, then someone must translate and interpret the data. This leads to the potential of not interpreting the response correctly and it is time consuming. If it is difficult to meet the requested delivery time, negotiate a change as early as possible. Calling a few hours before the due date is unacceptable. These may appear to be minor details but to the buyer they are of great importance. It is the first indication of the potential vendor's operating style. We all know how important the first impression can be.

Who is the thought leader of the business relationship? Will it be the customer or the seller that takes the role? Do we really care? The importance of these questions is the ability to reach a consensus. The objective should be an arrangement that benefits each party, adding value to both businesses. Debating for the purpose of establishing domination is not productive. A healthy push back is a positive addition to the relationship. Ideally the position of thought leader will be jointly shared. If that is not possible then one will lead and the other will follow. Vigilant listening is essential in order to overcome this potential dilemma.

Designing the first impression could be done with the RFP response. More likely it was done with an earlier sales call or some other interface. The challenge for the vendor is to determine what the buyer wants. One approach is to immediately lay out what you consider to be your strengths and hope that they will meet the needs of the buyer. Another approach is to ask the buyer to take the lead and describe what they want. Listen carefully and develop an understanding of the buyer.

The objective of both the buyer and the seller should be to join the supply chain. One of the ways to prosper in any business is to create an alliance that is focused on the growth of the market share of the end product. This can be accomplished by spending time and effort focused on the product's competition and not on the various parties within the supply chain. Building this type of relationship will provide an open door to all of the members of the chain. Aggregating a supply chain is a strategy that many providers are planning to use to build their business in the deregulated market. It makes sense to take advantage of the natural supply chain from the raw material through to the end user—created by the supply chain for a particular product. Build the relationship with one and then expand to the full chain. Listening and working together will be the glue that holds a group like this together.

Maintaining the relationship between the buyer and seller is an often-neglected aspect of the energy business. Customers have spent years in an adversarial relationship with the other member of the relationship. To lower all of our costs we must change this behavior. It is expensive to change suppliers or recruit new customers. If two parties have seen enough value in each other to start a business relationship, then it should make sense to attempt to maintain the relationship. We all are guilty of turning off our listening skills after the relationship has some age to it. If we have asked the question, "Is anyone listening?" then we must also ask the question, "What caused us to stop listening?"

The execution of the first contract is not the objective, it is the first step in a long journey of building a strong business relationship. Are you listening? I hope you are.

CHAPTER 13

Creating Economic Value through Risk-Based Pricing

Stefan M. Brown
Douglas W. Caves
Ahmad Faruqui

With the introduction of retail competition in previously franchised markets, incumbent electric suppliers are facing increased pressure on revenues and profits. Only those incumbent providers that successfully can create economic value for their shareholders and customers will survive in this competitive environment. The experience of other competitive industries suggests that one of the key ingredients to creating economic value is differentiation of products, markets, and customers. That is, incumbents need to differentiate themselves from the competition. At first blush, this appears to be an insurmountable task, since all providers are selling the same commodity, electricity. Not surprisingly, many analysts have concluded that such a commodity market is likely to be characterized by ongoing price wars with firms surviving only through progressive cost cutting.

This chapter argues that there is another way: risk-based pricing (RBP). The opportunity for RBP arises as a consequence of the relatively unique underlying properties of the basic electricity commodity—it is nonstorable with highly variable demand. With deregulation, these properties will cause power prices at the wholesale level to become highly volatile. Retail customers will have very different preferences regarding their willingness to be exposed to such volatile prices. Some customers actually can profit from volatility by shifting usage to low-cost periods; others will find the volatility to be risky and will want stability. Given these almost certain preference differences there is a rich opportunity for product differentiation along the lines that best match customer's risk preferences. RBP involves offering a menu of different electricity sale contracts on terms that best match the distribution of customers' risk preferences, while at the same time providing an attractive return to the retail merchant who develops the products and services the customers.

The purpose of this chapter is to introduce and define RBP. The chapter provides a discussion of the nature of RBP and its place in the emerging market place. It then draws upon quantitative results from a market simulation of the profitability of RBP to demonstrate its relevance to utilities that are moving into the competitive arena. Firms may approach the competitive arena with substantially different market strategies; thus the quantitative results highlight the role and impact of RBP within several disparate strategies. The results, therefore, provide insight not only into the profitability of RBP but into the profitability of differing market strategies.

Although there are practically endless variations on the theme of RBP, the following examples illustrate the major product categories.

- Spot pricing of electricity, similar to one-part real-time pricing.
- Forward pricing of electricity, i.e., the sale of a fixed quantity of electricity today for delivery at a future time, at a pre-specified price.
- Combination of spot and forward pricing of electricity, similar to two-part real-time pricing, in which a predetermined portion of the customer's load is covered by a forward contract, and the balance is traded at spot.
- Guaranteed (or "flip the switch") pricing of electricity that protects the customer against the uncertainties of future spot prices for any quantity the customer might take.
- Guaranteed bill for electricity, which insulates the customer against the uncertainties of some types of usage variation (such as that due to weather) as well as the uncertainty of spot prices.

Similar RBP products have been offered in other industries. The most obvious example is the banking industry. Prior to the deregulation of banking, home mortgages were only offered on a fixed-rate basis. After deregulation, adjustable rate mortgages were introduced to better manage the risk of operating in a competitive market. The adjustable rate allowed banks to offer lower average rates to customers who would bear the uncertainty of rate adjustments usually tied to underlying financial markets. Pricing of Internet services is being widely implemented through a fixed bill mode, at $19.99 a month for unlimited usage. In the airline industry, customers who are willing to buy a nonrefundable fare 21 days in advance represent a lower risk to the airline, and are offered a lower fare. In comparison, customers who buy their tickets less than a week in advance, and want to have the flexibility of changing their minds, generally pay more. Similar concepts exist in the natural gas industry.

MARKET STRATEGIES

RBP may be introduced by firms that are pursuing several different market strategies, and the quantitative benefits of RBP will depend upon the market strategy selected by the firm. A firm that is determined to be the first on the market with RBP products, for example, will see the largest market share impact from introducing the products, if the products are well-designed. Whether such an approach benefits the supplier depends upon the pricing of the product and the cost of such rapid innovation. In order to place the introduction of RBP in the proper context it is necessary to spell out the major market strategies that might accompany its introduction.

The first strategy is market leadership. A market leader would continually develop and introduce forms of RBP in advance of other market players. For the purpose of this chapter, "market leadership" is used to describe innovation with RBP even in those cases where the innovation may occur within markets that are still protected by traditional franchise arrangements. Transmission and distribution firms, for example, are likely to maintain substantial elements of franchise arrangements even in the emerging deregulated markets. Yet such firms still can demonstrate leadership in the introduction of products that manage customers' risk.

A second strategy is market passivity. In adopting such a strategy, a firm simply maintains its existing product line regardless of what its competitors do or what its regulator may urge it to do.

Third, the firm might engage in market imitation; that is, the firm would adopt innovative RBP products, but only after competitors do so. A company might be an imitator because it lacks the creative talent to be a leader, or it might intentionally adopt the imitation strategy so that its competitors absorb the major costs of innovation and of working the "bugs" out of innovative products.

Finally, a firm might integrate RBP with much broader product strategies that involve non-kilowatt-hour products. Such products might involve, for example, the provision of an end-use product like HVAC on a pay-by-season or pay-by-cubic-foot basis. Such a product might substantially alter the risks faced by both the customer and the supplier, and may prove to be highly fertile ground for product differentiation. For the purpose of this chapter, however, we confine the discussion to the kilowatt-hour, commodity-based products.

CUSTOMER MERCHANT PREFERENCES

The effectiveness of RBP in all of the above strategic contexts depends upon how the potential customers view the traditional electricity supplier in comparison to new market entrants. It is likely that incumbent electricity suppliers have a certain amount of "brand recognition," or customer loyalty. This brand recognition may be a potentially powerful tool that the incumbent merchant can use to introduce RBP on price terms that are more favorable than would be available to an entrant. In all of the analyses of RBP in this chapter the computations are carried out for two alternative states of customer loyalty—one in which the customers show a substantial degree of loyalty to the incumbent, and a second in which the customers are indifferent among suppliers and make their choices on the basis of price and product characteristics only.

Types of Risk-Based Pricing

The differences in the risks products engender on providers and customers can be shown best by examining several product types that may be offered in a competitive electricity market. The products we shall consider are spot, spot plus forward, spot plus a price cap, spot plus a collar, time of use, guaranteed price, and fixed bill service. A spot plus a price cap product includes an upper limit on the spot price to the customer. A spot plus a collar includes a price cap and a price floor. The discussion will begin with the least risky product from the merchant's perspective and progress to riskier products. In this context, a merchant is a reseller without either generation facilities or long-term purchase agreements.

Retail prices under a spot-based product are based on wholesale market prices. Since changes in wholesale spot prices are passed directly to the customer, a spot-based product does not impose price risk on a merchant that purchases power on the spot wholesale market. Rather, the customer assumes all price risk with a spot-based product. As a result, from a merchant's perspective, a spot price pass-through product may be the least risky product it can offer to customers. However, the merchant still faces load risk since the customers' usage is uncertain. A fixed quantity, spot-based product would be even less risky to a merchant since it would eliminate the merchant's quantity risk. Such a contract would require either a balancing product such as spot or guaranteed price, or self-generation by the customer.

A spot plus collar product reduces the price risk to customers and increases price risk for the supplier compared to a straight spot-based contract. A spot plus collar product can reduce price uncertainty to a customer for their entire load, or for a prespecified amount of the load, depending on how the contract is written. Lowering the price cap increases the supplier's price risk and decreases customers' price risk. Conversely, raising the price floor decreases supplier's price risk and increases customers' price risk.

A forward plus spot product gives customers some measure of protection from changes in the wholesale market by letting customers purchase fixed quantities of electricity at fixed prices. Customers retain the ability to adjust load to changes in the wholesale price of electricity since deviations from contract quantities are priced at (or near) spot. The merchant faces price uncertainty on the forward contract quantity. This means that a merchant needs to charge more for a forward plus spot product than for a spot-based product. Fortunately, customers should be willing to pay more for a forward plus spot contract since they face less price risk than with a spot-based contract.

Compared to a spot plus collar product, a spot plus price cap product increases price risk to a supplier and reduces price risk to customers than spot plus collars. Price risk to the supplier is increased because without a price floor customers can benefit from low wholesale prices while being protected from high wholesale prices by the price cap. Conversely, a spot plus floor product would reduce supplier price risk and increase customer price risk compared to a spot product.

A time of use (TOU), or time differentiated, guaranteed price product fixes the prices that customers pay for electricity where the price differs with the hour in which the energy is used. As such, a TOU product insulates customers from uncertainty in wholesale spot prices. Customers typically still face higher prices during heavy use hours and low prices during light use hours, but hourly prices are known well in advance and do not fluctuate with the wholesale spot price. As a result, compared to the products that have been described earlier, a TOU product transfers price risk for customers' entire load from customers to the producer. (Unlike a supplier that purchases power on the spot wholesale market, a supplier with either generation facilities or long-term purchase contracts may not view guaranteed price contracts as risky, since long-term generation commitments effectively hedge retail guaranteed price products.)

A guaranteed price product with a flat price—sometimes called a 'flip the switch' (FTS) product—transfers even more price risk from the customer to the supplier than a TOU guaranteed price product. With a flat, guaranteed price product, customers face the same price each hour, irrespective of wholesale price or system load. A guaranteed price typically would be higher than the expected load weighted retail spot price to compensate the supplier for the additional risk associated with it. This is analogous to the "risk premium" built into fixed rate mortgages, when compared with adjustable rate mortgages.

The last type of product we consider is a fixed bill product. With a fixed bill product, customers pay a flat fee for a given time period in which electricity usage does not affect a customer's bill. A flat fee product likely would have several restrictions in addition to a restriction on reselling, which typically is found on all retail products. For example, a possible restriction is that only inelastic customers would be eligible for the rate. Since the marginal cost of additional usage is zero to the customer on a fixed bill product, an elastic customer theoretically would purchase an infinite amount of electricity. This unlikely event can be avoided by making the customer's future fixed bill dependent upon past bills. With a flat fee product, customers have no bill uncertainty, but they do have price uncertainty since the average price per unit depends on usage. A supplier faces both price and quantity uncertainty with a fixed bill product. For example, suppose that the customer's usage is positively correlated with wholesale spot prices and that the customer is on a fixed bill product. When wholesale prices are high, customer demand is high and the retail price per unit is low. As a result, suppliers likely would require a premium for a fixed bill product over a guaranteed price product with more elastic customers being charged a larger premium than inelastic customers. In practice, suppliers are charging a premium of 2 percent to 5 percent in the electricity market.

Economics of RBP

As the above listing of RBP products indicates, both electricity suppliers and customers may face price and load uncertainty. Because the contract terms and customer usage characteristics determine the price and load risks borne by the parties, each combination of electricity product and customer segment can alter the level and allocation of these risks between the supplier and the customer.

Customers will tend to pay more for products that are less risky. This tendency will not be uniform across customers and, depending on their usage characteristics, some customers will have very different views as to the riskiness of different RBP products.

The owner of a vacation home that is used only on weekends, for example, might view a spot product as virtually risk free and would pay little extra to have a guaranteed price. In contrast, the operator of a dairy, who must run milking and processing equipment according to strict schedules every day might find the spot product to be unacceptably risky. Such a customer would be willing to pay extra for a product with more price certainty.

In order to supply less risky products, suppliers must absorb some of the risk, and they may undertake a variety of activities to lessen the impact of absorbing these risks. Such activities may include investment in fixed assets, such as generation assets, or the purchase of hedge contracts. But in general, all of these activities have costs of their own and may in fact generate additional kinds of business risk. For these reasons, suppliers will charge more, in general, for supplying less risky products.

The interaction of customer risk preferences and supplier willingness to provide risk management services in the form of RBP products generates a buyer/supplier risk frontier. Figure 13–1 illustrates the trade-off between the risk a product imposes on the supplier and the risk it imposes on a customer. The risk trade-offs of each product will depend on the wholesale market, customer characteristics, product features, and the set of purchase and sales contracts held by the supplier.

A customer's product selection depends not only on the individual product's features, but on the customer's attitude to risk. If each product is priced only

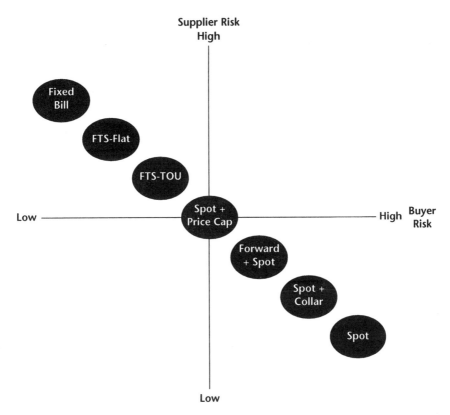

Figure 13–1   *Supplier Risk/Buyer Risk*

to compensate the supplier and/or the marginal customer for accepting the risk inherent in that product, a customer that is more risk-averse than the marginal customer is more likely to select a product from the upper left of the chart, such as a guaranteed price product. Conversely, a customer that is less risk-averse than the marginal customer will be more likely to select a product from the lower right of the chart, such as a spot product.

Benefits of Risk-Based Pricing for the Retail Electricity Merchant

Why should a power supplier be concerned with RBP? The introduction of competition in other industries such as railroads, airlines, and telecommunications, has demonstrated that competition puts substantial and ongoing downward pressure on revenues and profits. In these industries, successful companies responded with product differentiation to gain market advantage. It would be very surprising if these kinds of events do not take place in the power industry as retail competition takes hold. Furthermore, the electric power industry in many areas is particularly vulnerable to profit erosion from competition because plentiful capacity will make it difficult for firms to hold prices at current levels.

Under these circumstances, RBP offers a means for incumbent utilities to increase margins and offset some of the profit erosion that may follow the introduction of retail competition. This is analogous to developments in the airline industry following deregulation. The introduction of frequent flyer programs, advance purchase discounts, and single carrier service through hub cities, were all important tools that airlines used to bolster margins and volume in the face of fierce competition.

Market Simulation

To illustrate the benefits of RBP, we present results developed using EPRI's market simulation model, *Product Mix*. This model has been developed as a tool to help utilities assess the consequences of offering retail products with a variety of risk profiles. The model uses detailed information on wholesale market costs, customer loads, customer response to price variation, and customer choice of products and suppliers.

In this application, the model is applied to a market that is evolving from a traditional franchise market to a market with retail competition. To keep the example simple, we focus on a single customer segment comprising 20 customers with an average load of 50 megawatts. Both the incumbent supplier and the new entrant are pure merchant firms that buy power at spot prices from a competitive wholesale market, and make the power available to retail customers under different RBP arrangements. The application covers a time period of one year, during which the wholesale spot prices range from $8.80 to $26.60, with an expected price of $20 per megawatt-hour.

The model is first used to determine the impact on the incumbent utility of losing the franchise. This value provides a benchmark for assessing the subsequent evaluations of RBP. We then use the model to determine the value of RBP within the franchise context. Next we turn to how RBP can impact firm profitability under the three competitive market strategies: market leadership, market passivity, and market imitation. In all cases the results are presented for two levels of customer loyalty—one in which the customers have some preference for remaining with the incumbent, and the other in which customers are indifferent between the incumbent and its competitors.

Franchise Value

The magnitude of the profits at risk from losing the franchise can be examined by comparing monopoly profits to profits when there is a competitor. In this example, we make the comparison when both the incumbent and the challenger offer only the single guaranteed price product. The cost of losing the franchise also provides a reference for the benefits of the RBP scenarios examined.

Monopoly profits are calculated with the monopoly supplier offering a guaranteed price product to a single captive customer segment. That is, the customers are not allowed a choice of either product or supplier. The supplier's expected profits from the guaranteed price product are $20.5 million with a standard deviation of $30 million. This variation in profits arises because wholesale prices and customer loads are both volatile, while the price paid by the customers is fixed. This result, and other results from subsequent simulations, are shown in Figure 13–2.

The cost to the monopoly supplier of losing the franchise is calculated for two scenarios: one in which customers prefer the incumbent, and a second in which customers are indifferent between the suppliers. For both scenarios we specify that the challenger offers the product at a slight discount relative to the incumbent.

Even when customers prefer the incumbent supplier, and the former monopoly retains 61 percent of its customers, its profits decrease by $8 million from $20.5 million to $12.5 million. The challenger supplier's expected profits are $7.1 million, and the incumbent's cost of losing the franchise is $8 million.

If customers do not have a preference for either supplier, the incumbent's expected profits are $9.6 million and the challenger's expected profits are $9.7 million. In the absence of customer preference for the incumbent provider, the challenger attracts 53 percent of the market, leaving the incumbent only 47 percent of its once-captive customers! Without the benefit of customers' good will, the cost to the incumbent supplier of losing the franchise is nearly $11 million. This example highlights the importance of developing good estimates of customer data such as customer's preference for different suppliers. Such estimates will have a large impact on the incumbent's overall pricing strategy.

Customer Loyalty	Company	Market Structure						
		One Franchise Supplier			Competitive Market			
		RBP Strategy			RBP Strategy			
		Market Leadership			One Product	Market Leadership 2 RBP Products	Market Passivity One Product	Market Imitator 2 RBP Products
		One Product	2 RBP Products	3 RBP Products				
Prefer Incumbent	Incumbent	$20.5 million	$21.8 million	$22.6 million	$12.5 million	$16 million	$8.7 million	$13 million
	Competitor	NA	NA	NA	$7.1 million	$10.7 million	$5 million	$7.5 million
Indifferent	Incumbent	$20.5 million	$21.8 million	$22.6 million	$9.6 million	$13.3 million	$6 million	$9.9 million
	Competitor	NA	NA	NA	$9.7 million	$13.1 million	$7.4 million	$10.2 million

Figure 13–2 *Profits Under Various RBP Strategies*

Market Leadership

The value of market leadership can be determined by modeling profits when products with different risk profiles are offered. The value of market leadership is calculated when customers cannot select their supplier (monopoly), and when customers can select their provider (competition).

If the monopoly supplier offers customers a choice by introducing a spot-based product, its expected profits increase to $21.8 million—an increase of $1.3 million. At the same time, the standard deviation of profits is reduced dramatically—by 50 percent—to $15 million. Customers are split, with 50 percent remaining on the guaranteed price product and 50 percent switching to the spot product. In this case, expanding the product offerings and introducing product choice increases profits by more than 6 percent.

The benefit of further expanding RBP is examined by adding a forward plus spot product. This further increases profits by $770,000. The standard deviation of profits also decreases from $15 million to $12 million. Again, customers are evenly split with one-third of the customers selecting each product. Although adding a third product did increase supplier profits, most of the gains in this example were achieved by introducing the second product. The key is to add products only if the benefits outweigh the costs, including administrative costs, of adding the product.

The benefits to the incumbent supplier of being the market leader once challenger suppliers have entered the market are calculated in this example by specifying that the incumbent supplier offers an innovative RBP product, which in this case is the spot product. Both suppliers continue to offer the guaranteed price product. The benefits of market leadership are calculated both with and without an incumbency bias.

When the incumbent is the market leader and also benefits from customer preference, as shown in Figure 13–2, its expected profits are $16 million and the challenger's profits are $5 million. That is, the value to the incumbent of being the market leader is $3.5 million. The incumbent retains 73 percent of the market versus the 61 percent garnered by offering only one product. In this case, market leadership has paid off with both increased profits and increased market share.

If the incumbent is the market leader, but customers do not have a provider preference, the incumbent's expected profits are $13.3 million, and the challenger's profits are $7.4 million. (See Figure 13–2). Market leadership increases the incumbent's profits by $3.65 million versus offering only one product to customers who have no supplier preference. Commensurately, the incumbent's market share increases from 47 percent to 61 percent.

Market Passivity

The cost of being a passive supplier, of only offering the original product while the challenger is a market leader, is estimated in this example by creating a scenario in which the challenger offers two products, a guaranteed price and a spot product, while the incumbent offers only a guaranteed price product. If customers prefer the incumbent, and the incumbent continues to offer one product while the challenger offers two products, the incumbent's profits decrease by $3.8 million to $8.7 million and its market share drops from 61 percent to 42 percent. Conversely, being the market leader increases the challenger's profits by $3.6 million to $10.7 million, and increases its market share from 39 percent to 58 percent.

If customers do not prefer either provider, the cost to the incumbent of being a market follower is $3.6 million and its market share decreases from 47 percent to 29 percent. In this example, customer supplier preferences impact the overall profits from being a follower or a leader, but they do not appreciably alter the incremental benefits and costs of being a market leader or a market follower.

Market Imitation

The above results show the impact of being a market follower who continues to offer only the original product. It is possible that the incumbent supplier would adopt the new product as well—after the competitors have shown the way. In this example, we calculate the benefit to the incumbent supplier of catching up to the innovative challenger supplier by comparing the scenario in which the incumbent supplier is the follower to the scenario in which both firms offer a guaranteed price and a spot product. This calculation is made for two cases: one in which the customers have a preference for the incumbent and one in which the customers are indifferent among suppliers. A similar analysis could be done to calculate the value to the challenger of catching up to a market leading incumbent supplier.

When the incumbent supplier imitates the challenger supplier such that both suppliers offer two products, and the incumbent is the preferred supplier, the incumbent's profits increase from $8.7 million to $13 million, and its market share increases from 42 percent to 60 percent. That is, the benefit of imitating a market leading challenger is $4.3 million. If customers do not have a supplier preference, the incumbent's profits increase by $3.9 million, and its market share will decrease to 45 percent. As with the benefit of being the market leader and the cost of being the market follower, the presence of provider preference by customers affects the level of profits for copying the challenger's product offering, but does not substantially affect the incremental benefit of copying the challenger's product offering. If a competitor has a market share slightly greater than that of the incumbent supplier, this does not mean a supplier preference on the part of the customer; rather, it shows a preference for lower prices, for the competitor underprices the incumbent in all cases.

In summary, in this example, losing the franchise reduced the incumbent supplier's profits by 39 percent to 53 percent, depending upon customer's provider preference. By being the market leader, the incumbent is able to recover 56 percent to 66 percent of its lost profits. Being a market follower further decreased the incumbent's profits relative to its monopoly profits by 18 percent to 19 percent. Catching up to a market leading challenger supplier increases the incumbent's profits 50 percent to 65 percent relative to being a market follower. Although these results are from a simple example, the results clearly demonstrate the importance of RBP.

The expected profits and incremental impacts (in millions of dollars) of the different RBP strategies are presented in Figures 13–3 and 13–4. The former includes the effects of market leadership for the incumbent supplier when the customers prefer the incumbent, both with and without the franchise. Figure 13–4 presents the incremental impacts on the incumbent's expected profits for RBP strategies under competition when customers are indifferent between suppliers. In each bar chart, the left hand bar represents the incremental effects on expected profit, and the right hand bar is expected profit for the incumbent supplier for each strategy.

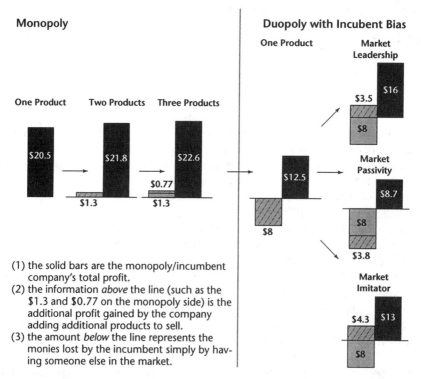

Figure 13–3   *Incremental Impacts of RBP Strategies: Customer Prefers Incumbent*

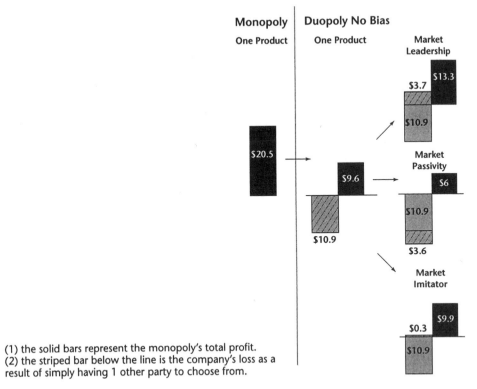

(1) the solid bars represent the monopoly's total profit.
(2) the striped bar below the line is the company's loss as a result of simply having 1 other party to choose from.

Figure 13–4 *Incremental Impacts of RBP Strategies: Customers Indifferent to Suppliers*

Pricing Tactics

The above discussion has been concerned with the broad strategies of RBP. No less important are key tactical questions regarding the detailed structure and pricing of RBP products. The resolution of these tactical issues can have profitability impacts that are as important as the impacts of the strategic issues examined above.

As an example of a tactical issue, consider the proper pricing of a guaranteed price product in a competitive market. Utilities have long viewed customer load factor as a dominant determinant of the cost of serving a customer. However, in recent EPRI research, we have shown that the load factor can be a very misleading indicator of costs. Instead, it is the correlation between a customer's load and the uncertain wholesale price that is the underlying determinant of the cost of offering a guaranteed price product in the context of a competitive wholesale market.

To demonstrate this, Figure 13–5 shows the relative cost of offering a guaranteed price product to three customers. The customer represented by the left-

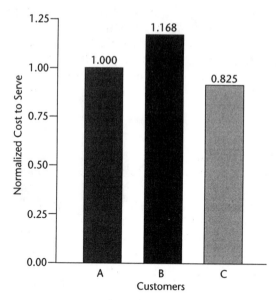

Figure 13–5 *Normalized Cost to Serve Customers*

hand bar (A) has a load factor of 100 percent and therefore has no correlation with wholesale market prices. The customers represented by the next two bars both have load factors of 60 percent, but one customer's load is positively correlated with wholesale market prices while the other customer's load is negatively correlated with wholesale market prices. Costs have been normalized to one for the 100 percent load factor customer. Using the conventional rule of thumb, costs for customers B and C would be expected to be 1.67 (100 percent load factor divided by 60 percent load factor) times those of customer A. However, as the chart shows, relative costs of the guaranteed price product are 1.17 for the customer with positive correlation, customer B, and only 0.83 for the customer with negative correlation, customer C. It is actually cheaper to serve a customer with a load factor of 60 percent, rather than a load factor of 100 percent.

Data Requirements

Strictly speaking, RBP requires only that the supplier knows or can estimate the characteristics of actual and potential customers, and understands the wholesale market in which it operates. That is, RBP is entirely appropriate for a firm operating in a regulated industry. In the absence of a wholesale market, the wholesale "market" could be a supplier's own generation facilities or its purchase contracts.

Competition among suppliers for customers adds the requirement that suppliers consider the behavior of competitive suppliers, and how customers select suppliers whether or not RBP is used. A consequence of the shift in the electric industry from a regulated environment to a competitive market is that profit margins likely will decrease. This suggests that RBP will gain in importance as competition enters the wholesale and retail electric markets. Since RBP is less important to monopoly suppliers than competitive suppliers, we will assume that suppliers compete in a competitive retail market. Naturally, both RBP and a shift to a competitive market increase supplier's data needs.

Customer Data

Electric utilities already collect data on their current customers, such as demand elasticity, flexibility parameters, load shape, and the correlation of load with wholesale price, in order to design and price retail products in a regulated environment. In addition to data currently being collected, RBP (with product and supplier choice) uses customers' provider and product preferences, inertia bias, ability to compare alternative products, load volatility, and aversion to risk. Much of the customer characteristics currently collected and used can be determined by examining actual customer data such as loads and prices. The additional customer data needed for RBP likely will require using data from foreign electricity markets, other products such as natural gas or telephone service, and surveys of U.S. customers, such as those described in Chapters 3, 4, and 5 of this book. The importance of getting accurate values for these additional customer characteristics can be seen by comparing the empirical simulations in which the customer prefers the incumbent supplier to those in which the customer is indifferent between suppliers.

The importance of gaining accurate customer data is brought out clearly in the example of three customers in the previous section. If customers B and C were offered the same guaranteed price product, the cost of serving negatively correlated customer C would be lower than that for serving positively correlated customer B. Even though the customers may appear identical in many respects, the cost of serving the customers may be markedly different. The supplier can lose money (or customers) unless it knows what customer information is important and takes that information into account when it prices its products.

Wholesale Market Data

The principle notion in RBP is that uncertain flows are discounted for risk. This means that suppliers need to evaluate products using forward prices and loads rather than spot prices and loads. That is, suppliers need to convert expected wholesale spot prices to forward prices. Published forward price data may be used if a liquid forward market exists for the commodity. Forward prices will need to be calculated for each relevant commodity that does not have a forward market. Even if a forward market exists, forward prices will need to be calculated for any analysis that extends beyond the time frame covered by the forward market. Calculating forward prices requires the current spot price, spot price forecasts, the rate at which current differences from typical conditions revert to normal, and the expected growth in spot prices.

Competitor Data

As previously mentioned, RBP does not require knowledge of competitor's product offerings or their reaction to changes in the market place. RBP can be used, however, to calculate the break-even price for each product offered. Although RBP provides a price floor for products, the price customers will be willing to pay depends on their alternatives, as well as their demand. This means that information on competitors products and competitors' likely response to your offerings is crucial to profitably offering products that customers will purchase.

CONCLUSIONS

RBP is a powerful business strategy that electric service providers can use to differentiate themselves from their myriad competitors in commodity electricity markets. It has been tested and widely used in other industries. In this chapter, we have discussed several different types of RBP, and shown that they can be used to improve supplier profitability. In addition, by providing customers with more choices, they also enhance customer satisfaction, and can ultimately lead to higher rates of customer retention and acquisition.

CHAPTER 14

A Market-Oriented Approach to Electric Utility Strategy

Kenneth R. Bartkus

For over seven decades, financial and engineering considerations have been the primary driving forces in the development of electric utility strategies. As the deregulation process evolves, however, utilities are beginning to realize that they will have to abandon "the luxury of selling a bulk commodity to a captive consumer at regulated monopoly price."[1] Specifically, when consumers are provided with choice, electric utilities will be pressured to adopt a market-orientated approach to strategic decision-making. At a minimum, a market orientation requires commitment to the following general precepts: (1) satisfying both the consumers' and the utility's objectives, (2) systematically examining consumer tastes and preferences through market research, (3) targeting customer groups using principles of segmentation analysis, and (4) utilizing all elements of the marketing mix in the development of a competitive strategy.

Although the electric utility industry has yet to fully embrace the concept of a market orientation, the experiences of other recently deregulated industries provide some evidence of the competitive implications involved. Who could have envisioned, for example, the concept of frequent flyer programs under a regulated airline industry?[2] Similarly, who could have believed that such innovations as wireless phone technology, phone cards, and caller identification would have been developed under a regulated telecommunications industry?[3] These and other changes occurred not simply as a result of deregulation, but because firms throughout the affected industries adopted a market orientation. Those that did not became casualties of deregulation. In the airline industry, "mergers, takeovers and bankruptcies shrunk a 13-company industry to five major players, whose share of market value rose from 55 percent to 91 percent."[4] If electric utilities are to remain competitive, they will need to learn the lessons of the airline, telecom, and other industries who were forced to develop new strategic philosophies.

The purpose of this chapter, therefore, is to comment on the general principles of a market orientation as they relate to the electric utility industry. Central to this objective is an analysis of what can be done to prepare for deregulation. First let's discuss the concept of consumer satisfaction and how it relates to the attainment of such organizational objectives as customer loyalty and profitability. Next is a discussion on how the principles of market research and segmentation analysis can be used to target customer groups, followed by an analysis of the marketing mix as it relates to a deregulated electric utility industry. Specifically, the chapter comments on how pricing, product, marketing communications, and distribution considerations can affect the ability of a company to compete successfully in a changing environment.

THE SATISFIED CUSTOMER AND COMPANY PROFITABILITY

There is little doubt that customer satisfaction is an important asset to firms operating in a competitive environment. Satisfaction is an asset because it tends to influence loyalty to the company and its offerings and, consequently, profitability. In the regulated environment of the electric utility industry, however, the concept of customer satisfaction often has been overlooked. When consumers have little or no choice, the costs associated with providing higher levels of service become prohibitive. Since utilities have enjoyed the luxury of a captive audience, all that has been required is for the utility to satisfy the marginal standards and guidelines established by the regulatory agency. To be sure, customers complained, but the regulators made the call on whether or not service levels were satisfactory. It is not surprising, therefore, that utilities have been less than responsive to the individual needs of the consumer.

In all likelihood, utilities will become more responsive as the industry moves toward deregulation. As this process evolves, utilities will find that additional resources will need to be allocated towards ensuring that consumers are satisfied at a level that prevents them from switching to a competitor. This, of course, is a complicated challenge. While company resources are finite, consumer needs are often infinite. So the strategic issue is not merely one of customer satisfaction, but satisfaction at a profit. But what exactly is customer satisfaction and how is it related to company profitability? To answer this question, we should start with a formal definition of customer satisfaction and then progress to its relationship with profitability.

Definition of Consumer Satisfaction

Consumer satisfaction can be described as a "person's feelings of pleasure or disappointment resulting from comparing a product's perceived performance (or outcome) in relation to his or her expectations."[5] Three statements should be made about this definition. First, satisfaction is defined as a function of both perceived performance and expectations. If a person's expectations are low, then satisfaction can be attained with lower levels of performance and vice versa. Therefore, utilities are faced with the task of ensuring that perceived performance and expectations are well understood. Second, the definition implies that satisfaction at the margin is sufficient to retain customers, an implication that is largely unsupportable by survey data. Third, one might draw the conclusion that satisfaction is directly related to profitability when, in fact, the relationship is often mediated by customer loyalty. Since each of these statements has important implications for the development of a competitive strategy, some elaboration is warranted.

Customer Satisfaction, Perceived Performance, and Expectations

It should be obvious that both perceived performance and expectations about performance can have a dramatic affect on customer satisfaction. With regard to perceived performance, it may be important to note that consumer perception may be contrary to the facts. Service may be adequate, even exemplary. However, if the customer thinks it is sub-par, satisfaction will suffer. The important lesson is that companies need to develop a better understanding of both perceived and actual performance.

On the one hand, if perceived performance is lower than actual performance, the utility has an image problem. The solution is simple: the company needs to develop and implement a marketing communications strategy that can effectively inform consumers that the company's offering has valued benefits. Apple Computer ran into this problem when it launched the MacIntosh personal computer. Since it didn't look like any other personal computer at the time, many consumers thought it was simply an over-priced toy. Apple responded by implementing an aggressive marketing campaign to overcome consumer misperceptions. It worked, for a while.

An equally troubling condition is when perceived performance is greater than actual performance. In this case, the consumer is getting "ripped-off." While some companies may revel in the thought of having this "problem," they should remember that it is a very real one indeed. In today's sophisticated market, it won't take long for a consumer to develop post-purchase dissonance. In this scenario, the consumer is likely to switch providers, and never return.

Consumer expectations of performance also can present problems for the firm. This is due to the fact that the formation of consumer expectations are not under the complete control of the company. Instead, external factors, including competitive marketing communications, adverse public relations, and word-of-mouth communications, may influence expectations.

Despite the relative lack of control, companies should be cognizant of such factors and attempt to neutralize those that provide the greatest threat. The classic battle with regard to neutralizing tactics involves MCI and AT&T long-distance communication services. Each company has attempted, in its own way, to neutralize the competitive communications of its rival. As distasteful as such advertising may be, there is little doubt that it has become necessary to counterbalance competitive claims.

Aside from responding to external sources of influence, companies also must plan and execute their own strategic influences. In doing so, they should provide a balance between what can be realistically delivered in the way of performance and what can be realistically promised. If a utility promises more performance than it can deliver, expectations will rise beyond reasonable levels and customers are likely to be disappointed. Conversely, if a utility promises too little, consumer expectations will decline and the company will have difficulty attracting new customers.[6] In either case, the company has committed a blunder by not coordinating the promises with consumer expectations. To help alleviate such problems, utilities should adhere to the following age-old maxim: promise only what you can deliver and deliver more than you promise.

With regard to influencing expectations, some companies have been relatively successful. Enron, a power marketing and generation company, spent $30 million to $40 million in 1997 to advertise its name. It must be working. Studies have shown that Enron's recognition rate in California alone is somewhere between 20 percent and 30 percent.[7] And although the evidence is still out on whether this will translate to brand switching, anecdotal evidence is mounting. Insiders at Puget Sound Energy joke that PSE actually stands for 'Pretty Soon Enron.'[8]

Marginal Satisfaction, Customer Loyalty, and Profitability

The definition of consumer satisfaction may lead one to conclude that if consumers can be satisfied at the margin, they will remain loyal to the company and its offerings. Unfortunately, this notion is often contrary to the facts. The trouble with the marginal satisfaction approach is that a satisfied customer may be more likely to switch to a competitor than one who is *highly* satisfied. With regard to highly satisfied customers, one study found that 75 percent of Toyota buyers were highly satisfied with their purchase and about 75 percent of these said they intended to buy a Toyota again.[9] In this case, high levels of satisfaction translated to customer loyalty.

But that is not always the case. Consider a survey by Coopers and Lybrand involving the recently deregulated U.K. electric utility industry. The statistics show that while "almost 60 percent of consumers declare a high level of satisfaction with the service they receive, their contentment does not translate into loyalty, with just 39 percent admitting to any sense of loyalty. But even this loyalty is fragile in the face of attractive alternatives – only 29 percent pledge not to switch allegiances."[10]

Clearly, the relationship between satisfaction and customer retention can be nebulous. This is due to the fact that the process of determining what motivates consumers to act in a certain way is relatively complex. One thing is for certain: utilities must be able to retain as well as attract customers. Lessons from the telecommunications industry provide some evidence to this effect. In particular, a recent study asked consumers their behavioral intent with regard to changing long-distance providers. The percent that actually changed providers (after saying they would) ranged from a low of 36 percent to a high of 64 percent.[11] More important is the fact that not all telecom companies have the same retention rates. AT&T, for example, has the highest retention rate in the industry.

Another important point is that the automobile industry has a long history of consumer choice. As such, consumers always had the right to exercise choice. In doing so, they established patterns of behavior that were consistent with the satisfaction-loyalty relationship. In the telecom industry, however, choice was a newly acquired right with all the privileges to experiment with alternative providers, hence the higher rate of switching. The electric utility industry should prepare itself for a similar phenomenon. As choice is introduced to the market, utility consumers likely will be driven by the need to "exercise choice and the feeling that the electric utilities are not meeting their needs. They may not have their needs met by switching, but they will leave in order to find out."[12]

In short, satisfaction must be translated into loyalty if utilities are to attract and retain consumers. This can be accomplished by providing exceptional value vis a vis competitors and by creating an environment where the consumer views the utility as a partner rather than an adversary. But a newly deregulated electric utility industry likely will experience a period of uncontrollable consumer volatility. Therefore, utilities that focus on longer-term objectives and provide quality service will be the ones most likely to succeed in attracting and retaining customers.

At this point, the relevant questions are: (1) How do we know what the customer wants? and (2) What if different consumers want different products and services? To answer these questions, utilities must rely on information provided by market research and segmentation analysis.

MARKET RESEARCH AND SEGMENTATION ANALYSIS

The adoption of a marketing orientation requires a commitment not only to understand customer needs and wants, but ability to effectively target consumer audiences.

Market Research

Market research of consumer behaviors can be dichotomized as either qualitative or quantitative. Qualitative research assists in the development of tentative information about the market. Techniques such as focus group interviews, observation studies, and juries of executive opinions help strategists develop an understanding of general phenomenon. Qualitative research cannot develop, however, an understanding of the consumer based on probabilistic outcomes. Quantitative research is required for that. Quantitative research is concerned with analyzing data for the purpose of confirming, refuting, or otherwise describing some marketing phenomenon.

When examining markets, most researchers suggest that it is desirable to utilize qualitative research first in order to "gain insight into both population whose opinion will be sampled and the subject matter itself. Usually this approach is thought to result in a better designed, more informed survey instrument."[13]

The Importance of Qualitative Research

Qualitative research has emerged over the last decade as an increasingly important method for analyzing consumer behaviors. It typically is equated with focus groups, although there are many other approaches to qualitative research. Focus groups tend to be the most common form of qualitative research, however, so the following discussion will be restricted to the advantages and disadvantages of this method.

At its most fundamental level, the focus group interview is defined as an "unstructured, free-flowing interview with a small group of people," typically in the range of six to 10 persons.[14] As with most qualitative methods, there are advantages and disadvantages to the focus group interview method.[15] The primary advantages are that they can be conducted quickly, they are easy to conduct, and they are inexpensive. The most significant disadvantage is probably the fact that, given the small sample size, the results cannot be generalized to the population with any degree of confidence. Nonetheless, focus group interviews can provide important insights into the company, its offerings, and its image.

Perhaps the most famous example is from the Ford Motor Company. In the mid-1950s, corporate executives were interested in developing an automobile to be named after one of Henry Ford's sons: Edsel. When market researchers conducted interviews, they were met with responses such as "diesel," "Ethel," and "pretzel"; hardly the type of associations one would embrace. Whether or not the name contributed to the failure of the Edsel is debatable. The point is that qualitative research provided rather concrete evidence that consumers were not receptive to the brand name Edsel, but the Ford executives didn't listen. This illustrates that misuse of research can contribute to deleterious effects. In essence, marketing research "is not worthwhile when the potential research users have already decided on a course of action or are unlikely to consider research-generated insights with an open mind."[16]

Utilities that opt to utilize qualitative research also should be careful not to generalize. This occurs when the qualitative evidence is so intuitively compelling that the company decides to generalize the results of a few focus group interviews to millions of current and potential customers. While this may turn out to be true, conventional wisdom dictates that focus group interviews typically are followed with quantitative evidence obtained from representative samples. Qualitative research is most useful when it serves to clarify important market information. By refining the issues to be considered for further study, the benefits and costs of quantitative verification are better realized.

The Importance of Quantitative Research

While qualitative research is based primarily on the exploratory side of a market investigation, quantitative research focuses primarily on verification. The vast majority of consumer-based quantitative research is based on describing the characteristics of a particular market or on the statistical testing of a market relationship. With regard to descriptive studies, much of the research has been devoted to providing information on the attitudes and behavioral intent of consumers. For example, a considerable amount of research has focused on describing the "green" energy market. In virtually every study, consumers have expressed a willingness to pay more for electricity that was "environmentally sound." Unfortunately, what consumers say they will do is often not the same as what they actually will do, at least in the short term. In other words, the rate of adoption for a "new" product is often contingent on a host of factors in addition to behavioral intent.

For example, some consumers may be hesitant to commit to "green energy" until they have competent and reliable evidence that the energy is indeed "environmentally friendly." To help consumers determine what sources do (or don't) live up to their claims, "[g]reen power certification is progressing on several different fronts. Commissions and legislatures are mandating resource disclosure of generation sources. Legislation at both the state and federal level are investigating the use of renewable portfolio standards. The Federal Trade Commission is considering establishing voluntary guidelines for electricity advertising like those already established for other retail products in the FTC's 1992 "Guides for the Use of Environmental Marketing Claims' (16 CER 260)."[17] Given consumers' predisposition to favor environmentally friendly products, these certification efforts are likely to increase the rate of adoption of green energy in the future.

Most environmental researchers now agree that descriptive studies previously overestimated the size of the market for green energy. Therefore, inferences derived from descriptive research should be applied with discretion unless there is additional evidence to support the conclusions.

Perhaps the most important application of descriptive research is in the area of segmentation analysis. The ability of segmentation analysis to accurately describe the demographic, geographic, psychographic, and lifestyle profiles of consumer audiences facilitates effective target marketing. This goes beyond the typical application of descriptive research of using usage rates or the consumer/business dichotomy. While these applications undoubtedly are important, they represent only the initial exposure to understanding the consumer.

With regard to statistical testing of market relationships, a utility might want to examine the extent to which a consumer might consider switching to another provider, given a change in the marketing mix. Typically, this involves price-sensitivity issues, but it needn't be so restricted. Consumer surveys dealing with other aspects of the marketing mix are also important. As a result, an increasing number of studies are devoted to understanding the extent to which other factors might influence a purchase decision. Specifically, while price still appears to be the most important factor in the purchase decision, studies also have shown that when prices are fairly compatible, "reliability, customer service, and special services become deciding factors to many consumers."[18]

This discussion suggests that both qualitative and quantitative research can provide useful information about consumers and the marketplace. Used intelligently, market research can provide a strategic advantage to the firm.

Segmentation Analysis and Target Marketing

The process of segmenting markets so that the most desirable ones are targeted is an important consideration in the strategic development process. Historically, electric utilities have taken a relatively passive approach to segmentation. In some cases, the product is viewed as a homogeneous product with little potential for differentiation. Perhaps the most typical segmentation process is to segment markets according to usage levels. As deregulation takes hold, utilities will find that more comprehensive methods of analysis and targeting are necessary to remain competitive.

In general, effective segmentation has four basic requirements. First, the market segment must be measurable in terms of purchasing power and size. There is a reason that geographic and demographic segmentation are the most commonly used bases for segmentation; they are easily measurable. Psychological and lifestyle bases are used less frequently as the primary basis because they are not as tangible. On the downside, geographic and demographic bases provide little insight into the lifestyle and personalities of the segment. As a result, most segmentation analysis starts with the bases that are most tangible and complements them with psychological and lifestyle analysis.

Second, the market segment must be economically accessible. A segment may be defined, but if it cannot be economically accessed it has little strategic value to the utility. This is another reason why demographics and geographic bases of segmentation are often utilized as the primary factors in targeting a specific audience.

Third, the market segment must be large enough to be profitable. A segment may be measurable and accessible, but if the segment is not large enough to sustain as separate strategy, it has little value. Consider, for example, why so few G-rated movies were produced during the 1960s. The most logical reason is that there were not enough children. It was not until the baby boomers married and had children of their own that the segment for G-rated movies once again became profitable.

Today, some utilities are betting that the green energy market segment is large enough to be economically profitable. This may be so, but at this time it is unclear whether or not the segment is large enough to support more than a few utilities. If the cost of supplying such energy is reduced over time, the segment base undoubtedly will increase in size and allow for the entry of more competitors.

Fourth, the market segment must match the capabilities of the company. This requirement is directly linked to the concept of meeting customers' expectations. If a utility decides to enter a market segment for which it does not have appropriate capabilities, it will be unlikely to meet its customer obligations. In essence, while it might be able to attract some customers by promoting a quality of service expected from that consumer base, the company will not be able to retain them over time. The long-term implication should be obvious: erosion of brand image.

Once these requirements are well understood, the company can begin the process of segmentation. Although there are numerous methods for analyzing consumer segments, the one most consistent with a market orientation is benefit segmentation. As the name implies, benefit segmentation delineates the market according to the attributes or benefits sought by consumers. The underlying assumption is that consumers do not purchase products or services, they purchase benefits. The most famous example is in the toothpaste market. A 1963 study[19] identified five benefits sought by consumers: low price, decay prevention, bright teeth, and good taste. Each segment had a distinct consumer profile related to demographics, lifestyle, and psychographics. Consumers that were likely to seek the benefit of bright teeth, for example, tended to be teens and young adults (demographic), smokers (lifestyle), and/or highly sociable and active (psychographic).

A review of the electric utility market indicates that some utilities already are moving toward benefit segmentation. The marketing of green energy is a clear example of providing a benefit rather than a commodity. Consumers in this segment often are willing to pay more for the product because it provides the benefit of social responsibility. While not all consumers value this benefit, there is at least a portion of the market that is responsive to it. Others may respond to more personal customer service or lower prices. The point is that the market can be divided according to a variety of benefits. The challenge for the utility is to examine those benefit segments that match the offering of the firm. In doing so, it not only relies on its competitive strength (the benefit), but it ensures that the market is effectively and efficiently segmented.

MARKETING MIX ELEMENTS

The development of an effective competitive strategy requires a thoughtful consideration of the marketing mix elements. They include price considerations, product and service considerations, communications considerations, and distribution considerations. Each of these has important implications for the development of an effective utility strategy.

Price Considerations

There is little doubt that price will remain an important factor in the consumer decision process. Time and again studies have shown that consumers are generally receptive to lower prices. But this is not really news. When reliability and service are held constant, sophisticated shoppers almost always will go with the low-cost provider. The problem is that reliability and service may not always be held constant. As such, one utility may be able to charge a higher rate and still attract and retain consumers who fit their "value" profile. Therefore, utilities will need to become creative in their pricing strategies. One emerging trend is competitive bidding.

As the name implies, competitive bidding starts when the user issues a request for proposal (RFP) and invites providers to submit competitive bids. The result is a customized, negotiated contract that usually saves money for the user. While utilities may wince at the idea of having to submit competitive bids, bidding is becoming commonplace in the telecom industry.[20]

What types of clauses can be expected in an RFP? Evidence from the telecom industry suggests that RFPs will ask for clauses involving business downturn or divestiture, technological advances, rate stability and indexation, account and service support, billing format requirement, and commitment levels and exclusivity. Utilities probably will want to add an exit fee to cover stranded costs and this may be where semantics plays a part. These charges are likely to be called "competitive transaction charges" because a "stranded cost charge" sounds more ominous.

The bottom line is that utilities likely will be facing increasing competition for large customers and it appears that the bidding process will become more commonplace. Utilities that offer value pricing will be the most likely winners in the bidding wars. Low price alone will be insufficient. As one telecom analyst has argued, "If you always market yourself on price, the only thing you're going to do is create a lot of fickle customers."[21]

Product and Service Considerations

Product and service considerations will become more important as consumers are provided with a greater choice of providers. As a result, utilities may seek competitive advantage by differentiating the physical characteristics of the product or service. Although the industry has not fully embraced the concept of green energy, Green Mountain Energy Resources sees the potential for 20 percent to 30 percent of California customers opting for a green brand over time.[22] Puget Sound Energy announced plans to provide automated meter-reading in its service area,[23] and San Diego Gas and Electric Company has installed Smart Meters with radio links that monitor energy use in real time. One benefit of this innovation is that unusually high rates can be spotted, the customer notified,[24] and the net result is better quality service.

How companies respond to customer concerns is also an indication of the service quality. It has been noted that

> [of those] customers who register a complaint, between 54 and 70 percent will do business again with the organization if their complaint is resolved. The figure goes up to a staggering 95 percent if the customer feels that the complaint was resolved quickly. Customers who have complained to an organization and had their complaints satisfactorily resolved tell an average of five people about the treatment they received.[25]

The problem is that consumers are unlikely to complain. Studies show that less than 5 percent of dissatisfied customers complain, and that most people simply will buy less or switch to another supplier.[26] Or they could decide to opt-out of utility service and build their own stand-alone unit. Walgreen's in Peoria, Illinois is moving in that direction right now.[27] How should a utility address these dilemmas?

There are a variety of tools to help monitor product and service quality.[28] First, companies should develop and implement a formal complaint and suggestion system. By making it convenient for customers to provide feedback, the company is able to gain potentially important information and act more rapidly to resolve problems. Second, companies should conduct regular customer satisfaction surveys as a means of auditing service levels. These surveys should be considered an investment in market research rather than an expense.

Communications Considerations

Communication considerations also will become increasingly important as consumers' need for information increases. In its most fundamental sense, marketing communications should inform consumers about the company, its offerings, and most importantly, the benefits that can be derived from doing business with the firm.

Although most people think of advertising when they think of marketing communications, in reality there is a multitude of tools. Public relations, for example, can play an important role in a utility's communication strategy. A Commonwealth Edison Company program provides energy discounts to schools that participate in a pricing experiment that teaches children in grades kindergarten through 12 about conservation and energy issues. In return, ComEd learns about energy usage and how it might change as deregulation takes hold.

Brand naming is another important communication consideration. Although the process of selecting a brand name may appear intuitively obvious, there are a few basic rules that should be followed. First, the brand name should distinguish the product from competitive brands. Second, the brand name should describe the product or its benefits (e.g., 3M's Post-it Notes). In this regard, brand names sometimes are derived from common words (e.g., Acura, Infiniti). Third, the brand names should be compatible with the desired image of the company. For example, the name "Santa Clara Electric Department" did not connote the image that city officials desired. As a result, the name was changed in 1998 to a more captivating name, "Silicon Valley Power."[29] Fourth, the brand name should be memorable and easy to pronounce (e.g., Enron).

Aside from creating a good brand name, companies also need to develop creative communication channels. One of the more interesting trends is the use of sponsorships. Cinergy Corporation, for example, is paying $6 million over five years for the rights to rename Cincinnati's Riverfront Stadium to Cinergy field. Edison International followed by sponsoring Anaheim Stadium. Its new name? Edison International Field of Anaheim. The Pittsburgh Civic Arena, home to the Penguins Hockey team is now the Allegeny Energy Dome.[30] Touchstone Energy, the brand launched by the National Rural Electric Cooperative Association in 1998, won a 10-year naming rights contract to the River Centre convention center complex in St. Paul, Minnesota, and is sponsoring both a NASCAR race and a pro golf tournament.[31] While such sponsorships may appear expensive, the rewards can be far reaching. Cinergy spokesperson Steve Brash, for example, maintains that their sponsorship of Riverfront stadium is an "effective way to build brand-name recognition on a national basis."[32]

An additional trend in utility communications involves the use of sales promotions. Perhaps the most creative one was developed by Green Mountain Energy Resources of Vermont. After recognizing the overall market trend toward green products, the company began promoting use of its environmentally friendly power sources and offered customers "Eco-Credits" for energy

saving deeds such as dining by candlelight.[33] Not all promotions are successful, however. Enron, for example, offered California residential consumers an
incentive of two free weeks of electricity. The promotion was discontinued
when Enron pulled out of the market, citing too much consumer confusion
and too little opportunity for profit.[34]

DISTRIBUTION CONSIDERATIONS

Consideration of distribution strategies will intensify as traditional channels
become more complex in terms of its participants. For example, multi-level
marketing plans already crowd the Internet with promises of fortunes as a
utility distributor. While many of these may turn out to be illegitimate business operations, the effect of their presence has important implications for
the industry as a whole. Similarly, utilities may look for competitive advantage by forming strategic alliances, joint ventures, or cooperatives with other
firms in the industry. Enron Power Marketing and Amway, for example, have
joined forces to pursue retailing activities.[35]

CONCLUSIONS

The adoption of a market orientation requires a substantial shift in the organizational practices of electric utilities. The lessons of other deregulated industries has provided an indication of the kinds of changes that can be
expected as the electric utility industry adjusts to market competition. While
many utilities already have displayed a willingness to adopt a market orientation, others have taken a more passive approach. This may be because they
lack the initiative, or maybe they believe that the industry is not really in a
fundamental transition. In all fairness, the railroads did not have the historical evidence to guide them. The electric utility industry has no such excuse.
The evidence of competitive change within the airline and telecom industries, post-regulation, is sufficient to direct utilities in the right direction.
Successful utilities of the future will be the ones that "find benefits to position against that are valued by customers, are unique as compared to the
benefits of other companies, are credible in the minds of customers and have
benefits that fit with the overall strategy of the company."[36]

NOTES

1. Peter Navarro, "Electric Utilities: The Argument for Radical Deregulation," *Harvard Business Review*, January-February 1996, 122.

2. See, Robert L. Thornton and M.K. Thornton, "Who Ambushed Airline Deregulation?" *Business Horizons*, January-February 1997, 41–46.

3. Lindsay Audin, "Techniques for Navigating the Utility Landscape," *Energy User News,* June 1998, 16.

4. Lori M. Rodgers, "Advertising and Branding: Are Utilities Getting it Right?" *Public Utilities Fortnightly*, 15 January 1998, 57.

5. Philip Kotler, *Marketing Management: Analysis, Planning, Implementation, and Control*, 9th ed. (New Jersey: Prentice-Hall, 1997), 40.

6. Kotler, *Marketing Management.*

7. John R. Hall, "Deregulation Marketing Blitz Doesn't Seem to Impress Consumers," *Air Conditioning, Heating, and Refrigeration News*, 8 June 1998, 8.

8. Nancy J. Kim, "Utilities Warming up to the Power of Marketing," *Puget Sound Business Journal*, 13 February 1998, 8.

9. Kotler, *Marketing Management*, 41.

10. Statistics are from a Coopers and Lybrand study, as referenced in "Shock for the Electricity Market," *International Journal of Retail & Distribution Management*, 26(1), 60.

11. Kerry Diehl and Rich Gillman, "Why Your Customers Switch," *Public Utilities Fortnightly, 15* April 1997, 37–40.

12. Kerry N Diehl, "Utilities Can Avoid Telecom Deja Vu," *Electric Light and Power*, August 1997, 36–37.

13. Michael L. Garee, and Thomas R. Schori, "Focus Groups Illuminate Quantitative Research," *Marketing News*, 23 September 1996, 41.

14. William G. Zikmund, *Exploring Marketing Research*, 3rd. Ed. (Orlando, FL: The Dryden Press, 1989).

15. For a comprehensive review, see, Pamela L. Alreck and Robert B. Settle, *The Survey Research Handbook*, 2nd ed. (Chicago, IL: Richard D. Irwin, Inc., 1995).

16. A. Parasuraman, *Marketing Research* (Addison-Wesley Publishing Company, 1986).

17. Lori M. Rodgers, "Green Electricity: It's in the Eye of the Beholder," *Public Utilities Fortnightly*, 15 February 1998, 40.

18. Heath, Rebecca Piirto, "The Marketing of Power," *American Demographics,* September 1997, 60.

19. Russell J. Haley, "Benefit Segmentation: A Decision Oriented Research Tool," *Journal of Marketing Research*, July 1963, 30–35.

20. Tom Hug, "How to Select a Long-distance Carrier and Save Big Bucks Too!" *Communications News*, June 1996, 10–11.

21. Robin Ceurvorst, vice president of Market Facts, quoted in Heath, "The Marketing of Power," 62.

22. Kristen Bole, "Company Plugs 'Green' Power: Green Mountain Sells Energy at Macy's," *San Francisco Business Times*, 19 December 1997, 1–2.

23. Peter Neurath, "Puget Sound Energy Plans Automated Meter-Reading," *Puget Sound Business Journal*, 13 March 1998, 10.

24. Joshua Levine, "What Brand Is Your Electron? Worried about Deregulation, Electric Utilities Are Starting to Pay Attention to their Customers," *Forbes*, 28 August 1995, 108–9.

25. Karl Albrecht, and Ron Zemke, *Service America!* (Homewood, IL: Dow Jones—Irwin, 1985), 6–7.

26. Kotler, *Marketing Management*.

27. Toni Mack, "Do-it-Yourself Power," *Forbes*, 9 March 1998, 54–55.

28. See, Kotler, *Marketing Management*.

29. Peter Delevett and Alastair Goldfisher, "Deregulation Sparks Ad Drive," *The Business Journal*, 6 April 1998, 1–2.

30. Rodgers, "Advertising and Branding," 53.

31. See various press releases on http://www.nreca.org.

32. Heath, "The Marketing of Power," 60.

33. Delevett and Goldfisher, "Deregulation Sparks Ad Drive."

34. L. Patrick Briody, "Notes on the California Experiment," *Electrical World*, July 1998, 47.

35. See, "A Major Endorsement? Amway and Enron Link Up, Downplay Their Venture," *Public Utilities Fortnightly*, 1 May 1998, 30.

36. Erika Seamon of Kuczmarski & Associates, quoted in Denise Warkentin, "Developing, Managing Brand Image Becomes Necessary Business Practice," *Electric Light and Power*, June 1997, 21–23.

CHAPTER 15

Managing Customer Choice: Implications from Market Segmentation Research, Theory, and Practice

David C. Lineweber, Ph.D.
Kerry N. Diehl
Patricia B. Garber, Ph.D.

The onset of retail competition in energy markets has engendered a good deal of wonder, and speculation, about how customers will choose among energy providers. In lieu of a better strategy, many energy services providers (ESPs) have said that they will follow the mantra of the recent business press and focus on customer loyalty. On its face, this seems to make good business sense: the costs of acquiring a new customer are generally higher than the costs of retaining a satisfied customer. Serving a loyal customer base seems intuitively more appealing than acquiring and retaining fickle, unruly, or short-lived customers.

Pursuing this strategy is difficult, however. It means, as a starting point, having a sparkling understanding of what customer loyalty really is, and how to create it. Even more importantly, it means being able to—and being committed to—doing the things necessary to create the very customer loyalty that is the object of so much desire. Unfortunately, while much of the business press seems to suggest otherwise, customer loyalty is neither a simple concept to define, nor a simple thing to create. As this chapter will argue, understanding customers' choices and decisions—and the role that customer loyalty plays, and can be made to play, in those decisions—is complex, and requires an understanding of a multifaceted set of dynamics between the customers, competitive providers, product features, and the nature of the product "offers" available on the market.

This chapter will explore the concept of customer choice and the way that customer choice in energy markets is likely to be affected by customer loyalty. Our goal is to provide a "how-to" guide that will assist ESPs as they attempt to acquire and retain market share and create and leverage customer loyalty in that process. Understanding these issues should help ESPs answer "big" business questions, such as:

- What offers should I make, and to whom (in order to be profitable)?
- Should I lead with brand, or products, or offers?
- What can I do now that will help me build customer loyalty, market share, and margin later?

This chapter is organized into two sections. The first section reviews customer choice theory and provides a framework for understanding customer acquisition, retention, and loyalty, both from the point of view of the customer and from the point of view of the supplier. Exploring these issues from the point of view of the customer will show why customers make the choices they do in a competitive energy market, while exploring these issues from the point of view of the supplier will show the outcomes that suppliers might face as they choose one marketing strategy over another.

The second section of this chapter lays out an action strategy based on the practical implications of this framework. It describes the information needs of the marketing manager in a competitive energy service organization and uses 'green power' as a case study in exploring how the value that customers attach to different product attributes can be used in implementing an actionable market segmentation strategy and customer loyalty program.

CUSTOMER LOYALTY AND CUSTOMER CHOICE

Customers, whether they are business people or residential homeowners, are constantly bombarded with a wide array of choices of products and services that represent endless combinations of product features offered by providers of varying reputation and quality. In order to deal with this complexity, customers typically develop buying rules that help them to simplify their purchases. Some of those "rules" might result in customer loyalty (when the rule is, "I always buy Pepsi," for example). But talk to customers in more detail about this process, and one finds that rarely are the "loyalty"-inducing rules that customers use this simple. Different buying "rules" that customers have articulated which might be thought of as engendering loyalty include the following:

- "I buy from Company X every chance I get, but sometimes I buy from Company Y when it is more convenient."
- "I buy most of my service from Company X, and they're great, but there's this one little problem that only Company Y can fix, so I call Company Y occasionally when I need them."
- "I usually buy from Company X because they carry my favorite top-quality brands and give great service. However, our warehouse doesn't really need top-quality or immediate service, so when I need something for the warehouse, I go to Company Y because they're cheaper."
- "I usually buy from Company X because I like the person I deal with, but when she is unavailable, I turn to Company Y to get immediate service."
- "I almost always buy from Company X—I'm stuck with them because I get a pretty deep discount through their frequent buyer program. But I've had problems with them and as soon as Company Y decides to accept my 'enrollment points' I'll never do business with Company X again!"

Each of these customers is, in fact, "loyal" in one way or another, and each is "loyal" to a specific component or "attribute" of the product, service, or offer available to them from Company X. Ultimately, from the customer's point of view, loyalty is not a simple matter of "repeat purchase" or "sole source," but rather is a result of a set of complex practical, emotional, and idiosyncratic decisions. Loyalty, in other words, is simply a label that can be placed on some set of customer choices, but whether or not to call some choices "loyalty" or not can be a tricky business.

From the supplier's side of the equation, the issue is no less simple. Companies generally see customers only at one point of contact—theirs. Not unexpectedly, most companies measure what they can see. For this reason, suppliers tend to track customers' purchases and repurchases. They measure continuity of customer purchases—and with varying degrees of accuracy—and measure the relative penetration of their products and services within their target market. Since suppliers only see the tangible outcome of customer decisionmaking, they generally measure only the end result. As a result, most companies do not know how customers made their the purchase decisions, nor what other considerations were dominant factors in the customers' decisions. For this reason, suppliers are really no more clear about which customer choices they should call customer loyalty than are customers. Just because a customer buys from you twice, for example, does this make them a loyal customer? If a customer stops buying from you after having bought from you 10 times, have you "lost" a loyal customer? In each case, there are clearly situations in which the answers to these questions are

"No." A customer may buy multiple times out of some random confluence of events (because the road he or she typically uses to go home is under construction, for example).

Calling some choices examples of customer loyalty, in other words, ends up being an important and somewhat tricky decision. Not all repeat purchases represent customer loyalty and not all loyal customers regularly repeat their purchases. Considering both the customer side of this issue and the supplier side, it is clear that what is to most customers a complex, textured, and potentially idiosyncratic set of preferences and behaviors (customer choices and customer loyalty) is typically measured and tracked in ways that do not recognize the complexity of customer loyalty processes.

Up to this point, the energy industry has been no exception. Most ESPs have enjoyed a captive market in a high-margin, protected market. As the industry deregulates, the needs of ESPs are driving them to reexamine their conceptualizations of customers' needs, concerns, and loyalties. As ESPs begin to function as full-fledged competitive businesses, however, they face questions about how they should measure and manage customer loyalty as well. The real question, however, is whether or not ESPs can and should simply borrow the customer choice and customer loyalty measurement tools used in other industries, or whether it is both viable and appropriate to develop new methods of understanding and measuring customer choice that should provide better insight for business decisionmaking.

EVERYONE MAKES CHOICES

Decisions Customers Make in a Competitive Environment

Applying existing tools to the prediction of customer choice in energy markets is difficult for a variety of reasons, but one important reason is that it is almost impossible to describe with any certainty the choices that energy services customers will face over the next one to two years. In some states, such as in California or Rhode Island, customers will have options to select a "standard offer" with their incumbent ESP, or choose a competing provider offering one of a variety of new or different products. In this situation, customers actually can "choose" the standard offer for at least two quite different reasons. For some customers, the decision to stay with their current provider is an active decision. They think about the decision and determine that they want to stay with their current provider because the standard offer is the best offer available, at least from their perspective. For other customers, the decision to accept the standard offer is really a "non-decision;" the end result of simply not being able, or not wanting, to decide what offer to select.

While the standard offer environment creates decision problems for customers, this is not the only context that customers will likely face. In other regulatory environments, the options that customers will face may be quite different. Some states may not require a standard offer, for example, or they may bid out the standard offer so that the provider of the standard offer may not be the "home" utility. Additionally, in some states there may be a relatively small number of competitive suppliers making a limited number of offers for service, while in other states there may be many ESPs making a wide variety of offers. (Some will offer "green power" under multiple specifications and prices, some will offer greater or lesser amounts of customer service or reliability for a greater or lesser price, and some will offer more or less in the way of bundled additional services).

If ESPs can understand how customers will choose from among competitive options, then they will know how to capture and retain customers most effectively. Some of those choices will involve "switching" from one supplier to another. Similarly, some of these choices might be called "loyalty" (when a customer actively chooses to remain with a current provider).

Understanding Customer Decisions

At some level, it's not very helpful to say that predicting customer choice in competitive energy markets is difficult. Certainly there are competing theories these days about what customers will do, which lead to different strategic conclusions about what ESPs should do to win market share and build loyalty. But absent "gut feelings," are there any analyses of customer choice behavior that can provide some insight about what customers will do, and as a result, suggest appropriate actions that an ESP should take to ensure their success over the long-term? We believe there are.

Bradley Gale, in his book *Managing Customer Value*,[1] lays out an approach to customer acquisition that provides an excellent starting point for this analysis. Gale argues that customers buy value. Products that are successful in the marketplace, in other words, are viewed by the market as a better value than the competitors' within a given product class. In assessing value, customers effectively trade off two fundamental features: quality and price. Given a fixed price, the greater the quality of a product and the greater its value, the more successful that product should be versus its competitors'. Any supplier should be able to assess the market position of its products by developing a map that shows how each product performs against its competitors, both in overall value assessment, and in terms of the two key component measures, quality and price.

Marketers, when designing products and evaluating product success, clearly need to understand both how customers make quality assessments and how customers make relative price assessments. On the quality side of the analysis, marketers need to understand how customers attribute qualities to products and assign weights to each of these attributes. A new laptop computer, for example, might excel on three of the four attributes that customers use to determine the "quality" of such a product, but if it fails on one (say, battery life), which customers think is very important (and therefore assign a very high weight), then the overall quality assessment of the product will suffer. By exploring the way that customers define product attributes, the way they weigh those attributes, and the way they rate a given company's product—and its competitors—on each attribute, a provider can understand the quality strengths and weaknesses of its product, its value proposition for customers, and the likelihood that customers will choose the product in a competitive market.

Using this logic, an aspiring retail ESP should explore the attributes that customers would use to evaluate the quality of an energy service offer and the relative weights customers assign to each attribute. This is difficult, of course, because today's energy customers do not know much about how energy service offers could differ and it is difficult for customers to assign "importance" to attributes they have never experienced before (at least within the context of energy services).

However, it is critical to recognize that within any marketplace, customers differ. Even in the energy services marketplace, there is a great deal of variability in terms of the way in which customers calculate the value equation. Some customers, for example, place far more importance on quality, while others place far greater importance on price. Even within the quality domain, some customers place far greater importance on some attributes (having "high-touch" service, for example, or having local offices, or being involved in the local community, or having no outages) than do other customers. Contrary to public opinion, price is only one among a diverse set of product attributes to which customers assign significant weight. As a result, ESPs can offer customers superior value, even at higher prices, as long as the quality part of the equation is appropriately higher.

The end result is that there is no single product that will represent "the best value" for all customers. Different energy products will represent the "best value" for customers depending on how they calculate the value equation for those products. Some customers, for example, may rate quality as critical over price and believe that high-touch service is the key element in quality. For these customers, the best value will be provided by an energy service product that promises direct contact to a customer service representative, personal service, and the like. Other groups of customers will calculate the value equation in different ways, meaning that, for them, a still different product would be best. Realistically, most markets can be segmented into at least five to 10 different submarkets, and many markets can be segmented into more than 20 different submarkets.

Systematic differences between customers mean that there is almost always the opportunity to provide a tailored product to one or more market niches. Since customers do not, cannot, and probably should not all calculate the value equation in the same way, the product that yields the greatest value will not be the same product in each case. This insight yields several important implications for energy marketers:

- Products developed for a broad class of customers may fail ultimately because they are suboptimal for almost everyone. (Teaching to the "average kid" in the class, for example, sounds like a fine strategy until you realize that there's only one average kid, so the 10 fast ones are bored and the 10 slower ones are confused.)

- Customers may buy suboptimal products for a time (they may be better than the available alternatives), but they are not "loyal customers" in any sense of the word; they will leave as soon as a product comes along that better meets their needs.

- Over time, capturing large shares of a gross market, such as 70 percent of residential customers in a geographic region may be impossible with "mass" products; alternatively, large market shares will be acquired by capturing a "stack" of submarkets with a portfolio of products (i.e., capturing 60 percent of the "green" market with two products and capturing 55 percent of the "high-touch service" market with another product. The high-touch market are those people with whom the company has direct contact, rather than via interactive devices.)

- The number of segments, or submarkets, that might be served in energy markets is unknown at this point. Little differentiation has been created yet, but this is constrained only by the creativity and insight of energy product marketers.

- Not all providers will want to—or can afford to—serve all possible submarkets. In fact, it is probably impossible for any one provider to serve all possible sub-markets. (Should the same brand that sells "premium quality product" also carry deep discounted off-brand support products?)

Customer loyalty—where this means repeat buying and customer commitment—is created when a vendor provides superior value on an ongoing basis, recognizing the ways that different customer segments assign value. For example, convenience-oriented customers are loyal to convenience. They will seek it, and stay with it when they find it, regardless of who offers it. One way, and possibly the most important way, that energy suppliers can acquire loyal customers is to ensure that they are making energy service offers that customers buy because those offers provide truly superior value, given the way that customer group determines value. If customers are buying because the company's offer is "no worse than the others," "not great, but the safe choice," or "the best of a bad lot," then it is hard to imagine how customer loyalty could result.

Developing a Branding Strategy

While much of the current discourse focuses on the importance of creating value for customers as the basis on which market share and customer loyalty can be bounded, the best ways to create and leverage corporate branding in this effort are still unclear. While many energy providers have begun to invest in brand development, the best ways to accomplish this, and the best ways to link these enhanced brand images to specific increased product sales, are still a challenge.

Across the U.S., and even globally, ESPs have been investing substantial sums of money in advertising, sponsorship, and promotional activities designed to create brand awareness and build brand equity. While there have been some number of fits and starts in these activities, it is still clear that brand equity is viewed by many as one of the mechanisms by which customer loyalty can be created, extended, and leveraged. To put this in a broader perspective, brands can work in one of two ways. They can:

- **Serve as a "marker" for the presence of key attributes.** If, for example, a customer—in her value calculation—attaches critical importance to the attributes of reliability and safety, then Volvo has served itself well with this customer by attaching those two attributes to its brand ("When you think Volvo, think safety and reliability," says much of their advertising). The value of the brand in this case is not that it makes the Volvo product "worth more," in some vague way, but rather that the brand market makes it easy for a customer to say, "Right, if that's what I want, I should go to that brand because I know they will deliver on the promise I am looking for." From this perspective, the equity of a brand lies in its ability to communicate its basic value proposition to customers. Products from a trusted brand are worth more because customers give them a higher quality rating based on the fact that customers can assume the presence of key attributes.

- **Fundamentally transform the consumption experience.** Some brands do more than simply serve as markers for the presence of product attributes. Some brands, by their very presence, fundamentally transform the experience of using or buying a product in a way that makes that brand unique and fundamentally irreplaceable. Coca-Cola is possibly the best example of this brand role. Drinking Coca-Cola is not the same as drinking another beverage for most Americans (indeed, for most of the people on the globe). For loyal Coca-Cola drinkers, the fact that it is Coca-Cola changes the experience of the beverage in such a way that a generic cola or other soft drink simply would not be a viable substitute. For brands like these, buying or consuming a product is intimately tied to the character of the brand; the brand is as much about how customers feel when they buy or use the product as it is about how the product actually works or tastes.

At some level, the second of these states is the goal of much corporate brand advertising. The Coca-Cola franchise is worth what it is because of the transformative character of the brand. That type of brand equity is the sort of benefit that many ESPs are hoping for in implementing their own branding strategy.

Coca-Cola, however, spends several billion dollars each year on advertising and promotional expenditures and has been spending substantially on that enterprise for more than 100 years. It is the rare energy supplier that has a fraction of that to spend on brand development activities. Given the very real limitations of time and money, attempts to create brand equity at a level that will create the sort of transformative consumption experience described above—only for energy markets—is probably unrealistic.

This is not to say that brand investments are inappropriate as a component of customer loyalty strategy. On the contrary, brand investments can be enormously important in creating the "brand marker" effect described above. As long as the organization is clear about what the brand should mean to customers (and "everything" is probably inappropriate), then brand advertising should attempt to build connections for customers between the brand name and the attributes that the company wants to attach to the name. The goal of this type of communication, however, is not to create simple awareness; nor is it to encourage customers to have a "good feeling" about the company, or to experience a specific type of emotional response when they encounter the company. Rather, the goal of this type of brand advertising is to encourage customers to attach a limited set of attributes to the brand—preferably those attributes that should be critical to the assessment of product value for the targeted market segments.

Developing a Customer Retention Strategy

If acquiring customers means developing targeted value propositions, supported by an appropriate branding strategy, then retaining customers means ensuring that the provider continues to offer a superior product for each of the targeted value propositions it seeks to serve. In Gale's language, the value assessment for a product must continue to exceed that of its competitors. Note, however, that the value proposition must reflect those unique sets of characteristics that are considered important to specific, targeted customer segments. This can be done either by creating a product with higher perceived quality for a given perceived price, or by having a better-perceived price for a product with a given level of perceived quality.

A provider may acquire new customers by promising to provide a certain level of quality and price, but customer loyalty—where this means that customers will tend and want to buy a given company's products over time—is best ensured when the company continues to provide superior value for those products over time. "You really want to ensure customer loyalty," this argument would suggest. "You need to provide products to customers that *continually* provide superior value relative to competitors." If an organization does not do this, then its market share will fall over time, regardless of how customers "feel" about the product or the company.

An ESP can improve its position by ensuring that the assessment that customers make of the value associated with its offers is superior to those of other providers. At some level, however, the value that customers assign to a product is based in part on the emotional connection that customers sometimes create between themselves and the products they buy. When customers create this emotional bond, they come to feel that they should do business with a company, or buy a product, even though they explicitly recognize that it may not be "the best." Customers who feel an emotional connection to a product (what some call a "loyalty bond"), tend to buy that product even if one or more of the values they may assign to the product is inferior to others ("They stood by me when I needed them; I'm a customer for life," or "I know he's not the best plumber, but I've been using him for 30 years and I'm not going to change now.")

The result of these emotional ties is that, in some situations, customers will assign a value to a product that is based on more than just the quality and price of the product. In these cases, there is also an emotional weight assigned to a product, which can increase the value enormously. Ultimately, these emotional ties can serve to accomplish several ends:

- When they exist, these emotional ties can separate a given product from a group of others that otherwise would be rated comparably.

- When they exist, these ties can move an inferior product into a category of otherwise comparable, but objectively superior, products.

Creating emotional ties, in other words, is one critical means by which customer retention can be enhanced. Having a prior relationship with customers means that the opportunity exists for the company to develop emotional connections with those customers. However, the down side of this perspective is that the strength and effectiveness of emotional ties is both difficult to measure and very difficult to build systematically. Because emotional states are more qualitative, it is a more trying task to develop consistent and reliable measures of those states. Even more importantly, however, it is difficult to create mechanisms that can be trusted to create consistently the emotional bonds that are the fundamental element of these relationships.

PRACTICAL IMPLICATIONS FOR ESPS

Several conclusions can be drawn from this model of decisionmaking in competitive choice markets. We have identified several lessons learned, which are summarized here:

- Gale's fundamental argument about the role of customer value management (CVM) in customer choice is a reasonable one: customers buy products based on which one provides the greatest value. That determination is itself a result of the comparison of the perceived quality of the product with its perceived price. What Gale does not highlight, however, is that customer segments exist that are differentiated in terms of what creates value for them; segments differ systematically in terms of how they assign value, how they perceive price, and how they evaluate the differences between these two items.

- Customer acquisition depends primarily on the ability of a provider to identify differentiated customer markets and provide solutions that will be evaluated as providing superior value within each of those submarkets (or at least among those which the company wants to serve).

- The acquisition of larger shares of energy markets likely will depend more on a company's ability to stack shares across customer submarkets than on a company's ability to attract large market shares to a single product, or small set of products.

- The creation of customer loyalty is best insured by making certain that separate customer submarkets find products that provide them with superior value according to the way each customer segment defines and trades off price and quality. In other words, there is no single set of actions that will enhance customer loyalty. What makes one submarket of customers loyal is different from what will make other customer sub-markets loyal, and understanding of this process presumes a significant market segmentation effort.

- Customer retention is best ensured by making sure that the product provided is superior (and continues to maintain that superiority over a period of time) in the ways that target customers care about.

- Brands can serve two functions—either as markers for the presence of desired product attributes, or as an umbrella attribute that serves to transform the consumption experience for a product. As attractive as the latter option might be, it is rare, and expensive to create; a more expeditious option is to pursue a brand strategy that focuses on ensuring that the brand "means" a specific set of things to its targeted customers.

- Brand loyalty, in the traditional sense in which this phrase is used, or emotional types of connections, can be very effective in creating additional customer "stickiness" over time; they are, however, difficult and expensive to create, and potentially idiosyncratic in nature. As a result, they may not be the most appropriate near-term goals for retail ESPs wishing to create a cadre of loyal customers.

Most important, however, this analysis leads directly to the conclusion that customer loyalty—where this is defined as regular repeat purchases—is probably best created in energy markets by ensuring that the company provides differentiated energy service offers that are targeted to customer groups that evaluate product value differently. By developing, targeting, and continuing to service these offers in consistent ways retail ESPs can acquire and retain attractive customers consistently.

So what should the marketing manager pay attention to? How can the marketing manager for a competitive ESP make use of this framework that explains the way that customer choice will likely operate in a competitive energy market? As expanded customer choice becomes available in the energy arena, ESPs who wish to remain competitive will need to develop methods by which they can predict and manage customer choice. The obvious response is to "build customer loyalty" through some combination of improving service and satisfaction, developing and offering value-added products and services, or creating other "hooks" into their customer base. However, as we have argued in this chapter, the obvious response is not always the best response.

Managers who fail to pay attention to "invisible" customer choices can make less-than-optimum decisions regarding customer retention strategies, program design, or product attributes. Some of these errors include:

1. **Creating products that are designed to capture large portions of market share or that are intended to serve the needs of large heterogeneous populations.** The incorrect assumption behind this decision is that the market is sufficiently homogeneous with respect to its need for, perception of, or value attributed to the program or product. The result of this error is that the product or offer will fail to capture sufficient market share, will not result in customer loyalty, and will appeal either to only a small number of market segments or to market segments that are not in the company's best strategic interests. The more accurate view is that within a large market there almost always exists a number of customer segments, each of which will require products, programs, services, or offers that are tailored to its needs, perceptions, and purchase values.

2. **Assuming that brand investments should be oriented toward brands that try to change the consumption experience rather than brands that function as effective markers.** This assumption is incorrect because such a branding strategy is very difficult and costly to achieve (costs can soar into the range of billions of dollars). The result of this error is that the company can invest huge amounts of money for little outcome; the strategy can backfire if it does not match the needs and perceptions of the targeted market segments. The more accurate view is to develop a brand marker strategy that appeals to the specific value propositions that represent the needs of your target market segments.

3. **Assuming that customers are loyal to brands or that brand loyalty is one of the most important forms of customer loyalty.** This assumption ignores the fact that customers may be more

loyal to the "offer" rather than the brand. A more accurate assumption would be that the most consistent way to maintain market share may be to create and maintain "offer" loyalty (in which customers buy from you because they trust that you have and will continue to provide the value proposition that they want to buy). The result of this error is an incorrect brand strategy. The more accurate view is to recognize that the best way to capture and retain market share in emerging markets may be to develop attractive offers for specific targeted segments and "mark" those offers with an appropriate branding strategy.

4. **Spending valuable time and effort trying to create emotional attachments to customers.** This incorrect action is based on the assumption that since emotional ties "feel good," they will attract loyal customers. This can be a dangerous strategy. Emotional appeals can backfire if the attachment is created to a person rather than to the company or to the offer. Such a strategy can be subject to idiosyncratic shifts in customer perceptions, experiences with specific customer service people, and can serve to undermine an otherwise stable market position if customer perceptions of value (or rather lack of value) do not match the specific emotional appeal. The easier approach is to focus less on emotional appeals and more on matching value propositions to segment needs and perceptions. Market success is more likely to arise from developing targeted offers that meet the specific product/service needs that solve real problems from real segments.

5. **Concluding that there is no way to make money in mass markets without developing a low-price or discount product.** This all-too-common error occurs when ESPs develop a market strategy that unnecessarily and prematurely commoditizes the product. This occurs as a result of the incorrect assumption that the only attributes that differentiate value for customers in this product category are price related—a common but costly mistake. Instead, the marketing manager needs to examine the market opportunities represented by specific customer groups who calculate the value proposition in such a way as to highly value the product, service, and offer. These are the customers who would not only buy, but would prefer a higher priced option if it were available.

In order to optimize decisions and avoid these mistakes, the successful marketing manager needs information that will help develop a usable understanding of the target market. That is, one must understand how the market can be categorized into segments most relevant for the value propositions carried by the products and services of the company as defined in its specific offers, brands, and suppliers. The market research protocols that support such information needs will allow the marketing manager to define attributes for product category which drive different value assessments by customers, and assess the weights each customer segment attributes to products and suppliers.

Product Attributes

Understanding product attributes is key to understanding market segmentation and ultimately, customer choice. In the simplest terms, products attributes are distinctively different 'things' that customers value. (Product attributes also are frequently referred to simply as "benefits." This convention will also be adopted here to reduce the potential for confusion between the supplier and product benefits.) The most important element about attributes is that they exist as customer perceptions and, as such, may be accurate or inaccurate pictures of reality. Like other symbolic systems such as language or mathematics, the concept of "attributes" allows us to talk about what may be very complex concepts in simple shorthand. The reason that particular products or services are purchased is that that they satisfy customers' expressed and unexpressed needs better than any other product available. One way to think about market segmentation is that different market segments exist because consumers place different values on attributes presented by the specific product or service, and, as a result, make different purchase choices.

Marketers frequently speak in terms of the features of a product or service. Features mirror the benefits valued by the segment(s). The more effectively features are aimed at a segment, the better the results. The choice faced by marketers is whether to develop products and services by seat-of-the-pants experimentation ("trying things") or to conduct systematic research, map customer values, and develop packages with benefits that match those of high value groups.

Supplier Attributes

Supplier attributes help determine whether—and how seriously—a buyer will consider a given supplier during their search. Buyers exclude potential suppliers from their search if they don't expect that supplier to be associated with a given set of products or to offer products that fall within an acceptable range of product quality.

Consider the illustration of a Rolls Royce automotive dealership expanding by opening up a convenience store under the same roof. If the dealership maintained the exclusive ambiance that is characteristic of its image (a supplier attribute that is clearly an important attribute to its customers), is the convenience store likely to attract lots of traffic? Would the Rolls Royce dealership suffer erosion of business if the supplier attributes desirable for a convenience store began to bleed over to its floor? Large difference in supplier attributes would affect consumer search and selection behavior in both businesses. While we suspect that few Rolls Royce dealerships are actually considering using their valuable retail showroom space for this purpose, the example illustrates a problem with all product line extensions: how should a supplier extend its product lines and attract new customers without cannibalizing its current product mix or offending current customers? These will be important questions to ESPs as they extend their product lines into new and uncharted territories.

Energy companies, like packaged goods, automobile, or financial products suppliers, have to understand how they are perceived by their customers. One EPRI report, "Residential Customers: Perceptions About Utility Providers and Electric Switching Intentions,"[2] compares supplier attributes of utilities (electric, gas, local telephone, long distance, cable, and cellular). The results suggest that customers use multiple attributes to assess their utility providers. Price, product, responsiveness, customer service, size of organization, and ability to solve problems were some of the top-priority supplier attributes assigned to ESPs, although each market segment demonstrates a unique cluster of weights assigned to each attribute.

Well, this is interesting at some theoretical level, but what does it have to do with customer choice? Essentially, the choice framework mentioned above indicates that customers choose products and services depending on their assessment of value trade-offs between those products. Those value trade-offs are driven by the way in which customers define, and weight the importance of, a variety of attributes that might be attached to a product or service. Understanding the nature of those attributes and the way in which different customer groups respond to each of them is critical in building products that customers will see as a "good value," and which ultimately will engender customer loyalty. The section below outlines one example of the attributes that can be linked to an energy product—in this case 'green power'—and the way that different customer segments might attach different amounts of importance to those attributes.

CUSTOMER ASSESSMENT OF GREEN POWER PRODUCTS

In green power programs, customers are offered the opportunity to buy energy from environmentally friendly sources—usually at a higher price. Participants justify their purchase of a higher price product when they perceive an alignment of the program attributes with their own personal values. If people responding positively to this concept were asked to construct or list the benefits of green power, their list might look something like the one in Figure 15–1.

Allows user to be socially responsible
Allows user to belong to a group helping the environment
Price is within X dollars of prior bill (or represents a premium of Y percent).

Figure 15–1 *Key Green Power Benefits*

How Can Market Segmentation Be Applied in the Case of Green Power?

A savvy marketer, before trying to market a Green Power program, would look for other products or services whose attributes also appealed to these segments and bundle them into a package to offer to the target market. Actual implementation of the initial research requires that the marketer make translations between the benefits a market segment desires and the product features that supply (or reinforce) each of the benefits. An actual green power package might include the benefits and features listed in Figure 15–2. The package then would be developed with these features. The seller is simply acquiring energy from environmentally benign sources and delivering it to a transmission and distribution system. The buyer selects green power because the offer meets other needs—dealing with social responsibility, environmental responsibility, public awareness, and price.

The differences between products and services can be understood as differences in their attributes. Buyers make purchase decisions based on attributes they perceive in specific products or services. When choosing between alternatives, the alternative with the highest value wins. The buyer achieves that value by trading off value gained in one attribute against value lacking in other attributes.

Markets can be segmented effectively because products and services have multiple attributes and the package of attributes can be modified and molded to meet the value propositions desired by its target segments. Effective market segmentation is based on assessing and segmenting the market using customer perceptions of benefits, and then translating and integrating those benefits into the systematic development of product or service concepts that match the needs of the target market segments.

The following example provides an illustration of this systematic development process, starting with the Green Power example.

Key Green Power Benefits	Product Features
Social responsibility, group membership	Monthly Green Newsletter
Reinforces participant identity	Eco-power window plaque
Addresses price concerns	Rate guarantee
Additional revenue source, group membership	Green Power catalogue
Reinforces social responsibility	Reduction of air pollution due to program participation placed on bill

Figure 15–2 *Potential Green Power Package*

WHAT WOULD YOU DO?

The CEO of a new ESP affiliated with a regulated utility distribution company wants the vice president of marketing to consider launching a green power program. A list of potential benefits is compiled, using published materials.

In the first meeting, discussion turns on whether the research also should determine whether a green power offer would be acceptable if coming from the investor-owned utility. Discussion quickly migrates to how important supplier attributes are and whether there is an "ideal" set of supplier attributes for green power offerings. At the end of the meeting, the marketing vice president is challenged to respond to the following questions:

- How important are supplier attributes when compared with benefits?
- Does the IOU possess the necessary supplier attributes directly?
- What are the characteristics of the market segment that values these attributes?
- Which potential providers possess the necessary supplier attributes?
- Which potential providers will offer the stiffest competition?
- If the IOU lacks the necessary supplier attributes, can they be gained by joint offerings with another entity, or can the IOU develop another identity for offering this service?
- What is the best mix of benefits?
- Can the IOU attract enough customers, at an adequate price premium to make money on this effort?
- How will this product help the IOU meet its strategic marketing goals?

Table 15–1 outlines a systematic cost-effective approach to developing this program. Note that it's impossible to effectively answer all of the outlined questions in a single round of research—and some of the answers will reflect corporate strategy goals rather than market research goals. However, typically the most effective market research is performed in several ongoing stages in order for the information to effectively iterate between marketing, product design, and strategic planning. What is outlined below only reflects the marketing portions of this iterative strategy-research-marketing process.

Step 1: Feasibility Screens: Many firms regularly perform feasibility screens to segment their markets, providing a base of customer information that will assist in developing a number of potential offerings, spreading the cost of the initial market analysis across multiple products.

Step 2: Feasibility Analysis: Promising clusters of benefits identified in the first round then are carried to a second round, where they are analyzed to determine the likelihood that they either can be profitable or add value in specifically targeted customer segments.

Step 3: Concept Testing: Most research costs are incurred in these later stages, as concepts are tested and response models developed. The outcome from this stage generally includes price points, features, and even key advertising copy points.

Step 4: Targeting Analysis: Further analysis is conducted to refine profitability projections. As part of this effort, direct response models are often formulated to increase the cost-effectiveness of the targeted marketing campaigns.

Steps 5 & 6: Pilot Offer and Rollout: At this point, the refined offer is presented to a limited audience as a final test for features, pricing, and acceptance levels.

The cost of Stages 1 through 5 increases steadily, reflecting the need for increased precision. The research costs mirror the increased financial exposure

TABLE 15-1 A Systematic Approach to Developing Market Research		
Step	**Description**	**Goals**
1. Feasibility Screen	Focus on how customers value product benefits and "product attributes." Include psychographic (lifestyle and values) questions to provide insights into key ways to communicate offer. Include customer demographic or firmographic questions	Estimate overall market size and revenue potential. Identify the key benefits needed in the offer, an understanding of who prospects are, and what messages they might respond to.
2. Feasibility Analysis	Project survey results onto the region surveyed.	Determine potential market size.
3. Concept Testing	Assemble one or more concrete offers and test. This may include allowing respondents to choose between several "offers" or to assemble their own "best offer" from a list of options. Test Supplier attributes. Collect demographics and psychographics.	Develop refined market potential estimates for specific products at definite prices. Identify the "best" offers.
4. Targeting Analysis	Develop costs to deliver defined offers. Determine participants and probable cost of acquiring customers. Develop profitability model. Develop model to target high-potential customer segments	Refine financial estimates of program costs, revenue, and projected profitability. Determine which offering (features and price point) is best to offer.
5. Pilot Offer	Test actual product or service offering to a limited audience.	Refine offer, targeting, and actual execution of program
6. Roll-out	Full scale marketing campaign begins with best chance at success.	Financially successful program

that occurs as the company proceeds through product development, testing and introduction.

Market segmentation almost always is a difficult and expensive business. Even more important, it is ultimately unsuccessful at least as often as it is successful. Without it, however, the ability to provide value to customers—and the ability to win significant market shares consistently over time—will be constrained. If companies don't pursue market segmentation in order to provide customers with greater value, they run the risk of losing market share piece by piece over time. While the first 5 percent of customers lost to a new provider offering a targeted product with a better value proposition may not hurt too much, the next 5 percent, or the next 5 percent, or the next 5 percent, ultimately will hurt. This is probably reason enough to explore niche product opportunities before your competitors do.

CONCLUSIONS

Predicting customer choices in competitive energy markets is a daunting task. For a variety of reasons, it is difficult to predict how customers will choose, and as a result, it is difficult to predict the actions that ESPs should take if they are to effectively win and retain customers over time. While this is difficult, however, some conclusions can be drawn from a disciplined examination of how customers choose products across a wide variety of product categories. Importantly, this analysis shows that customer acquisition and retention likely will be most effective based on a segmented understanding of the customer markets, which recognizes the different ways that customer groups evaluate energy-related products. Even more importantly, this analysis suggests that customer loyalty is critical, but difficult to create using commonly recommended techniques. In contrast, the approach to customer choice outlined here suggests that value assessments—and not the creation of emotional ties—are the keys to the creation of customer loyalty and long-term profitability. Large choices loom ahead as competitive energy providers attempt to determine how best to acquire and retain market shares. By focusing on assessing the value of offers and the way in which significant groups of customers make those assessments differently, ESPs have the greatest hope of success.

NOTES

1. Bradley T. Gale, *Managing Customer Value: Creating Quality and Service that Customers Can See* (New York: The Free Press, 1994).

2. Electric Power Research Institute Report [December 1998] TR-108465.

BIBLIOGRAPHY

Aaker, David, *Building Strong Brands* (New York: The Free Press, 1996).

Berrigan, John, & Carl Finkbeiner, *Segmentation Marketing* (Harper Business Press, 1992).

Blankenship, A.B., and George Edward Breen, *State of the Art Marketing Research* (NTC Business Books, 1996).

Crainer, Stuart, *The Real Power of Brands* (FT Pitman Publishing, 1995).

Dick, Alan & Kunai Basu, "Customer Loyalty: Toward an Integrated Conceptual Framework," *Journal of the Academy of Marketing Science*, Spring 1994, 99–113.

Fournier, Susan, Susan Dobscha & David Mick, "Preventing the Premature Death of Relationship Marketing," *Harvard Business Review*, January/February 1998, 43–51.

Gosden, Freeman, *Direct Marketing Success* (New York: John Wiley & Sons, 1985).

Gruenwald, George, *New Product Development*, 2nd ed. (NTC Business Books, 1992).

Heskett, James, et al., "Putting the Service-Profit Chain to Work," *Harvard Business Review*, March/April 1994.

Hughes, Arthur, *Strategic Database Marketing* (Probus Publishing, 1994).

Jackson, Rob & Paul Wang, *Strategic Database Marketing* (NTC Business Books, 1997).

Jones, Thomas & W. Earl Sasser, Jr., "Why Satisfied Customers Defect," *Harvard Business Review*, November/December 1995.

Kasper, Hans, "On Problem Perception, Dissatisfaction, and Brand Loyalty," *Journal of Economic Psychology*, 9 (1988), 387–397.

Keaveney, Susan, "Customer Switching Behavior in Service Industries," *Journal of Marketing*, (April 1995), 71–82.

Marder, Eric, *The Laws of Choice: Predicting Customer Behavior* (New York: The Free Press, 1997).

Maruca, Regina Fazio & Amy Halliday, "When New Products and Customer Loyalty Collide," *Harvard Business Review*, November/December 1993.

Mazursky, David, Priscilla LaBarbera, & Al Aiello, "When Consumers Switch Brands," *Psychology & Marketing*, 1997 (Vol. 4), 17–30.

Morgan, Michael, and Chekitan Dev, "An Empirical Study of Brand Switching for a Retail Service," *Journal of Retailing*, (Vol. 70, No.3), 267–282.

O'Brien, Louise & Charles Jones, "Do Rewards Really Create Loyalty?" *Harvard Business Review*, May/June 1995.

Quelch, John, "Brands vs. Private Labels: Fighting to Win," *Harvard Business Review*, January/February 1996, 99–109.

Reichheld, Frederick, and W. Earl Sasser, Jr., "Zero Defections: Quality Comes to Services," *Harvard Business Review*, September/October 1990.

Reichheld, Frederick, "Loyalty-Based Management," *Harvard Business Review*, March/April 1993.

Reichheld, Frederick, *The Loyalty Effect* (Harvard Business School Press, 1996).

Rust, Roland & Anthony Zahorik, "Customer Satisfaction, Customer Retention, and Market Share," *Journal of Retailing*, Summer 1993, 193–215.

Wayland, Robert & Paul Cole, *Customer Connections* (Harvard Business School Press, 1997).

Weinstein, Art, *Market Segmentation* (Probus Publishing, 1994).

Relevant EPRI Reports:

"Residential Customers: Perceptions About Energy Providers and Electric Switching Intentions" TR-108465, July 1997.

"Green Power Guidelines, Volume 1: Assessing Residential Market Segments" TR-109192-V1, December 1997.

"An Assessment of the Residential Remodeling Market" TR-109491, November 1997.

"Customer Choice: A Framework for Leveraging Loyalty" TR-109242, December 1997.

"Energy Market Profiles, Volume 2: 1995 Residential Buildings, Appliances, and Energy Use" TR-107676-v2, December 1996.

"ReQuest II: An Investigation of Consumer Attitudes Towards Telecommunications and Electric Services" TR-106166, June 1996.

"ReQuest III: Assessing Changes in the Residential Telecommunications and Electric Marketplace" TR-107631, December 1996.

CHAPTER 16

Implications of Retail Customer Choice for Generation Companies

Daniel M. Violette
Michael King

The changes taking place in the utility industry have received wide coverage by the media, with much of the focus on the changes that will occur in the end-use retail market. Coverage has focused on retail pilot programs and on how customers are selecting new retail suppliers in states that have opened all or a portion of their markets.[1] Research and articles abound on customer switching behavior, the types of offers being made to end users by retail energy companies, and announcements about contracts with new suppliers in those markets where choice is now available.

While the media is focused on retail choice, the impetus for restructuring the energy industry has arisen from forces unleashed in the production segment of the industry. In the electric industry, the Public Utility Regulatory Policies Act of 1978 (PURPA) provided clout to a new player in the industry—the qualifying facility. PURPA created the environment whereby private capital not directly tied to the local electric utility could invest in power generation. Much of the impetus for restructuring the electric industry can be attributed to the commercial interest of private investors who have demonstrated that they could build plants less expensively and operate them more flexibly and efficiently than the incumbent utilities. Further, with the radical changes in generation technology cost and performance, the stage has been set for a re-capitalization of the electric industry.

Private investors have grown weary of the battles with local utilities to qualify private investment in power plants. They believe that the "single buyer" of power, in the form of a utility, effectively limits their market, and constrains their ability to grow their business. Thus, some private investors in generation have become one of the primary forces in reshaping the electric power industry.

As large (predominantly industrial) customers eyed the possible cost reductions associated with a more competitive wholesale market, they became strong advocates for restructuring and retail choice. Thus, competition in generation now is tied inextricably to competition at the retail level. Similarly, for retail choice to thrive, competition in generation must also thrive.[2]

This chapter discusses changes that are likely to occur in electricity generation. The economical operation of power plants has been viewed for some time as dependent on engineers and sophisticated operational technology. While this will remain a prerequisite for success in generation, considerable innovation will be required by the companies that choose to participate in the competitive power production market. Being a good technical operator of power plants is likely to be insufficient for success in the new marketplace. The business model that generation companies operated under for the past 80 years is being torn down. Many competitors will be innovative in their business model, and which business models will succeed is in part an issue of timing and in part an issue of vision and execution. Those companies that pick an appropriate business model, develop operating characteristics consistent with that business model, and lead the industry in innovation and become knowledge-based companies with respect to dispatching, bidding into exchanges, developing structured generation offerings to customers, and managing risk will be those that are more likely to attain above average margins.

A stylized version of the pre-retail choice value chain is shown in Figure 16–1. This value chain is now changing, with new industry players entering niche markets and companies providing specialized services (see Figure 16–2). The competitive focal points of the industry are on the two ends of the value chain, that is, the creation of electricity through generation (GENCOs) and the selling of electricity through retail companies. These are the enterprises that are being deregulated. The middle of the value chain—transmission (TRANSCOs) and distribution (DISCOs)—will, in general, continue to be subject to regulation. Substantial modifications and structural changes that will be required by TRANSCOs and DISCOs to support the information and transactions flow required by competitive markets and the effort required by these entities should not be underestimated, but they will remain regulated entities.

Given that the competitive portions of the market are occurring at the two ends of the value chain, where is the action today and in the future? Retail companies selling electricity to end-use customers may or may not own gen-

Figure 16–1   *The Old, Linear Electricity Value Chain*

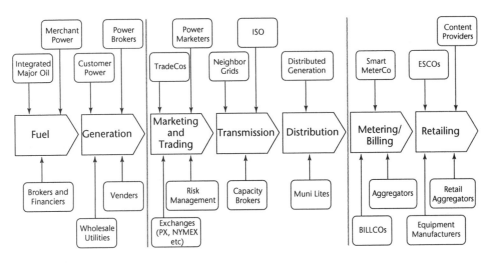

Figure 16–2   *The New Value Chain: Potential Allies, Competitors, and Partners Exist throughout the Chain*

eration, yet they have contracts with customers to provide electricity and will have to arrange to obtain these supplies of electricity. The GENCO portion of the value chain will produce electricity, and electricity will reach end-use customers through market arrangements with power marketers, brokers, traders, and retail organizations.

What does this mean for the generation portion of the industry? It means that the entire framework of the generation sector will change dramatically, as it becomes the first competitive step in a value chain resulting in retail choice for end-use customers. The structure of GENCOs, the value propositions they pursue and the business processes they will employ are now changing and will continue to change in ways barely imagined just two years ago.

GENCOS—THE HISTORICAL PERSPECTIVE

Most utilities in the United States remain vertically regulated monopolies. Every utility that owns generation assets is hard at work determining how to maximize the value of those assets under a competitive framework. The electric utility industry is one of the most capital-intensive industries in the United States. The restructuring of this industry into functionally independent GENCOs that operate in competitive markets with many sellers and buyers will require a shift in investment and a movement of capital that is unprecedented in U.S. history. From a deal-making perspective, this will be an exciting industry for at least the next decade.

The historical regulatory context under which utilities operated meant that utilities generally built generation to meet the needs of customers located in their own service territories. Due to the lead time required for plant construction and varying rates of demand growth, there were often opportunities to sell or buy excess power from neighboring utilities, although the incentives to maximize the benefits of trade were muted, to say the least. All new plants and major plant modifications had to be approved by regulators, usually through granting a certificate of public necessity and convenience. This certification process often involved public hearings examining the need for the plant and the economics of the specific proposal, including technology and location. Building generating plants was not an easy task in the 1970s and 1980s as many projects went over budget and were completed well beyond the target completion date. Nuclear plant construction was particularly susceptible to these cost and scheduling problems. Once approval was granted and the plant built, the utility was to run the plant as efficiently as possible and sell most of the output in its regulated service territory, and resell any available excess to neighboring utilities. Most plants were run essentially as separate entities, and often were viewed as separate cost/profit centers. In this environment, regulation served as a hedge against the plant being unprofitable by requiring customers to pay for embedded plant costs, purchase the output of the plant, and pay a regulated price of the electricity purchased.[3]

This environment used the regulator and the regulatory process to condition the decisions of utilities in their selection of generation technology, level of investment, construction management, operations management, and timing of investment. While all involved (utility, regulator, intervenors) were well-intentioned, there is little doubt that regulation has been an imperfect dispenser of discipline. Many now believe that markets are better able to ruthlessly discipline the decisions of generation owners.

GENCOS—THE COMPETITIVE CHALLENGE

All the rules for GENCOs change in the competitive environment. There is no longer a regulatory hedge that protects the generating plant. Now, the generator must assume the risks of the market. There are no "captive" customers to whom plant costs can be allocated and sales from the plant are not guaranteed at a regulated price. In fact, the key difference between the regulatory environment and the competitive environment is the imposition of market risk. A premise that will be developed throughout his chapter is that those firms best able to accurately identify market risks for their generation asset portfolio and appropriately manage those risks will be the firms that earn above average returns.

Two basic assumptions underlie the competitive restructuring of the electric industry:

- The first is the premise that began the drive to a competitive market place, that is, the view that generation is no longer a monopoly activity. Rather than customers leading the drive toward competition, it is other firms that see an opportunity to provide lower cost electricity and related services.[4]
- The second premise is that the industry must be restructured by breaking the existing utility into independently functioning business units that are at least comprised of the GENCO, the TRANSCO, the DISCO, and unregulated affiliates such as a retail power marketing affiliate or RETAILCO. Each of these is to operate as a fully independent entity.[5]

In the competitive era, the formerly regulated GENCO has no regulatory hedges. With each entity now independent, it can expect no favors from its corporate or holding company partners that purchase electricity. The GENCO has no guaranteed cost recovery and no guaranteed customers. If a competitor can provide electricity better suited to a customer's needs at a lower cost, then that customer will take its business to the competitor. Each GENCO must assume all risks common to competitive markets, develop business plans and customer acquisition plans based on value propositions that reflect the characteristics of its assets, corporate strengths, and market position.

This competitive challenge is compounded by the claim that electricity is a commodity, that one kilowatt-hour is exactly like another. As a commodity, all competition will be based on price, thereby marginalizing returns on assets. Electricity does have more commodity-like characteristics than do many other products; however, a company in the generation business cannot afford to accept the premise that its product is a commodity and cannot be differentiated. Each GENCO will need to strive with every fiber of its corporate being to achieve whatever differentiation is possible. In fact, this is one of the basic values of competitive markets. The drive for differentiation will unleash a wave of innovation that ultimately will redefine the industry, create new products and services, and ultimately create the value expected from competitive markets.

GENCOs will seek product differentiation in the same ways as do other firms in commodity-like industries, such as computer chips, steel, chemicals, or credit cards. Many of these companies have etched out footholds in product differentiation through brand, place, price, convenience, customer service, bundling information with the product, and customer-specific tailored products.[6] In fact, the first step toward failure for a GENCO may be accepting the idea that its product is a commodity and thus not striving for every possible edge or differentiation that might enhance margins.

In summary, GENCOs will have all the same challenges and opportunities facing companies in other commodity-like industries. Every generation plant owner is now assessing the future of its physical assets, whether it is an investor-owned utility, an independent power company, or a member of the public power community. There is not a single strategy that will be best for all participants. Different choices are being made today in anticipation of the restructured market. Several example strategies are shown below.

FUTURE OF THE GENCO INDUSTRY—BUSINESS PROCESSES

The existing business model and value chain in the electric industry are rapidly becoming obsolete and the components of what will make up the new value chain have not yet been defined. Competitive markets create constant change and the role of the GENCO in the energy value chain will not be static and instead will continuously evolve. Because the old value chain contained regulatory risk hedges that will not exist in the future, the structure of GENCOs and the business processes needed to identify and manage market risks largely do not exist and must be developed. GENCOs will need to build, buy, or align with providers to help manage risk. In addition, GENCOs will need to develop processes that quantify market uncertainties and risks. This will include estimating the volatility of electricity prices in different regional markets, defining risks associated with pricing terms for structured generation products, setting prices that appropriately reflect the sharing of risk between buyers and sellers, and managing the basic demand risk associated with the customer acquisition processes. Each GENCO faces many challenging choices in defining its business model and supporting business processes.

GENCOs that are offshoots of formerly regulated firms will need to change their focus from plant-by-plant economics to a comprehensive portfolio view. A coherent GENCO strategy will involve portfolio analyses of generation assets and the increased application of financial strategies to maximize the value of the portfolio. A sale of electricity no longer will be viewed as a sale from a given plant but, instead, will be viewed as a slice from a regional portfolio that, in part, self-hedges some of the risk of the sale.

Minimizing operating and maintenance costs will be as critical as ever to providing a portfolio of low-cost electricity-producing assets. Innovation and creative plant management will be important. However, the GENCO operation will be confronted with two choices regarding sales: (1) the GENCO can sell its product into an exchange where its output will be treated as an undifferentiated commodity; or (2) the GENCO can transact in the bilateral market, but will be forced to meet the price terms, take provisions, and load shape requirements of its customers.

Those GENCOs seeking a differentiation strategy must know their customers. Some GENCOs will develop certain structured offerings based on their assets and risk management procedures to meet the needs of power marketers and retail organizations.[7] These offerings will be tailored on relatively complex financial theory, with the portfolio and its assets treated as an option to provide electricity at times in the future.[8] Other GENCOs may specialize in ancillary services, providing a niche product such as voltage support or regulation. Decisions regarding unit commitment, dispatch, and maintenance will be conducted within a portfolio framework.

The development of a coherent business strategy will have to address the why, what, where, and how questions shown in Figure 16–3. These must be addressed in the context of portfolio management of these assets and they illustrate the initial types of decisions that will make up a business strategy. Figure 16–4 illustrates some of the more obvious GENCO business models that might emerge. Three models are illustrated: (1) the "regional dominator" that picks a specific fuel and type of plant with the objective of becoming a dominant provider of power in a region; (2) an "asset collector" that attempts to create mass by building a diversified portfolio of assets across several regions; and (3) the "differentiator/speculator" that looks for special situations, possibly in a region they know well, where they can provide arbitrage opportunities and hedges for other entities in uncertain markets. These

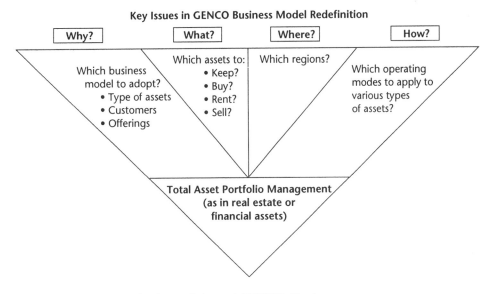

Figure 16–3   *Developing a Coherent GENCO Strategy*

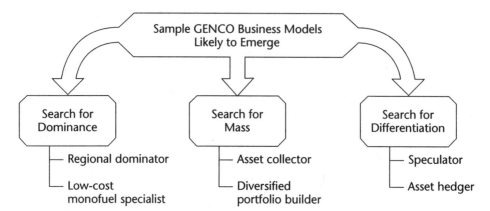

Figure 16–4   *GENCO Business Models*

three business model views illustrate a few of the choices to be made by GENCOs, and demonstrated that there will be more than one way to make money in the GENCO side of the industry.

There are a number of business processes that a self-standing GENCO business will need. Efficiency in running power plants is viewed as an area of expertise by at least some regulated utilities. Although efficient plant operation will continue to be important, it represents only one aspect of the GENCO business. One of the basic premises of the restructured industry is that each entity must operate as an independent entity. This means that the GENCO must have the full range of business processes usually possessed by a player in a competitive commodity-like market. It cannot expect to simply produce power and have customers come to its door asking to buy all the power that unit can produce. Figure 16–5 presents six basic business process areas that a GENCO will need. These include the following:

1. **Asset operations and management** includes the traditional plant management functions (fuels procurement, operations and maintenance, and plant accounting).

2. **Supply, logistics, and trading (SL&T)** are needed to manage the portfolio of assets and contracts that have been entered into by the GENCO. This involves using financial tools and market forecasts to engage in transactions (both buying and selling), scheduling and dispatching of assets,[9] and making the transactions necessary to develop an appropriately diversified portfolio of assets, asset rights, and contracts.

3. **Risk management** is related to SL&T but it is a separate process that is critical to the success of the GENCO enterprise. This function monitors trading limits, executes hedging strategies, and performs overall risk assessment through stress testing or value-at-risk methods. Recall that the primary difference be-

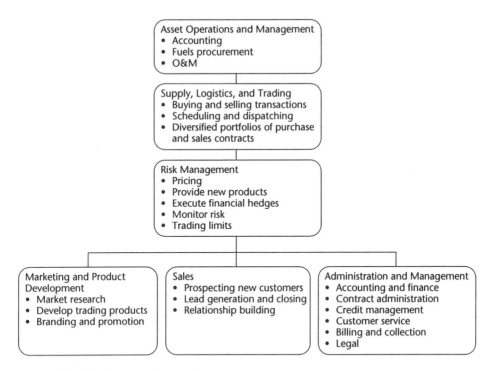

Figure 16–5   *GENCO Core Processes*

tween a regulated market and a competitive market is risk. This business function identifies and manages risk in a manner that maximizes the value of the portfolio of assets and the GENCO's book of business.

4. **Marketing and product development** are necessary for any self-standing business enterprise. Structured products can be based on price characteristics, quantity characteristics, quality characteristics, service characteristics, or time-differentiated characteristics. Any differentiation strategy first must spring from understanding the customer. Clearly, the product needs of a residential aggregator are likely to differ from those of whole-sale traders or independent system operators. Market and brand research are part of this new product development.

5. **A sales function** will be necessary to prospect for new cus-tomers whether they be municipals, industrial, power mar-keters, or retail entities that might benefit from a GENCO's uniquely structured offers (e.g., buying through another power marketer's interruptible contract). This will involve prospecting

for customers, lead generation and closing, as well as relationship building.

6. **An administration and management** function also is needed
   to handle customer service, accounting, contract administration,
   credit management, billing and collections, and legal issues.

A GENCO does much more than just provide for the efficient operation of its
generating assets. It will need business planning and all of the business
process functions outlined above to be successful as a self-standing functional entity in a competitive market.

BUSINESS STRATEGIES FOR SELECTED COMPANIES

Some entities in the electricity sector have announced specific strategies for
the generation sector that illustrate the diversity of market plays that are
under consideration. Several strategies are discussed below.

PECO Energy—Growth through Nuclear

The strategy for generation growth at PECO Energy of Philadelphia, Pennsylvania, is to expand its nuclear generation portfolio.[10] While the company
acknowledges that many companies expect better returns in trading electricity and in distribution, PECO's position is that someone has to produce
power and that it can do this best by operating nuclear power plants. This is
a contrarian strategy, in that few if any companies are looking to acquire nuclear plants, but owning and operating nuclear power plants is an activity
that PECO believes it does well. It has lowered operating costs at its nuclear
plants substantially through innovative cost-control techniques. PECO also
believes that grouping nuclear units can help build economies of scale. Many
units operating in the United States are stand-alone, and linking units in
geographic areas can reduce plant costs through the benefits of consolidation, elimination of overhead, exchange of best practices, utilization of a
skilled workforce, and strategic vendor partnering.

PECO has taken several steps along this path. It already has operation and
maintenance contracts for the Millstone and Clinton plants and in September
1997 PECO formed a joint venture with British Energy and created AmerGen
Energy Company to pursue opportunities to acquire and operate nuclear generating plants in North America. They are currently discussing operating
Ontario Hydro's 20 nuclear plants, as well as exploring other opportunities.

PECO Energy acknowledges that a strategy focused on nuclear plant ownership is not for the faint of heart, but PECO believes it can run nuclear units
safely and efficiently, and produce competitively priced electricity in an
open generation market.

GPU, Inc.—Counting on Transmission and Distribution with a Little Generation on the Side

GPU, Inc. of Morristown, New Jersey, is taking a different approach, in which it plans to concentrate on transmission and distribution, and essentially exit the bulk generation business. The first step along this path is auctioning its 26 fossil and hydropower plants. It also has been actively negotiating with potential buyers for its two operating nuclear power plants. GPU is exiting the merchant plant business as well. These generation assets are being auctioned, even though GPU could have decided to stay in generation under both Pennsylvania and New Jersey state law.

The company acknowledges that the infrastructure business it has chosen will continue to have regulatory oversight and that regulated rates of return will apply to much of its business. However, this provides a certain amount of predictability and stability that some investors will find attractive. GPU does plan to try to leverage the brand name through its wires services to sell unregulated revenue-cycle services to enhance overall return.

While GPU is auctioning off its "bulk power" plants, it still sees a role for itself in the commodity business with small- and medium-sized commercial customers. A few relatively large players will characterize the large-scale bulk power and trading game. However, GPU sees a second level for regional players serving niche clients and locations. GPU plans to exit the arena of market niches in which generation is now being converted from a contract business to a commodity business. GPU believes there are opportunities to pursue in having contracts with businesses or localized electric systems to take the output of a plant.

In summary, GPU is focusing on the transmission and distribution infrastructure aspects of the industry, but it will also take a role in regional generation for small and medium business customers on a contract basis.[11]

United Illuminating—Exiting the Generation Business

United Illuminating, of New Haven, Connecticut, has laid out a plan to exit the generation business and focus future activities on electricity distribution and its non-regulated services businesses. The company plans to sell 1,337 MW of generation assets and purchased power contracts by early 1999. At the 1998 annual stockholders' meeting, the outgoing UI chairman, Richard Grossi, indicated that the company decided that UI was not big enough to compete in the deregulated Northeastern market and will be selling its generation business, even though under state law it could bid to retain the assets. The newly appointed UI president, Nathaniel Woodson, stated that UI will become both a premier regulated distribution company and a leading provider of value-added energy services for key customer segments.

The three utilities discussed above—PECO Energy, GPU Inc., and United Illuminating—illustrate a broad range of approaches being taken toward participating in the generation segment of the market. They illustrate that utilities will have to make important decisions that will determine their companies' future. This can involve expanding in generation, playing a niche in the generation market, or exiting the generation business.

GENCO KEYS TO SUCCESS

Deciding to retain a presence of some sort in generation is a first step for a company, but many strategic decisions remain regarding where the business should position itself within the value chain. As was discussed in the "Future of the GENCO" section of this chapter, generation will be a functionally independent operating unit and investment in business processes will be needed to support a functionally separate organization. The eventual retail customer has much to gain from an efficient generation market that improves productivity through innovation and efficiency. The ability of the other players in the value chain to provide retail products and service offerings will depend upon the types of structured offerings and competitive organization of the generation segment of the industry.

The GENCO industry is just beginning to explore its options, and new types of operations not previously envisioned undoubtedly will characterize the competitive generation market five to 10 years from now.

The divestiture process that is characterizing the regulatory restructuring process is resulting in a large number of generation assets coming on the market. The next few years in the utility industry promise to bring the largest market transfer of assets experienced by any industry in the United States. Some utilities are selling the generation assets in their own service territories and acquiring assets elsewhere. Figure 16–6 outlines a path starting with what might be termed the initial "shock wave" of deregulation, where market players are establishing their strategies and testing them in the market. This is followed by a "consolidation" phase in which unsuccessful strategies begin to be weeded out. The final, "alignment" phase is when the strategies, product offerings, and scope and scale of market players begin to align with the actual economics of the market. This illustrates that experimentation and learning will occur as this market develops. Deciding when to enter a market may be as important as deciding what strategy to employ.

Reasons for buying generation assets, based on research of companies which have already done so, include the following:

Strategic:

- establish scale in merchant plant generation[12]
- gain national presence

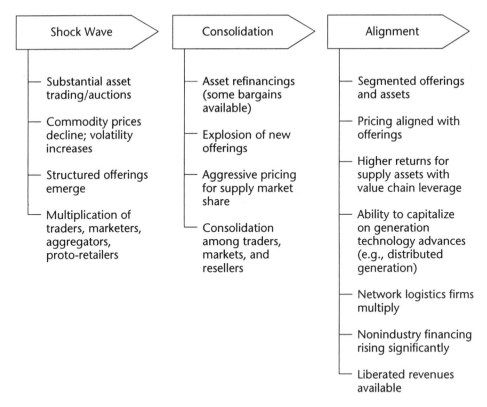

Figure 16–6 *Deregulation of the Energy Supply Sector*

- get in on the ground floor of deregulation
- complement other business lines and build value chain
- build regional platforms of supply, trading, and marketing business
- operate as a "converter" (i.e., fuel to electricity), letting other companies take the power marketing risks[13]

Tactical:

- create a portfolio with fuel diversity to allow for hedging
- improve trading margins with backing of hard generation assets
- grow through enhanced gas/electric arbitrage opportunities and trading volumes
- gain proximity to interstate pipeline

Technical:

- implement efficiency measures at plants and reduce costs
- increase output through capital additions—topping cycles or new units at the site
- repower old plants with clean and efficient technology.

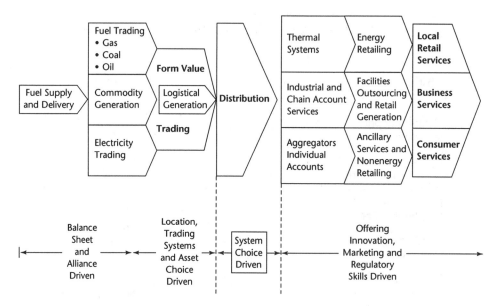

Figure 16–7 *Business Segmentation and Opportunities Along the Value Chain*

The reasons given for acquiring assets reflect different views of where a company wants to be placed in the overall electricity market value chain. Figure 16–7 illustrates the value chain from fuel supply to retail services. At the commodity generation end of the chain, success will be driven by strong balance sheets and appropriate alliances to facilitate fuel trading, electricity trading and form-value trading. The new business models will embody the traditional engineering aspects of producing power, but keys to success also will include adaptation to convergence, innovative approaches to financing, and the importance of intellectual capital and knowledge-based systems.

Successful generation companies in the new value chain must focus on portfolio management of their assets and continuous and intensive margin management. Factors that will drive success in attaining these portfolio management objectives and continual margin management are shown in Figure 16–8.

CONCLUSIONS

The ability of the restructured electric market to meet the needs of end-use retail customers will depend, in part, upon the development of a robust and competitive generation market. After all, the movement toward competition is predicated on the assumption that electric generation is no longer a natural monopoly. The type of generation market that will develop will reward those that recognize both the new opportunities and the risks of this competitive market.

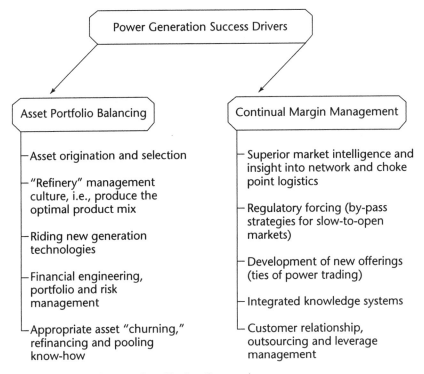

Figure 16–8 *The New Value Chain: Generation*

For retail customers to obtain the full benefits from competition, regulators will need to assure the development of a market that promotes competitive generation and provides for a level playing field for both current industry participants and new entrants.

The well-prepared companies will have opportunities to improve yields on generation assets. Historic tools such as production costing and traditional unit commitment models are not tailored for this new market, however. Margin management of portfolio generation assets will require a new view—a view that treats each element of the portfolio as a financial asset that must be managed through the use of hedges, options, and other financial approaches.

Generation companies will need to develop probabilistic price forecasts to be used to price exchange traded products and in the scheduling, bid strategy, and asset selection of generation companies. A generation company can increase the yield of its portfolio of assets by:

1. Focusing on yield across a portfolio of assets that includes fuels and plants.

2. Explicitly recognizing that prices are uncertain and that the bidding strategy must start from this premise.

3. Continually assessing price volatility in markets to price traded commodities in order to participate in both financial and physical markets.

4. Continually monitoring the position of the firm and making trading and operations decisions based on whether the firm is long or short in both physical and financial markets.

5. Retaining margins through superior identification of the risk the firm faces at any point in time, allowing the book of business to be optimally hedged through physical and financial instruments.

6. Developing structured products from the generation assets thereby allowing physical assets to be used in innovative ways.

7. Building the business processes appropriate for the implementation of its business model and execution of its business strategies.

Form-value trading, where the spread between the energy equivalent prices of gas, coal, and electricity in the same market is arbitraged, will become increasingly important. This is a direct consequence of convergence and competition. Single-commodity traders will be at a disadvantage to traders who have the ability to arbitrage multiple fuel forms. They will miss margins from the cross-commodity swaps. As a result, single commodity trading margins will fall. Without fuel trading, it will be extremely difficult to survive in the electricity business. The types of portfolios that provide these opportunities will be diversified in their fuel use, location, offerings, and contracts. Without a portion of the portfolio in merchant power, it will be difficult for a company to engage in physical arbitrage.

The opening of a competitive generation market will challenge the full range of entrepreneurial talents of those who choose to participate. The new generation market, far from being a stodgy market doomed to irreversible marginalization, provides opportunities for advanced knowledge-based market players who enhance the output of existing and developing generation technologies, identify their customers and differentiate products to meet their needs, and manage the risks associated with an extremely dynamic marketplace.

NOTES

1. Rhode Island began allowing retail choice in the summer of 1997, Massachusetts began full retail choice in March 1998, California opened all of its electricity market to retail choice in April 1998, and a number of states are expected to open their markets in late 1998, 1999, and 2000.

2. This assumption already is being borne out in the industry. The concept of a merchant power plant that would be built without any guaranteed customer contracts was unheard of two to three ago. Now, over 20 merchant power plants are being planned with the expectation that they will be able to produce power at below cur-

rent market prices, and thereby sell the output of the plants. Merchant power plants are being considered in most regions of the country, but regions with high prices and low reserve margins are attracting much of the attention. These areas include New England, California, New York, Texas, and the upper Midwest surrounding the Chicago load area.

3. This is an oversimplification of the regulatory environment that generating plants operated under, but it outlines the implicit agreement between the regulators and the utilities.

4. In telecommunications, much of the impetus for change came from equipment vendors who sought to break Western Electric's stranglehold as equipment supplier to the regional Bell operating companies. In fact, the initial impetus for change in telecommunications came from the Hush-A-Phone case, where the developer of a mechanical muting function for telephones successfully fought the denial of interconnection of "foreign" equipment to the telephone network. [*Hush-A-Phone Corp.*, 20 FCC 391, 12 PUR 3d 73 (1955), aff'd *Hush-A-Phone Corp.* v. *United States*, 238 F. 2d 266 [15PUR3d 467] (D.C. Cir. 1956).] Subsequently, MCI broke AT&T's monopoly on long distance through construction of private microwave networks. The advance of technology and its potential to dramatically increase productivity and expand service offerings resulted in the dramatic changes seen.

5. The authors believe that many competitive firms will seek to span the value chain by integrating retailing companies with generation companies. We have no quarrel with this business premise, but object to combinations of monopoly functions with competitive functions. The opportunities for cross-subsidies from the monopoly functions into the competitive functions are too great and are likely to stifle competition.

6. Hagler Bailly conducted a survey for a utility client of marketing executives at non-electric companies in commodity-like industries, and found that these companies strive for differentiation through providing information, customer service, bundling information with products, and tailoring of products to specific customer needs and profiles.

7. These structured offerings could leverage exchange-traded products such as New York Mercantile Exchange futures contracts. They also may be defined by the GENCO, with the structure provided by the needs of its customers. For example, some customers will be interested in imbalance service that allows them to contract forward for supplemental energy instead of relying on the real-time market with the extreme price volatility inherent in that market.

8. The appropriate pricing of structured offerings is largely an unsolved problem. There has not been enough experience in competitive markets in the United States to be able to predict future price volatility with any degree of confidence. Most structured electricity sales contracts that GENCOs are likely to enter into can be decomposed into a set of options and related financial derivatives. The pricing of these products will depend mostly on projections of future price volatility along with forward price curves or what the authors, in an earlier work (see below) term "forward price views," due to the lack of depth in current electricity financial markets. Several approaches for estimating and addressing price volatility, and the pricing of generation products are discussed in papers contained in the proceedings of the Electric Power Research Institute's 1998 "Pricing Energy in a Competitive Market" Conference held in Washington D.C. Select papers that address these issues include Hamden, K. "Electricity Spikes;" Shumway, N., "Stochastic Stress Testing;" and Violette, D. and M. King, "Electricity Price Forecasts and the Forward Price Curve for Electricity."

9. Scheduling and dispatching can be a complex task if the GENCO is bidding into a power exchange and competing with other entities in structured markets for power. Daily and monthly price volatility, and the probability of spikes in prices due to market events (e.g., plant outages and extreme temperatures) all must be factored into the dispatch and bidding decision process. The strategy selected might well determine whether the firm earns a return on its investment.

10. This is adapted from "PECO Forges Nuclear Excellence into New Business," *Power*, May/June 1998; and also from the PECO Annual Report for 1997.

11. For more detail, see "Inventing a Business in Wires & Pipes," *Public Utilities Fortnightly*, 15 June 1998.

12. Many utilities believed that economies of scale had been exhausted in utility ownership of generation. In the new environment, however, portfolio effects which hedge the risk of individual assets through the portfolio and cost of capital reductions resulting from ownership of a portfolio of diversified generation assets and balance sheet structuring may impart new economies of scale in merchant generation.

13. Some fuel companies see merchant generation as a refinery that provides choice of market for their raw fuel. These companies, and marketers associated with them, have been driving fuel-form convergence beyond gas and electricity to coal and oil. A portfolio of electric generation plants with their fuel stocks allows much greater leverage in fuel form arbitrage in which the company can take advantage of discrepancies between the prices of a multitude of fuels.

CHAPTER 17

Customer-Connected Strategies

Robert E. Wayland
*David E. Jones**

New industry structures beget new business designs. The restructuring of the utility industry is being driven by customer choice. Utility executives need to replace their product- and asset-driven strategies with a customer-based strategic framework. This chapter outlines and applies the Value Compass[sm] framework[1] that addresses four dimensions of customer and shareholder value creation:

- Selecting the right customers to serve.
- Determining the right value proposition(s) to offer those customers.
- Identifying the industry value-added role that best leverages the firm's capabilities.
- Creating appropriate reward and risk sharing mechanisms with customers, suppliers and employees.

Using this framework, we review the industry's current strategic initiatives and find that far too often they are based on internally perceived capabilities instead of a realistic sense of the marketplace. Far too little attention, in our opinion, has been given to the most fundamental strategic decision: determining which customers companies should focus on attracting, developing,

*The authors gratefully acknowledge the comments and suggestions of Ahmad Faruqui of EPRI, Alan Willer of James J. McEnroe & Associates (St. Louis), Mark Gerber of M.S. Gerber & Associates (Columbus, Ohio), Paul M. Cole of Ernst & Young, LLP (Boston) and Brian Ferrell of the Edison Electric Institute. Part of this chapter is based on a 1997 presentation to EEI's Strategic Planning Committee.

and retaining, and which to disengage from. Most utilities have focused on the second dimension of value creation and have expanded their value propositions, most commonly by buying or building energy service companies. Relatively few companies have developed a strategy for finding and securing their most profitable value-added role, despite the pressures to unbundle the vertically integrated utility business model and the efforts of many companies to move "upstream" on the industry value-added chain. Finally, there have been major innovations in pricing and risk management but they have come almost exclusively from companies outside the traditional industry boundaries.

Although a few companies have moved successfully to exploit the opportunities of industry restructuring, many more have found it difficult to create shareholder value outside of their traditional regulated operations. As the disappointing results of many early competitive initiatives accumulate, we believe that executives will reassess their product- and asset-driven strategies and focus more sharply on developing customer-based strategies.

THE STRATEGIC LANDSCAPE OF THE UTILITY INDUSTRY TODAY

As more states open their electricity markets, CEOs are asking: Where are tomorrow's profits going to be made? Even before the final shape of the industry is clear, a number of companies have moved aggressively into new businesses and geographies. Among those who are doing more than taking a wait-and-see position, we see five main strategic directions being taken. Let's call them the five "cons":

- **Consolidation.** Investment bankers looking at a map of utility service territories see a lot of undersized companies crying out for consolidation into a few regional giants. New mergers of more-or-less contiguous utilities seem to be announced daily.

- **Conglomeration.** Thin and dropping margins on commodities are prompting many to move into "value-added" energy services. The energy services company idea is so compelling that nearly every utility now has one. Several are aggressively buying up heating/ventilation/air-conditioning businesses, energy trading operations, or lighting and engineering firms. Some are buying not-so-related businesses such as home security and real estate brokers.

- **Convergence.** Some see natural gas and electricity, even oil and propane, coming together into a universal energy business. Duke's purchase of PanEnergy and Enron's acquisition of Portland General Electric are prominent examples of horizontal combinations. Platform convergence leverages competencies related to pipes and wires to justify companies offering electric, gas, telephone, water, cable television, and so forth.

- **Concentration.** The forced as well as voluntary unbundling of vertically integrated power companies is proceeding more rapidly than many thought possible a few years ago. A few companies, GPU for one, seem to be pursuing a policy of concentrating only on part of the industry value-added chain.
- **Confusion.** A fair number of companies seem to be pursuing two or more of the above strategies. This strategy is probably more testimony to the effect that free cash flow has on the imaginations of investment bankers than it is evidence of untapped management expertise in the utility industry.

In a few years some of these strategies will be hailed as acts of genius, while others will be seen as the energy sector's version of the Allegis, Peoples Express and AT&T acts of shareholder value destruction. How can we tell which of the emerging energy business strategies is likely to succeed and which to fail? Strategists almost always can rationalize their moves. Allegis had its fans, Peoples Express was hailed as a bold new model for its industry (and was for a while), and some people feared that an unleashed AT&T would dominate not only communications but also computing. Many of the words used then are being heard again today.

Customer Emancipation

We tend to think of deregulation as the freeing of firms to pursue new strategies but it is the freeing of customers to choose that really matters. Firms' strategies are driven by customer choice, not vice versa. All of the great deregulation initiatives came after customers (at least some of them) were able to breach the walls of the entrenched monopoly—often with the aid of aggressive entrepreneurs such as Freddie Laker (who founded, and for a time successfully ran, Laker Airlines, which undercut the majors on trans-Atlantic flights, thereby undermining arguments about fare levels and exposing the potential market opportunities to later entrants) and William McGowan (the founder of MCI Communications).

In the early days of restructuring, the voice of the customer is often silent as companies pursue strategies based on internal logic and attempt to leverage their self-assessed competencies. But the market is a powerful teacher and after awhile the need to truly understand customers becomes apparent. As we write this there is some evidence that leading companies are rethinking their initial strategies and assumptions about customers.

- Enron retreated from the California residential market citing the high cost of customer acquisition relative to profitability.
- UtiliCorp has pulled the plug on EnergyOne, an ambitious cross-sell and shared services play for which there was little demand.
- Consolidated Natural Gas, a well-regarded gas utility, has announced that it will exit the electricity marketing business.

CUSTOMER-BASED VERSUS PRODUCT-DRIVEN STRATEGIES

The language, objectives and metrics of people who approach strategy from a customer-based perspective are different from those who have a product-driven orientation and call for a different vocabulary, as illustrated in Table 17–1.

Despite what some of the more evangelical members of the customer service and total quality tribes say, profit is not the inevitable consequence of virtuous behavior, and being obsequious to customers is not a business strategy. First we will set out the basic connection between customer and shareholder value. Understanding this connection goes a long way toward illuminating the customer-based approach.

THE CONNECTION BETWEEN CUSTOMER AND SHAREHOLDER VALUE

Customer-based strategy is built on the insight that a firm's customer relationships are its most fundamental assets and the only sustainable source of shareholder value. In a customer-based strategy the starting point is the source of the revenue flows (customers), not the means (products). Figure 17–1 illustrates the connection between customer and shareholder value.

The creation of shareholder value starts with producing value for customers. At the margin, customer value created equals the price paid. A firm's revenue is its share of the total value it creates for customers.

Creating value for customers is a necessary but not sufficient condition for creating shareholder value. The customer cash flow generated by a buyer must cover not only product costs but also relationship-related costs and, of course, taxes. Chasing and acquiring customers whose revenues do not cover

TABLE 17–1 Product vs. Customer-Based Strategic Perspectives		
Factor	Product-Driven	Customer-Based
Basic Unit of Value:	Product	Customer Relationship
Primary Value Driver:	Product Sales	Customer Acquisition, Development and Retention
Competitive Index:	Category Share	Share of Customer Expenditure
Profitability Indicator:	Product Margin and ROI	Customer Equity and Shareholder Value
Customer Pulse	Satisfaction (historical)	Probability-of-Purchase (future)

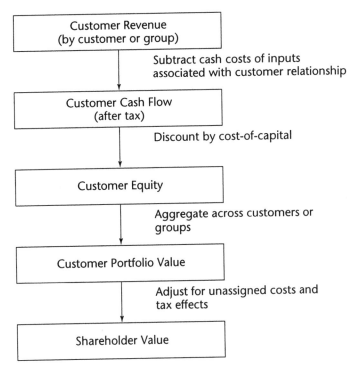

Figure 17–1 *The Customer-to-Shareholder Value Connection*

the costs of serving them destroys shareholder value and weakens the firm's ability to build profitable relationships with others.

Customer relationship value, sometimes called customer equity, is the present value of the net cash flows from the customer over time. Understanding this value is critical to allocating marketing and sales resources, evaluating mergers and acquisitions, or assessing customer-related technologies. The value of a customer relationship depends upon the volume of purchases per period, the margin on those purchases, the duration of the relationship and the costs of acquiring, developing, and retaining the relationship. All of these variables will change as competition takes hold in the electricity markets.

Customer portfolio value is simply the aggregate value of individual customer relationships. Portfolio value depends upon the quantity and quality of customer relationships. With some adjustments to account for tax factors and cost assignment issues, the customer portfolio value is equal to shareholder value.

SEARCHING FOR UNTAPPED CUSTOMER AND SHAREHOLDER VALUE

Now that we have established the linkage between customer and shareholder value, we can consider the various means of creating and capturing those values. There are really only four fundamental ways to create value: finding and keeping the right customers, offering them the right value proposition, doing the things that best leverage your company's capabilities, and sharing rewards and risks intelligently. Of course there are many ways to approach each of these and the relative importance of each may vary or change for many reasons.

The Value Compass[sm] is a useful framework for searching for untapped sources of value. Each axis of the compass represents one of the four dimensions along which a firm can move to increase customer and shareholder value.

Each of the four axes of the compass represents a potential means of value-creation.

- **Customer Portfolio Management.** A customer portfolio may be managed at successively more granular levels represented by the

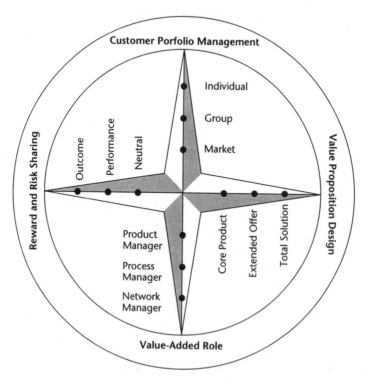

Figure 17–2   *The Value Compass[sm]*

market, group, and individual positions on the axis. Accustomed to universal service, very few utilities understand the value of their customer relationships and allocate resources accordingly. Minnesota Power & Light Company and NIPSCO Industries have developed the value of their relationships with their very largest customers through sophisticated key account management techniques.

- **Value Proposition Design.** The core offer, extended offer, and total solution positions represent successively more complete coverage levels of the customers' total requirements. In contrast to the slow pace of movement in customer portfolio management, nearly every utility has rushed to extend its value proposition beyond the core product of energy service.

- **Value-added Role.** The industry value-added chain is the sequence of steps performed to create a final product. Companies may play product, process, and network manager roles with respect to their customers and suppliers. With restructuring, the profitability of steps will change and new competitors will arise in every step. Large utilities traditionally have been vertically integrated product managers. There have been significant initiatives to move toward process management roles such as operating customer facilities and network management positions providing wholesale marketing and trading functions.

- **Reward and Risk Sharing.** Operating competitively will require kicking the cost-plus pricing habit and developing new arrangements for sharing the risks and rewards of value creation efforts with customers and suppliers. The neutral, performance, and outcomes-based risk/reward positions denote progressively more interdependent arrangements. Enron Capital and Trade Corporation has revolutionized commodity pricing with its risk management products. Emerging new products such as weather hedges are indicative of the creative possibilities available to companies.

Moving out on these axes increases the potential relationship value but it requires knowledge and the capability to execute against that knowledge. Since this knowledge and capability are often costly, it does not always pay to move out to the extremes on each dimension. A firm maximizes shareholder value by moving to the point along each dimension at which the increase in potential value is offset by the cost of achieving it. Table 17–2 provides summary definitions of representative positions on each axis.

MAPPING STRATEGIES WITH THE VALUE COMPASS^SM

The customer-based components of a firm's strategy can be mapped by connecting its positions on each axis of the compass to create a "spider-web" diagram. Using this technique we can analyze the shift in moving from a utility to an energy services business model, as illustrated in Figure 17–3.

TABLE 17-2 Representative Positions on the Value Compass^sm Axes			
Customer Portfolio Management	**Value Proposition Design**	**Value-Added Role**	**Reward & Risk Sharing**
Market or Index: Customer portfolio reflects the composition of the market.	**Core Product:** Product addresses one part of the customers' value chain.	**Product Manager:** Emphasis on supply-chain management; hand-off to customer.	**Neutral:** Parties take what the market gives them in terms of bargaining position.
Group: Portfolio management emphasizes one or more specific customer groups.	**Extended Offer:** Contributes to two or more elements of the customers' value chain.	**Process Manager:** Works within the customer's value chain.	**Performance-based:** One or both parties put something at risk.
Individual: Each relationship is managed on customer-specific basis.	**Total Solution:** Addresses the entire value chain, provides result rather than input.	**Network manager:** Takes a central position among suppliers and customers.	**Outcome-based:** Parties share in the result of a common undertaking.

The traditional "utility" pursues an index portfolio strategy treating customers uniformly, provides a basic core product, is a product manager focusing on its supply chain, and relies on cost-plus pricing to share the risks and rewards of its operations with its customers. This relationship model and business design is shown in the left-hand panel of Figure 17–3.

In contrast, the "energy service company," illustrated in the right-hand panel, pursues a selective customer portfolio and extends its value proposition to cover a substantial part of its customers' energy-related value chain. It may play process or network manager roles and often earns at least part of its reward on a performance basis such as shared savings.

Surprisingly, the most common new business model being pursued by utilities, the transition from a utility to an energy services company, involves some of the biggest changes imaginable. The utility must make substantial changes along each of the four value-creating axes. It should not be surprising that so many of these operations have been unprofitable.

The Grand ESCO Experiments

Given their conservative reputation, it is remarkable that utilities (or their parents) have so boldly engaged in strategies that demand whole new ways of dealing with customers, involve substantially expanded value propositions launched against established rivals, require them to perform unfamiliar industry functions, and involve dramatically different levels of risk.

Not surprisingly, very few of these new companies are profitable. This unfortunate condition is sometimes blamed on the short-sighted behavior of rivals who are accused of "buying business," the need to "invest" in a new market,

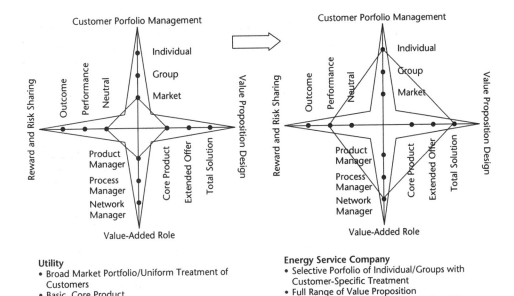

Utility
- Broad Market Portfolio/Uniform Treatment of Customers
- Basic, Core Product
- Product-Focused/Vertically Integrated
- Neutral, Market/Cost Based Pricing/Little Collaboration

Energy Service Company
- Selective Porfolio of Individual/Groups with Customer-Specific Treatment
- Full Range of Value Proposition
- Network-Focused; Aggregator, Broker, Alliances
- Performance-Based Reward and Risk Sharing

Figure 17-3   *Shift from Utility to Energy Services Model*

and the rationale most likely to cause shareholder anxiety—"these are strategic investments."

Because they involve supposedly "related business areas" these strategies are thought to be consistent with good strategic practice. But looked at with the Value Compass^sm, the real difficulty and stretch required is evident. Most companies will not be able to master the skills necessary to move out along all four dimensions of value creation.

There will be a shakeout. We are not arguing that the energy services business model will never prove successful. We suspect that it will for a few firms, but it won't work for nearly as many companies as are trying it. As we write this, there are signs that some companies are reevaluating the notion of chasing customers one at a time in their neighbors' territory with a value-proposition and brand name scarcely different than the pack of rivals.

Is there a better way to develop a customer-based strategy? Yes. It requires starting with the customer portfolio and working around the four dimensions of value creation in light of your own company's situation and resources and developing a unique strategy that fits you and your customers:

- Manage your customer portfolio effectively.
- Optimize your value proposition.
- Determine the right role on the value-added chain.
- Establish effective risk and reward sharing arrangements.

EFFECTIVE CUSTOMER PORTFOLIO MANAGEMENT

As customers gain the right to choose suppliers, firms need to learn how to select customers. Choosing which customers to compete for is a company's most important strategic decision. All else follows—the competitor set, the value proposition, even the corporate structure and organization.

Until recently, utilities enjoyed exclusive customer relationships within their service territories. By default, they pursued nonselective portfolio strategies (what we would describe as an index or market portfolio) knowing that the losses sustained in some relationships were averaged out by the high returns in others. For the most part, all customers (within tariff classes) were treated equally.

It will be increasingly possible for companies to shape their customer portfolio. Companies may change their customer portfolio by reconfiguring their service territory through swaps, sales, or acquisitions (as PacifiCorp has indicated it will do) and by aligning their customer acquisition, development, and retention efforts with the potential relationship values of their customers. Since the incumbent utility will remain the provider of last resort it is especially important that it know which relationships are inherently unprofitable and seek equitable regulatory treatment for the social costs it is bearing.

It is likely that some companies will pursue a group portfolio strategy by specializing in meeting the needs of particular types of customers. An energy marketer, for example, might aggregate only customers in a specific industry, such as health care. Some companies already appreciate the wisdom of managing on an individual basis their very best customers and seeking to serve them across all of the customer's locations.

In order to manage portfolios effectively, companies must understand:

- the distribution of customer relationship value;
- economic segmentation;
- the supply curve of customer relationships; and
- the customer acquisition, development, and retention cycle.

Given the importance of customer knowledge to succeeding in an environment of customer choice we will discuss each of these factors in some detail below.

The Distribution of Customer Relationship Value: All Customers Are Not Created Equal

Electric power companies that continue to treat all customers the same while competitors target and win the best customers will destroy shareholder value. Since the value of the firm equals the aggregate value of the firm's customer relationships, executives should think of themselves as managing a portfolio of customer relationships. Managing this portfolio means making choices about which relationships to acquire, develop, or retain (or to disengage from). Most managers realize that all customer relationships are not of equal value. But many utility executives, having grown up in a monopoly environment tempered by a strong service ethic, believe all customers deserve the same level of effort.

Portfolio management involves creating and using customer knowledge to allocate resources. This requires understanding the expected asset value of prospective and current customer relationships and allocating resources accordingly. If the relationship values are uniform, modest, and positive it may not pay to invest heavily in customer-specific knowledge since one customer is much like any other. If, on the other hand, the distribution of customer relationship values has a significant spread the return on investment in customer specific knowledge can be very high.

Figure 17–4 illustrates a representative bell-shaped curve of customer relationship values. Simply calculating the distribution of relationship value often prompts managers to appreciate better the need for selective investment in customer-related investments.

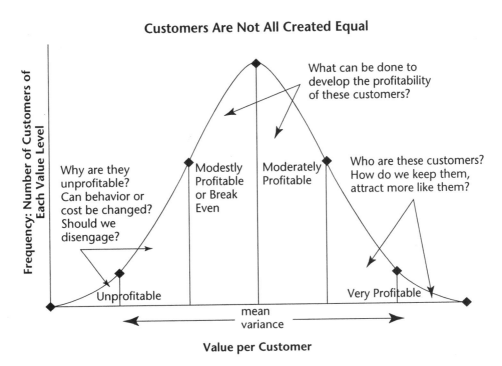

Figure 17–4   *Customer Relationship Value Distribution*

Economic Segmentation: The Willie Sutton Principle

Willie Sutton, a prominent member of the financial community in the 1920s, had two distinctive competencies: robbing banks and escaping from jail. In a famous but probably apocryphal story, Sutton, when asked why he robbed banks, replied, "Because that's where the money is." Saint Willie (as consultants refer to him) put his finger on what should be the fundamental objective of strategy.

In preparing for increased competition, many electric power companies have undertaken expensive segmentation studies to better understand and approach their customers. Millions of dollars have been spent on efforts to find the mathematical center of gravity for a mixture of customer characteristics and creating cute labels like "global gladiators" and "hassled homeowners." Unfortunately, these segmentation schemes are often poor predictors of customer relationship value.

Economic segmentation goes to the heart of the question, where is the money and what are our chances of getting it? More technically, it groups customers both in terms of their expected relationship value and their responsiveness to your value proposition. Figure 17–5 illustrates the dimensions of an economic segmentation scheme.

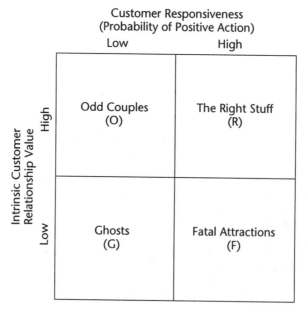

Figure 17–5 *Economic Segmentation*

Customers with the "Right Stuff" are both valuable and responsive and are likely to be the firm's primary targets. Although "Odd Couples" are valuable, they are not responsive and can be a black hole for fruitless marketing and sales efforts. For electric utilities, an example would be a large industrial plant that produces a commodity, competes only on price, and has a confrontational attitude toward vendors. "Ghosts" are seldom seen and are best left alone. Perhaps the worst are the "Fatal Attractions"—customers who are aggressively responsive to your value proposition but do not form valuable relationships. These people can drive your fulfillment and customer service costs through the roof.

The Customer Relationship Supply Curve

Like any other asset or economic good, the supply of customer relationships depends upon their value and the cost of "producing" them. Both the value and cost of relationships change over time and in response to various factors. The notion of a supply curve for customer relationships is a little hard for some managers to swallow but it's essential to effective portfolio management.

Many executives believe that maximizing the number of high-value customers in their portfolio maximizes shareholder value. But, when we take account of the relative levels of customer responsiveness and the acquisition, development, and retention costs, the fallacy of this notion is evident.

Figure 17–6 illustrates a situation where there are only two acquisition cost curves: the one closest to the y-axis for the two unresponsive customer segments [Odd Couples (O) and Ghosts (G)], and the other for the two responsive

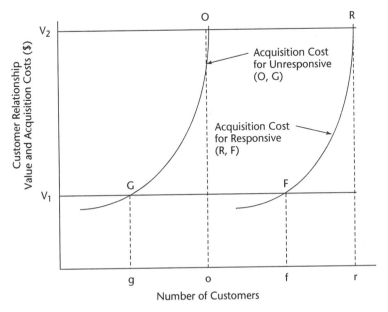

Figure 17–6 *Optimizing Customer Portfolio Weighting*

groups [Fatal Attractions (F) and Right Stuff (R)]. To keep it simple, we also have assumed that there are only two levels of relationship value, V_1 for the low-value groups and V_2 for the high-value groups.

The rule for maximizing the value of the customer portfolio is to invest in customer relationships up to the point where the expected value equals the cost of acquiring, developing or retaining the relationship. In our hypothetical case, this results in a portfolio with weights of g, o, f, and r for the four customer groups. Any other investment pattern would produce a lower value portfolio. Acquiring another 'r' would cost more than the value of the relationship, acquiring one less 'g' would forego a cash flow stream of greater value than the cost of acquiring it, and so forth.

People often are surprised that the optimal portfolio usually will be a mix of customers representing most or all of the economic segments. This is because at some point the acquisition, development and retention cost curves slope upward due to the increasing marginal cost of acquiring, developing, or retaining successively more customer relationships. The slopes of these curves are also one reason why average acquisition, development, and retention cost figures can be misleading guides for allocating resources.

The Customer Acquisition, Development, and Retention (ADR) Cycle

Customer relationships often go through a cycle from acquisition through development and, in many cases, retention or reacquisition. The concept of an ADR cycle can be combined with economic segmentation to plan and allocate resources.

Some people argue for an emphasis on retention, because "it costs less to keep a customer than to get a new one." Even if this statement were always true (and it's not) the relative costs of customer acquisition, development, and retention are not the proper basis for allocating resources. It's the value of the customer relationship relative to the cost of securing it that matters. So long as the expected value of the customer relationship exceeds the cost of acquisition, development or retention, it is worthwhile to make the investment.

Figure 17–7 shows how knowledge of customer relationship value and responsiveness across the ADR cycle can be used to acquire the most promising customers, focus development on those with high potential but low realized relationship value, and work to retain the most highly valued with the greatest likelihood of switching.

OPTIMIZING A COMPANY'S VALUE PROPOSITION

Customers' value chains represent the activities that they undertake to achieve some objective or to satisfy some need. The potential rewards of

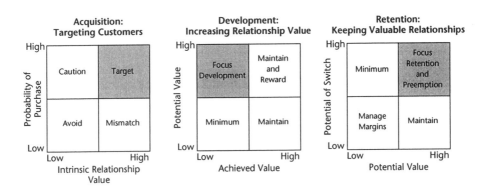

Figure 17–7 *Managing Across the ADR Cycle*

meeting these needs are a function of the expenditures on the chain (or across the "customer experience") and the competitive position of rivals serving the chain.

In general, value propositions may be characterized as core, extended, or total solution positions serving progressively more of the customers' needs and addressing more of their value chain, as illustrated in Figure 17–8. When the "process" view of the value chain is not representative of the way customers combine factors, it is often better to use a "total experience map." Nike, for example, has moved across the "athletic accessories experience" starting with shoes, moving on to apparel, and now is entering the sports equipment arena.

Most companies start out offering core products. Many excellent companies such as Coca-Cola and Stanley still do. Companies usually try to expand their presence on their customers' value chain because they realize that they are capturing only a portion of the money customers spend to meet their underlying needs. Thus is born the desire to cross-sell. Aggressive cross-selling is one reason many of us dislike banks, insurance companies, and brokers who, having sold us one thing, feel compelled to sell us anything else even remotely related.

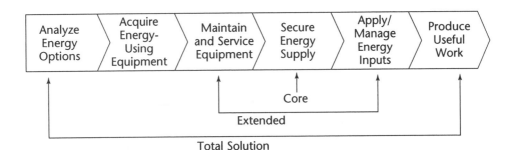

Figure 17–8 *Customer Value Chain for Energy Related Needs*

Nearly every electric power company is enthusiastically extending its value proposition to provide a full range of energy related services or in some cases to provide what is called an "integrated utility service," comprising every product in which wires are to some extent involved. A number of energy companies have in mind something like addressing the "total energy experience" as illustrated in Figure 17–9 below.

Given the substantial brand equity that many utilities have created over the years, moving "beyond the meter" with existing customers may be an effective strategy. The trick in providing a total experience is to have a credible basis for the customer to trust in your ability to bring the package together effectively. It may be very difficult to leverage this customer trust beyond your service territory where you may have little or no brand equity.

A variant on the total experience play is to position your company to meet an emotional need. In the case illustrated in Figure 17–10, an exceptionally sensitive electric power company undertakes to provide customers with a full suite of products that contribute to the customer's "peace of mind." This is the concept behind the "Worry Free" package offered by Public Service Electric and Gas and similar programs of other companies.

Leveraging brand equity and trust can be extremely effective in a period when customers are confronted with an unfamiliar array of choices. Recognizing the power of incumbency and brand equity, a number of new entrants have attempted to restrict the use of the utility's brand or to emasculate the utility by limiting it to providing only "pipes and wires" services.

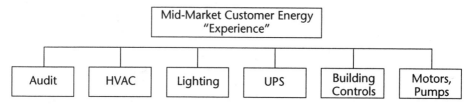

Figure 17–9   *Meeting the Total Energy Experience*

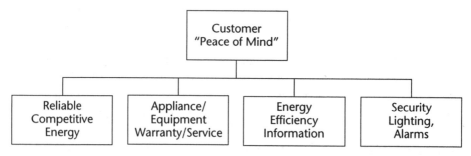

Figure 17–10   *Satisfying an Emotional Need*

Avoiding the Total Solution Virus

Many companies are infected with the total solution virus when their industry is restructuring. No one wants to be "just a bank"—they want to be your "financial partner"; they don't want to be "just a utility" they want to be your "total energy solutions source."

The lure of offering a total solution is very strong and has resulted in many a corporate near-death experience because efforts to offer total solutions often lead to acquiring companies that have little real reason to be joined together. As mentioned earlier, Allegis, an attempt to combine a hotel chain, a car rental franchise, an airline, and a reservations system into a one-stop travel solution nearly grounded United Airlines. AT&T, freed from having to stick to its knitting, found it necessary to acquire a computer company (NCR), start a credit card, enter the call center business and take stakes in satellite television and many other businesses in order to set a record for the destruction of shareholder value. (Their most recent purchase, TCI, at least offers the potential to provide local phone services.)

The risks of offering a total solution are similar to those of being a conglomerate. The odds that you will be very good at all of the different activities are against you. Just because two products are used jointly by the customer to satisfy a need doesn't mean that they share any supply-related characteristics, such as requiring the same competencies, using the same channels, and so forth. Few of us are interested in buying an "integrated food solution" comprising hardware (stove), software (recipes), and food buying and cooking services (facilities management). Moreover, customers often prefer to manage their own value chains and some are quite good at it. In addition, the incumbent suppliers already occupying the desired space rarely cede it gracefully. Efforts to offer total solutions to customers often turn into multi-front wars with rival suppliers.

Now that we've outlined the dangers of the total solution virus, we acknowledge that it does little good to be told to stick to your knitting when everything around you is coming unraveled. It is sometimes possible and profitable to extend the range of a company's value proposition but usually only when it has:

- a deep understanding of its customers' total value chain or total experience;
- earned the customers' trust and confidence in your ability to address a greater part of their value chain or total experience; and
- competencies that provide an advantage over the extended offer or when you can realistically acquire new required competencies.

Consider Nike's move into sports equipment. It has unparalleled knowledge of its customers, it is one of the strongest brands in the world, and it has exceptional marketing and supply chain competencies that it can probably leverage in the equipment market. Still, even Nike is not certain to succeed in the much larger and more fragmented equipment market populated by many strong brands. Many utilities have ventured farther afield with less ammunition.

DETERMINING THE RIGHT ROLE ON THE INDUSTRY VALUE-ADDED CHAIN

The industry value-added chain is the name given the series of steps undertaken from acquisition of raw material to delivery of a finished product to the customer, as illustrated for the retail electric power industry in Figure 17–11. An individual firm may perform only part of a value-added step or, in some cases, may perform several or all of the steps in its industry's chain. The relative profitability of value-added steps often varies significantly. The steps involved in providing the microprocessor and the operating system for personal computers generate a disproportionate share of the personal computer industry profits for Intel and Microsoft.

There are two strategic issues here. First, which step(s) of the industry value-added chain should a company compete in? Second, how should it perform that function?

In terms of sales, the predominant business model has been the vertically integrated utility company, although there are more distribution-only entities and a sizable number of generation-only companies. Whatever the historic merits are of the "natural monopoly" argument in justifying the vertically integrated utility structure, it has for some time been an artifact of regulation. In many states companies are unbundling into generation, transmission, and distribution operations and in some cases divesting generation assets.

Some people advocate unbundling even further and eliminating all direct relationships between customers and "pipes and wires" companies by encouraging the creation of independent meter reading and billing companies. It is hard to find models for this sort of sterilized role: telephone, cable, water, and sewage companies all deliver product and perform (or at least control who performs) the customer billing function.

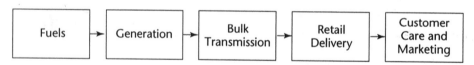

Figure 17–11   *Retail Electric Power Value-Added Chain*

As the institutional props of the vertically integrated utility business are re-moved, the industry will evolve toward its "natural" configuration or, more accurately, set of configurations. The current model won't disappear over-night; local market and political circumstances will determine the rate of evolution and it may persist as one of a range of business models. The trick is figuring out which steps in the new, restructured energy value-added chain are likely to be profitable and which of those require competencies that your company possesses or can reasonably acquire.

Most observers believe that the industry will consolidate. The question is whether it will consolidate into a few very large vertically integrated compa-nies, or whether consolidation will occur primarily within each element of the industry value-added chain, such as generation. Many managers seem to view companies and the functions they perform as a kind of Play-Doh that can be shaped at will and made to work by good management practice. In fact, the functions of the firm are determined to a great extent by the struc-ture and competitive intensity of the industry.

The conceptual framework for understanding the future structure of the indus-try, the functions performed by firms in the industry, and the relative levels of profitability were outlined over 220 years ago by Adam Smith in his theorem: "The division of labor is limited by the extent of the market." Smith's theorem and George Stigler's elaboration of it explain why, as a market grows, firms that once performed nearly every function necessary to produce and deliver their product are driven to concentrate on specific parts of the value-added chain. As the market grows, specialist firms arise to perform those functions for which there is sufficient demand to support an external market.[2]

The predictions and strategic implications of this theorem are:

- Vertical disintegration will take place in the energy business as mo-nopoly power is curbed and the "extent of the market" increases.
- Specialist firms probably will perform those functions that can be performed most economically on a scale greater than necessary to meet the needs of a single company.
- The returns that these new firms can earn will be limited by the cost of the firms who previously performed the function internally and that could re-enter the market if it becomes attractive.

A large number of utilities will find that their scale will not support competi-tive service and cost levels in many functions previously performed in-house. This is particularly true of many customer-related functions and processes.

Alternative Roles on the Industry Value-Added Chain

Once a company has decided which parts of the value-added chain to com-pete in, it's time to think about the type of operating role the company will play. By operating role, we mean the way that you interact with your suppliers and customers to create and deliver value. Are you going to be a product manager and add value largely by managing your supply chain? Or, will you work intimately with your customers and take responsibility for a part of their value chain as a process manager? Perhaps your most effective role is to bring together and manage customers and suppliers as a network manager. The general flow of value-adding activity for each of these representative positions is illustrated in Figure 17–12.

The Product Manager Role

The product manager creates value for customers by supplying a finished product such as electricity. The product manager captures value for shareholders primarily by managing the supply chain well enough to have something left over after paying for inputs. This is very much a competencies-driven strategy. The winner is likely to be the company that produces most efficiently.

The acquisition of out-of-territory generation assets (either domestically, as in Southern Company's purchase of units from Commonwealth Edison, Commonwealth Energy and Eastern Utility Associates, or the numerous international acquisitions by U.S. utilities) is an example of a product manager-based strategy to enter new markets. The acquirer is betting that their company can manage the units more effectively and produce a profit sufficient to offset the control premium paid. The high market-to-book ratios paid for domestic generation reflect the expected cash flows from deregulated operations including—in the case of gas units—the potential profits from arbitrage operations across gas and electricity price differentials. In some cases the acquisition premium may reflect an expectation that competition will cause nuclear units to be retired earlier than planned, thus tightening supplies and increasing mid- to peak-level clearing prices and raising the value of fossil plants.

Combinations of vertically integrated utilities, such as that of Delmarva Power and Atlantic Energy Inc. to form Conectiv, usually are justified on the basis of cost savings from the consolidation of general and administrative functions, increased procurement leverage, and so forth. Since the distribution and transmis-

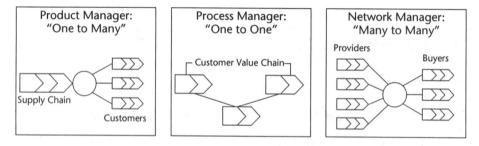

Figure 17–12   *Representative Value-Adding Operating Rules*

sion components of vertical mergers will remain shielded from competition, there will be relatively little pressure to improve performance in those areas. This accounts for the regulators' interest in performance-based regulation (PBR).

The Process Manager Role

The process manager is involved in an ongoing, interactive relationship with customers. In most cases the process manager assumes responsibility for some portion of the customer's production process or value chain. Outsourcing has been a major source of process manager relationships. However, not all outsourcing involves process management; in most cases it involves purchasing something previously produced in-house.

The facilities manager who operates a customer's generation, HVAC system, purchases their energy, and so forth, acts as a process manager by assuming responsibility for a function formerly performed in-house. To perform effectively, the process manager must be closely connected to the customer and be able to quickly respond to changes in the customer's activity levels.

The Network Manager Role

The network manager creates value by making it easier for people on both sides of a market to do business with one another. He captures value by operating his network efficiently and charging "tolls" that are high enough to generate profits but lower than the parties' cost of recreating or circumventing the network.

The network manager occupies a central location between at least two groups of buyers and sellers. These buyers and sellers would benefit by connecting directly to one another, but a number of factors may make it difficult or costly for them to establish one-on-one connections. The much maligned but economically effective HMO operating model is an example of a network manager.

The "system integrator" plays a network management role. In many cases, the integrator performs few, if any, of the functions or services that are brought to bear on the customer's need. As shown in Figure 17–13, the integrator's value-adding role is reducing the information, search and transaction costs of the buyers and sellers.

ESTABLISHING EFFECTIVE RISK AND REWARD SHARING ARRANGEMENTS

This dimension addresses issues usually covered in pricing and contracting but goes further to include all of the factors that determine how the rewards and risks associated with a relationship are shared. The degree of collaboration or interdependence between customers and suppliers ranges from simple transactions in which the customer's role is largely one of signaling and evaluating alternatives to intense interactive relationships involving the joint creation of value.

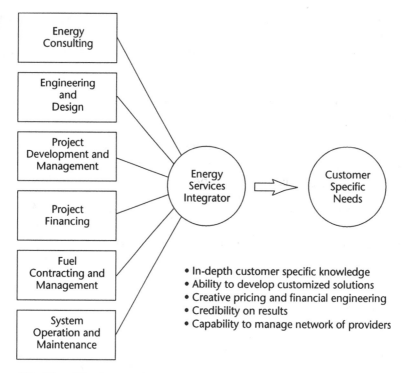

Figure 17–13   *The System Integrator*

The metaphor of a "game" is a useful way to explore the interactions between buyers and sellers and to examine their incentive to collaborate. A customer-supplier relationship can be thought of as a game in which each party seeks to maximize their reward from the exchange while bearing as little risk as possible.

Utilities have been conditioned to think that the game rewards reasonable behavior and that customers are obligated to share the risks of unforeseen outcomes. In a regulated world, the reward and risk sharing arrangements between the utility and its customers are mediated by a regulator who seeks to allow a reasonable return on the utility's investment, so long as that investment was prudent. Of course in a free market, rewards are determined by outcomes, rather than by purity of intention or diligence of execution.

Holding aside the ongoing effects of stranded cost recovery, the future range of reward and risk sharing arrangements can be illustrated by the three positions: market/neutral; performance-based; and outcomes-based.

Market/Neutral Arrangements: Taking What the Market Gives You

The most basic arrangement for sharing rewards and risks is for each party to simply take what its relative market power allows. Nothing promotes fuzzier thinking about customer relationships than the notion that there is always a "win/win" arrangement better than the market result.

A firm's ability to set terms and to capture rewards depends on its market power.

In a highly competitive market (many buyers and many sellers), neither the buyer nor seller has any incentive to form close or "intimate" relationships because the market produces a single price for all that is more efficient than they could produce through one-to-one negotiation. In other cases, buyers and sellers with substantial market power know that they are unlikely to secure better results through closer relationships (they are not interested in talking about "win-win" because they already have won). So the interesting cases and opportunities for relationship-based strategies are those in-between cases where there is some market power but a reasonable degree of choice. This is a fair description of most energy markets.

Price discrimination sounds awful but it is the key to leveraging customer knowledge in order to maximize profitability. Pricing (as opposed to ratemaking) is based on customer characteristics, not product costs. Energy companies are likely to find that they have varying degrees of market power and pricing discretion depending upon the particular customer's or group's opportunity costs and risk bearing capacity. This may lead to a proliferation of prices to different customers for essentially the same product.

Product differentiation also can provide some degree of market power or pricing discretion (what economists opaquely call monopolistic competition). The key is for some product attribute to be relatively scarce and therefore to command a premium. In many cases, particularly during the transition to full choice, the scarce factors tend to be familiarity and trust. This should provide all but the most despised incumbents an opportunity to charge a modest premium over what new entrants command.

The customers' opportunity costs include not only the price they are offered by other suppliers but also the costs of learning about and transacting with those suppliers. In general, incumbents should be able to capture a part of these switching costs. Increasing switching costs is an ancient customer retention strategy.

Performance-Based Arrangements: Putting Something at Risk

The first step beyond taking what the market gives you is to make the reward for one or both of the parties contingent upon some level of performance. For example, it is common for energy service companies to be compensated on a "shared-savings" basis: no savings, no payment.

It is likely that distribution and other regulated services will move to "performance-based regulation" (PBR). Although it takes only a few minutes for regulators to complexify the concept, PBR usually involves a cost-based rate with a "dead band" above and below the base cost level. Within the dead band the supplier assumes the risk of cost increases in return for the rewards of cost reductions. Obviously, the company goal is to have the upper range of the band as narrow as possible and the lower range as wide as possible and vice versa for the regulator.

The supplier—in this case the energy company—is usually in a better position than its customers to understand and bear energy-related risks and uncertainties because of its proximity to the market and its ability to hedge across a larger, more diverse population of buyers and sellers. The weather hedges offered by some energy marketers are examples of this. Thus, there often is an opportunity to assume a slightly greater degree of risk in return for a substantial gain in volume and margin.

Enron reinvented the natural gas business by providing customers a way to swap the volatility of the spot market for a certain long-term price. Enron introduced an innovation called the "gas bank," which enabled it to attract gas "deposits" from producers equal to the "withdrawals" from consumers. This was achieved by carefully calibrating the bid and offer prices in much the same way that a neighborhood bookie moves the odds or points until he attracts equal volumes on both sides of a bet. The premium commanded by the longer fixed contracts more than covered the costs of managing Enron's exposure to cost increases, and the volumes attracted by the vastly superior product propelled Enron to the top of its industry.

Almost any factor can be grist for financial engineering of the revenue stream and risk bearing load. The shared savings, PBR, and gas bank examples all deal with risks related to cost changes. The MCI Fund illustrates how rewards can be time-shifted to meet the preferences of customers for the immediate rewards of committing to a long-term contract. In this arrangement MCI capitalizes a part of its long-run savings and pays them to the customer for immediate use. As long as the present value of the contract is the same under both options, MCI is indifferent financially, but it is much better off from a marketing perspective. By reengineering the time pattern of savings for the customer, MCI has achieved higher sales and longer customer relationships.

Some energy services companies have employed a similar approach to overcome customer reluctance or uncertainty about the potential for savings. In some cases the energy service company "pays" the customer for the right to be the customer's energy manager by giving the customer a portion of the anticipated savings up front as a "signing bonus."

Outcomes-based Arrangements: Partnering for Dollars

In the outcomes-based arrangement, both parties have something at risk contingent on some outcome over which neither has complete control. This is the reward and risk sharing arrangement closest to what many people refer to as 'partnering.'

Here's an example. A large electric power company needed to refurbish several of its older fossil-fueled plants that had been neglected during the construction of nuclear units. The company, unable to secure reasonable project financing terms and reluctant to take on additional long-term debt, approached several equipment manufacturers and engineering firms and offered them a percentage of future cash flows from the plant. As a result, the plant owner reduced its cash outflows and capital requirements. But most interesting was the behavior change on the part of the engineering firm. Since its rewards no longer depended on the number of engineering hours it charged, it actively searched for ways to bring the plants on-line sooner and to operate more efficiently. Because their interests were aligned, both parties found that they could reduce the amount of resources devoted to checking on one another and arbitrating conflicts.

CONCLUSION

Any strategy depends in part on the initial conditions—the hand you are dealt and the resources and competencies you bring to the party. It is natural to want to start strategic planning with a consideration of existing capabilities and strengths and this is often appropriate in stable industries. But in times of profound changes, the options and the potential pay-off depend on recognizing emerging patterns in the evolution of the industry and understanding how new industry structures and technologies will affect the choices available to customers and the means of delivering them.

In this chapter we've provided a customer-based framework for evaluating strategies and identifying new sources of value. No one can be sure what the ultimate shape of the energy industry will be or which business designs will be most successful. We believe that the companies who know the value of their customer relationships, understand what those customers want, perform the functions that best leverage their capabilities and know how to manage the risk and reward sharing process will have a much better chance of surviving and prospering.

NOTES

1. The Value Compass[sm] framework was introduced in *Customer Connections: New Strategies for Growth* by Robert E. Wayland and Paul M. Cole (Harvard Business School Press, 1997).

2. George J. Stigler, "The Division of Labor is Limited by the Extent of the Market," *Journal of Political Economy*, June 1951. Reprinted in George J. Stigler, *The Organization of Industry* (Homewood, IL: Richard Irwin, Inc., 1968), 129–141.

SECTION V

WHAT ENABLING TECHNOLOGIES AND INFRASTRUCTURE ARE NEEDED?

CHAPTER 18

Infrastructure for Customer Choice: Managing the Risk and Cost of Implementation with a Sensible Business Model

Eric P. Cody*

The key design intentions of customer choice—a more efficient, market-driven means for driving energy costs down over the long run—depend to a great extent upon an infrastructure of processes and systems that is in place and works robustly. When customers express their choice of electricity supplier, those preferences must be recorded, verified objectively against key data, planned for execution on a schedule that meshes with other operational activities, and executed with predictability. Customer billing must continue to occur with the relentless efficiency that has become the hallmark of the utility industry over many decades. And the quantities of supply and load that ought to be roughly in balance every hour in the regional control area or power pool must be measured or estimated with adequate precision to enable a multitude of energy suppliers to function in the new market context. These requirements do not mirror those as traditionally understood and supported by utilities; moreover, some are counterintuitive.

---

*The opinions stated in this chapter are those of the author and are not necessarily the opinions of NEES Global Inc. or its affiliated companies.

The mechanics of competitive electricity supply—the so-called "competitive infrastructure" elements—are quite complex, and accordingly, costs to implement customer choice may be so large as to negate a year or more of competitively generated energy savings if they are approached halfheartedly by utilities, regulatory bodies, power retailers, and system operators. My first-hand observation of this industry in transition, through NEES Global's consulting engagements with utilities serving more than 25 million meters in 27 U.S. states and Canada, point to a wide diversity of approaches and levels of understanding of these monumental challenges.

This chapter is intended to uncover many of the mysteries of competitive infrastructure in order to clarify risks and point to approaches that moderate implementation costs. An overarching theme of this chapter is that there are many distinct business models for customer choice, and the trade-offs between and among their fundamental elements must be made in a reasoned and well-informed manner. A keen understanding of the issues, which can be gleaned from real implementation experiences, will raise the probability of implementation success and contribute to the achievement of value for ultimate customers. Utility executives I have worked with all agree on one point—the high stakes demand that implementation be done correctly the *first* time.

THE COMMON CHALLENGES OF 'CUSTOMER CHOICE'

What *is* 'customer choice'? In the simplest form, it is the enabling of end users to choose a competitive energy source provider (ESP), or supplier, not to be limited to affiliates of their local utility. Suppliers compete for customers, prices are set by the market as opposed to through cost-based rate recovery mechanisms, and most often the customer chooses the ESP directly.

When one considers the unprecedented aspects of customer choice infrastructure requirements, it becomes clear that the industry and its regulators are dealing with a project that is by its very nature high-risk:

- Performance of certain new activities in support of electricity market settlement fall upon the local distribution companies (LDCs).
- Large-volume customer transactions are involved.
- The most efficacious means for trading information between LDCs and ESPs is through electronic business transactions (EBT); however, many of the trading partners will not have prior experience with EBT as a matter of practice.
- Market-enabling organizations such as independent system operators and power exchanges are being created.
- All of these activities are occurring simultaneously, and must integrate seamlessly for successful market operation.

In fact, comparisons of risk between customer choice infrastructure and the Year 2000 (Y2K) computer problems are instructive and offer LDC management a general guide for how seriously to consider the challenges. (See Figure 18-1). The complexity of new business requirements, the uncertainty of underlying business rules, the potential impact on unaffiliated companies (who sometimes tend toward legal remedies when outcomes are not as anticipated), and unrealistic delivery timeframes appear more threatening for the customer choice systems project than for Y2K. Given that most companies initiated Y2K projects two or three years ago, it might be reasonable to ask the executives of LDCs that are opening retail competition within the next two years: "How long ago did your systems professionals begin working on the customer choice project?" Many information technology veterans consider the characteristics commonly found in competitive infrastructure to lend themselves to a rapid systems prototyping approach, as opposed to a business requirements-based development approach. That is, the business requirements are not well-defined because the processes are completely new. However, prototyping is less appropriate for large scale, production systems where lack of robustness may prove to be costly. In short, LDC managers who take the nuts and bolts of competitive infrastructure too lightly may suffer financial or legal consequences worthy of auditors' footnotes in an annual report.

The direct financial risks to LDCs of competitive infrastructure are routinely considered at the board level. Given the sizable dollar stakes, a level of success as low as 95 percent may result in a "problem" equivalent to perhaps as

Risk Factor	Y2K	Customer Choice
Complexity of New Business Requirement		□□
Pervasiveness of Systems Modifications	□□□	□
Degree of Systems Integration Testing Required	□□	□
Magnitude of Direct Financial Consequences of Failure	□□	□
Uncertainty of Underlying Business Rules		□□
Magnitude of Internal Retraining Requirements		□
Impact on Unaffiliated Entities	□	□□
Visibility of Failure	□□	□□
Implementation Timeframe	□□	□□

Figure 18–1 *Comparison of Risks: Year 2000 (Y2K) Project vs. Customer Choice Implementation Project*

much as one-half of a utility's annual net income. High implementation costs also could precipitate the need to file a rate recovery case, exposing other utility cost structure elements to scrutiny at a time when costs are being squeezed out, thereby jeopardizing the utility's ability to retain these savings in the form of improved earnings. Finally, the prospect of delays in retail choice for utilities that have reached major financial commitments such as asset divestiture agreements creates an awkward hiatus in management's planned transition to restructured operations. Competitive infrastructure, experience shows, is much more of a challenge than an opportunity for most utilities.

KEY LESSONS FROM EARLY IMPLEMENTATION EXPERIMENTS: THE "NUTS AND BOLTS" OF CUSTOMER CHOICE AND MARKET SETTLEMENT

Early Implementation Experiences

As of mid-1998, competitive electricity markets in California, the United Kingdom, and New England permitted choice of electricity supplier for some 13.5 million electricity customers, not including pilot programs. However, only about 100,000 of these market-eligible customers had elected to choose a competitive supplier of power, as lack of awareness and insignificant near-term savings for most of these customers led most consumers to "stay put" with their current utilities. While it is hard to attribute this lack of movement to specific underlying factors, it is undoubtedly the case that absence of a common market foundation has contributed to the confusion. Across the many regulatory jurisdictions I have visited, one thing has become apparent: no two customer choice implementation models are identical, and this has worked against the formation of a set of market mechanisms that invite competition to begin on the scale desired by choice architects. Moreover, there is fundamental debate over one of the key essentials: is the exercise about customer choice or a new commodities market? This debate is most easily understood by looking at the Massachusetts experience, where legislators and other public officials insisted upon across-the-board savings of 10 percent or more, even for those customers who "choose not to choose." This so-called "Standard Offer" generation service may constrain competition to some extent, at least for the years in which it will remain in effect; however, such elements provide real savings for all customers during the transition period when the market might otherwise be inclined to reward only large customers.

Other key lessons arising from early implementation experiences include the following:

1. Rules specifying the flows of market transactions and information have not been tested and confirmed through pilot programs in some jurisdictions. As a result, they may turn out to be grossly inefficient and expensive.

2. Lack of foresight by utilities as to what businesses they plan to pursue in the future inevitably leads to confusion over how new rules will affect them in the future.

3. There is often a lack of understanding on the part of LDCs of the risks and liabilities that are created by their performance of new activities such as load profiling, where LDCs generally have no "upside" profit potential and therefore should have a low tolerance for risk.

4. Politically expedient elements such as discounted standard offer services and default generation services interject nonmarket based rules and directly impact the free movement of customers to competitive suppliers.

5. Failure to recognize that customer bills must be unbundled into the new components (distribution, transmission, and generation) in advance of choice day makes it difficult for customers to comprehend the choices they are being empowered to make.

6. A tendency by some regulatory or legislative bodies toward further competitive unbundling of the traditional distribution services of metering, billing, and customer service (so-called 'call center') functions, adds significant complexity and risk to the already complicated challenge of delivering a competitive generation market, as was learned painfully in the U.K. in 1994.

7. Lack of understanding that 'exception handling' must be carefully considered and minimized in process design, given the high volumes of transactions that can be expected.

8. Superficial attention appears to have been paid by ESPs to the inherent accuracy and bias performance of most load profiling methodologies. However, when the financial flows through these systems finally become material, ESPs may find that regulatory rulemakings governing how this work is to be performed are long closed.

9. Finally, it is apparent that utilities almost never behave as a homogeneous group within any single jurisdiction, nor do ESPs, nor do customers. The design of market mechanics must take this market diversity into consideration. Few have thus far.

Building the Utility-Supplier Relationships

One of the primary challenges facing utilities in their efforts to deliver a workable competitive infrastructure is that unprecedented relationships must be built between the LDC and ESPs. Some form of electronic network must be put in place to facilitate the transfer of information and funds in a timely manner. Ultimately this electronic landscape must become standardized or suppliers will find the administrative burden of building the needed electronic interfaces with utilities to be overwhelming, particularly against the backdrop of what have turned out to be razor thin margins. Key processes must be created within this electronic information network, including, as a minimum:

- Customer enrollment with ESPs.
- Notification of successful transactions or errors, as applicable.
- Customer billing, generally supporting several optional models.
- Meter reading and passing of meter determinants for billing.
- Customer service support.
- Termination of supply by ESP.
- Service disconnection and supplier notification.
- Service restoration.
- Authorization and release of customer information.
- Installation of telemetering equipment and data collection/transfer.
- Determination of hourly supplier loads.

While generally accepted models for many of these components are now available via Internet sites for review, other procedures are not always subject to regulatory approval. For example, the detailed rules for handling of disputes between ESPs are often left to the discretion of the LDC.

Market Settlement: A Basic Prerequisite

Settlement activities—the detailed housekeeping of supplies of energy and customer demand, and the financial accounting within market control areas—have been approached jurisdictionally, as a general pattern. Because bulk delivery meters can no longer simply be read and the loads of monopoly generation suppliers reported to the control area or ISO, some form of simulation of supplier loads and associated losses is necessary for early choice markets to open. Even if LDCs were willing and able to invest the hundreds of millions of dollars to implement metering reading networks to collect hourly load data on a nightly basis, load profiling still would be essential as an insurance policy covering the contingency that network components fail or communications paths are disrupted during daily settlement processing.

Early implementation experience suggests that load profiling may have a life expectancy of three to five years, a period determined largely by the pattern of daily electricity prices in the market. While the accuracy of state-of-the-art profiling methodologies is quite good for energy (kilowatt-hours) when averaged over a typical month, the hourly errors may be quite large, especially for suppliers with small market shares. Thus, while fairly flat market prices would be somewhat forgiving of these error patterns, the converse is also true—that price spikes will combine with hourly error patterns to create volatility in settlement cash flows for some suppliers. Such volatility, should it occur early in the market evolution, would interact directly with thin supplier margins to threaten ESPs' financial viability. Experience in the Midwest during the hot summer of 1998 suggests that significant volatility will be present early on.

Nevertheless, most agree that load profiling is a basic prerequisite for the opening of competitive electricity supply markets until more cost-effective metering technology becomes generally available or the penalties of poor estimation become so onerous as to justify large scale technology investments.

Approaches That Simplify the Transition

Firsthand experiences with retail access and customer choice point to several cardinal rules for simplifying the transition and benefiting LDCs and ESPs alike:

1. Focus exclusively on making supply competition work at the outset; this is where the most significant savings are available to customers and the model is complex already.

2. Defer the further unbundling of metering and billing until the generation market launch is assured to be successful—this may involve only a period of months in some cases.

3. Maintain a design for the settlement process that minimizes the number of entities that must exchange critical information within operational timeframes (e.g., hours not days).

4. Build full-scale systems even when phasing in competition to confirm their integrity and the workability of the business rules that underlie them.

5. Organize implementation efforts without respect to current utility organization structures; also recognize that some elements of the design must be agreed upon by all affected parties and will be best resolved outside the regulatory hearing room.

6. Ensure that total utility franchise area load and other supply requirements are fully allocated in a balanced accounting sense to avoid the possibility that the distribution company will have to absorb costs of providing generation itself, other than as pass-through arrangements.

Experiences worldwide have demonstrated that individual utility implementation costs may exceed $100 million when competitive infrastructure decisions are reached in a manner that ignores such simplifying approaches. Current estimates of total implementation costs that California utilities will incur exceed $1 billion, and regulatory filings by the two largest investor-owned utilities in Michigan contain implementation budgets that exceed $100 million each. This adds further credibility to the comparison with Y2K project challenges; the difference in this case is that something can clearly be done to manage costs downward and mitigate risk.

DESIGNING A BUSINESS MODEL AND RELATIONSHIPS THAT WORK

Anyone who has attempted to specify workable rules for this new competitive energy market, and the new relationships that must be governed, quickly has realized that this is quite different from building a new utility billing system or implementing new operational procedures for handling customer calls. Other parties must coordinate their efforts with the LDC and they may not agree with the LDC's presumptions.

Two practical avenues offer some hope for timely resolution of such differences of opinion in matters where the outcomes are not controversial.[1] These avenues can be divided into guiding principles and consensus building approaches.

Guiding Principles

There is a set of guiding principles that may be applied when considering trade-offs between several detailed options for any customer choice process element. A typical set of principles might be to:

- Reduce the potential for customer confusion.
- Protect and facilitate the customer's right to choose a new supplier.
- Specify electronic solutions, not paper-based flows.
- Minimize the number and complexity of electronic transactions.
- Minimize the potential for exceptions that require manual intervention and troubleshooting.
- Control total cost, not solely that to be incurred by the utility or supplier.
- Streamline for efficiency, speed, and accuracy.
- Make standard background information such as profiles, meter reading schedules, and publicly available tariffs generally accessible, in anticipation of data requests.

Collaboration & Consensus Building to Get the New 'Rules of the Road'

Stakeholders should readily accept that formal regulatory proceedings are often a poor instrument for reaching consensus on technical implementation elements, such as electronic business transaction formats, validation of transactions, and specific utility-supplier service arrangements. Figure 18–2 illustrates the relationship between critical outcomes and approaches to resolving issues. There are numerous examples of such successful collaboratives and working groups in early implementation worldwide. It is up to each utility to determine how best to pursue closure; however, the one avenue not shown in the figure—litigation—already has been shown to be a highly unsatisfactory avenue for virtually all parties involved, creating customer confusion and uncertainty, as well as delays, which ultimately become costly for utilities, suppliers and customers alike.

Customer Enrollments & Transfers

Customer enrollments are the fundamental prerequisite to all other retail access business processes. In other words, if a customer's election of a new

Critical Success Factors	Regulatory Hearings	Technical Sessions	Collaborative Process	Legislative Committee	Bilateral Negotiations	Round-tables	Pilot Programs
Achieve market goals				✓			✓
Meet customer needs	✓	✓		✓	✓	✓	✓
Manage complexity		✓	✓				✓
Minimize implementation costs		✓	✓				✓
Enlighten decision makers		✓	✓			✓	✓
Educate new market entrants		✓	✓		✓	✓	✓
Create win-win situations			✓				✓
Recognize existing obligations	✓		✓	✓			
Base decision on facts and experience			✓				✓
Test feasibility of untried models	✓		✓	✓			✓
Minimize market barriers	✓			✓			✓
Protect consumers							

Figure 18–2   *Regulation, Legislation, or Collaboration? Critical Success Factors in Rules Development*

energy supplier is not acted upon correctly or in a timely manner, all other process steps are impeded. Accordingly, this process must be thoroughly designed. Some common design elements for enrollment have emerged:

- Customer selects a single supplier *per meter*.[2]
- The *supplier* accepts the customer's enrollment order and is responsible for maintaining proof that the customer requested the switch.
- The supplier enrolls the customer with the LDC *electronically*, within a specified number of days prior to the meter read date to allow for notification of old and new suppliers. (Note that the customer is not notified of the change by the utility, but rather sees confirmation of the supplier change on the next bill.)
- The LDC automatically changes the customer's supplier of record on the customer's regularly scheduled meter read date.

Other transactions, such as a supplier drop of a customer for nonpayment of the generation bill, may be patterned closely after this basic enrollment design.

Sharing Customer Information

Sharing customer information is critical to the successful operation of both the "back room" (technical processing) and "front room" (customer oriented) activities in the competitive market, although elements central to this activity often are considered fairly late in implementation planning by LDCs and their regulators. It generally is accepted in the utility industry that the customer retains rights over disclosure of his/her account information. However, procedures for release of this information must be clearly spelled out. Generally speaking, suppliers obtain some form of customer authorization, at which point the utility is expected to honor the request in a timely manner. Key questions remain however, such as: Are LDCs required to provide energy and demand data to suppliers when customers may already have the information? Can utilities charge for these reports since there may be thousands per year? If so, what information services should be provided at no cost?

Best practices in this area focus on "pushing" information out in such a manner that some of the demand or "pull" for detailed information that would exist otherwise is dissipated. This practice also promotes standard formats by its nature. Several LDCs have created customer information centers on their Web sites, for example, which enable customers, and their suppliers or aggregators, access to their individual usage information with appropriate confidentiality safeguards. Simpler forms of information push include more detailed usage history on all customer bills and periodic load reports sent directly via electronic mail to large customers who have interval data recorders installed.

Needless to say, the appetite for customer information that exists in the competitive energy market is large and the incremental work requirements associated with feeding it must be carefully considered, and addressed, particularly in the case of utilities subject to multiyear rate freezes.

Whose Customer Is This Anyway?

Even a cursory review of some utilities' retail access implementation plans on file with state utility commissions points to a philosophical gulf between utilities and competitive power suppliers, and that is, "Who owns the customer?" Customer handling processes often are articulated in such a way as to place the utility in the position of facilitating the customer's preferences and actions. And why not? Progressive utilities have striven for years to win improvement in levels of customer satisfaction and to achieve customer loyalty, and that effort spawned fundamental rethinking on how to answer customer calls, what options to offer for bill payment, and how to provide more accurate estimates of service restoration times. Many utilities seem to believe, then, that customer choice can be considered as just another choice within the basket of utility-sponsored offerings and to design processes that allow the utility to continuously affirm that things are working as intended.

However, those with retail access experience generally will disagree with this perception. One of the most essential realizations that occurs to practitioners is that once customer choice is offered, former customers "are no longer our customers for generation services." A key observation from early retail access experiences is that this change in thinking must occur for the necessary adaptations to occur that will produce even-handed and nondiscriminatory processes. To some observers, it appears that "customer loyalty" may become the banned phrase in the new utility lexicon, in much the same way that "marketing" once was.

MASTERING THE COMPLEXITIES OF COMPETITIVE METERING AND BILLING

A number of ESPs have argued before state utility commissions that consumers will gain additional economic benefits, beyond those offered by competitive generation, by immediately making metering, billing, and other customer services—The so-called 'revenue cycle' services in the California terminology—competitive also. The argument is that competition invites innovation and new technology; that it is the ESP who ought to be building the relationship with the customer, and that these services are more closely linked to generation than to the distribution or transmission functions.

One does not need to choose sides with the ESPs or the utilities to determine the practical answer to the question of immediate distribution services unbundling. Taking this step adds great complexity and risk to already complicated business processes and therefore one can conclude that temporary deferral of this step will be in the customer's, and perhaps even the supplier's, interests. As a component of the total cost of a kilowatt-hour, the cost of these services, and therefore the immediate savings to customers, is relatively small in comparison with generation; however, margins might well be more attractive to suppliers given the intensely competitive generation market.

Impact on the Business Process Model

A decision to open up metering and billing to competition creates a sizable impact for other processes that must be supported under customer choice. This added complexity transcends the contentious issue of who should be entitled to maintain the close relationship with the customer through the provision of these services. Some of the infrastructure-related questions, which must be answered in detail to understand how this expanded model of electricity competition could be made to work, include:

- How will outage calls be sent from suppliers to the LDC during major power disruptions? Will systems be directly interfaced?

- What entities will have authority to physically disconnect customers once the LDC has a limited or no part in billing or meter reading?

- How can (or why would) the LDC maintain even minimal call center capabilities without the economies of scale afforded by the requirement to provide metering and billing services also?

- How will metering information flow and what is the likelihood that this information will be available to those who need it in operational timeframes measured in hours?

- Who will maintain customer enrollment systems if the LDC considers suppliers its customers for billing purposes?[3] Who would fill the void since suppliers obviously cannot be permitted access to competitors' market share information?

- How will metering and billing services be arranged for customers on default generation service who draw their power from the Power Exchange? Will there be a revenue-cycle services provider of last resort, as there is for generation?

- What measures need to be taken to avoid inappropriate behavior by nonutility providers of metering and billing services, such as redlining of low-income customers in favor of those with good credit ratings or the ability to pay by credit card?

The debate over competition in metering, billing, and customer services often overlooks the practical requirements of operating the new electricity market so that all stakeholders enjoy the opportunity to participate and consumers derive the desired economic benefits.

Potential for Inappropriate Gaming

In order to make competitive metering and billing viable, the main problem to overcome may be that winners and losers are created. Winners will be those customers who live in densely populated areas where meter reading can be done for less than the utility's average cost, those with superior credit ratings, and those who have credit cards—in short, affluent urban dwellers and suburbanites. Losers might include customers living in remote areas, those with poor credit, and those still paying their utility bills in person at local retail stores—in other words, everyone else. Moreover, ESPs with the back room resources to provide large-scale metering, billing, and call center services will be presented with unique opportunities, clearly giving them a competitive advantage over their smaller rivals.

MARKET SETTLEMENTS: NECESSITY IS THE MOTHER OF INVENTION

Under the long-standing, vertically integrated utility model (shown in Figure 18–3), matching the generation quantities that flowed into the electricity grid every hour against the customer load and line losses they were contracted to serve was a technical challenge, but one that was readily accomplished. Every night, bulk electric meters, most often located at transmission delivery points, were called up and read over telephone lines. Interval load data were downloaded and adjusted for the presence of distributed generation and other factors to produce the utility's own load for balancing against generation. Generally speaking, all the kilowatt-hours flowing through the bulk meter were sourced from generating plants owned or contracted by the local utility, operating in a vertically integrated manner.

The new world of competitive electricity supply, shown in Figure 18–4, is somewhat problematic, by comparison. Now, many ESPs who are not affili-

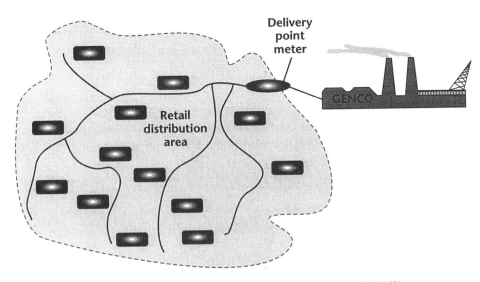

Figure 18–3 *Traditional Model: Integrated Retail-Wholesale Utility*

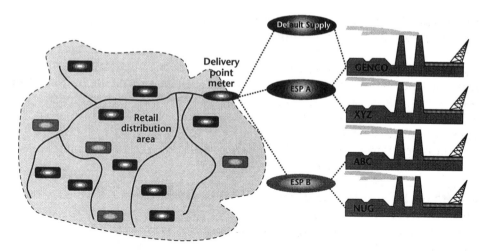

Figure 18–4   *Restructured Model: Competitive Electric Supply*

ated with the LDC supply energy through the same bulk meters to ultimate customers on the LDC's distribution lines. The bulk meter can no longer simply be read and reported as utility load. Moreover, even if hourly load recorders were to be placed on every individual customer load and read nightly, the previous level of accuracy still would not be achieved, since line losses are not reflected in the customers' meter readings and would have to be allocated to account for total load. As a result, some form of load estimation methodology had to be developed to enable load reporting in order for daily financial settlement to be performed by the control area authority or ISO.

Load Profiling & Reconciliation

"Load Profiling" is the generally accepted name for this new methodology, and it has been viewed worldwide as a necessary, and adequate, basis for the opening of the new electricity market. Load profiling, as it refers to the after-the-fact determination of hourly supplier loads, is a data-intensive technique combining customer class load profiles from the utility's load research sampling program or other source with customer monthly energy usage to produce preliminary estimates of the hourly loads served by each ESP on a given day. These preliminary estimates are adjusted to reflect line losses, then compared to bulk meter readings that represent the known load for those hours' in the geographic area covered. Any residual differences are allocated proportionately among ESPs so that all the metered energy is accounted for. This initial reconciliation to metered area loads is essential since the distribution company, by definition, may not have generation to cover any load that is not assigned to suppliers. It is important to note that all customers whose hourly loads are directly telemetered are excluded from the simulation, except for the loss adjustment step. These preliminary daily supplier load estimates are forwarded to the control area authority or ISO as a major input to the financial settlement process. Similar methods may be applied to produce ESPs' next-day energy schedules.

As more information, such as manually retrieved revenue meter readings and interval load data, becomes available to the LDC during each month, these too help improve the accuracy of the preliminary supplier estimates, offering a further reconciliation basis and improving the allocations of load. Most ESPs support load profiling, although most probably will wait to see how the technique performs when the financial consequences become material, and this may take several years.

Risks to the long-run market acceptance of load profiling include the technique's inherent accuracy and bias, potentials for inappropriate gaming, hour to hour variation in pool price, and degree of predictability of results. Many practitioners feel that it may be preferable to sacrifice some accuracy for improved predictability in the marketplace; in other words, load estimates may be more relevant than actual meter readings if they are the values that are used in financial settlements.

Telemetering versus Profiling

One of the key differences between retail access models around the world is the level at which customer telemetering is required for participation in the market. In the U.K., customers with demands exceeding 100 kilowatts must have a half-hourly recording device with modem and phone line for settlement purposes. The threshold is lower in California and there is no such requirement currently in Massachusetts and Rhode Island; however, in the latter jurisdictions, individual customer profiles are used in lieu of class average profiles to more accurately estimate the loads of large customers.

There is one important consequence to how large customers are treated in load profiling. Load profiling works best for large classes of relatively homogeneous customers—those which can be statistically sampled with smaller samples due to small within-class variability. Very large customers are known to be much less homogeneous and often are sampled at the 100 percent level for utility rate design purposes. If significant errors are introduced into the load profiling process by large customers whose load pattern is not well described by a class average curve, then other customers, and their respective suppliers, also will be impacted since it is a zero-sum game. Hence, requiring telemetering for some large customers is not a trivial decision.

Regional Concepts & Solutions

Suppliers face a patchwork quilt of regulatory rules and local conditions in the piecemeal, early market openings across the United States. State commissions and legislatures set specific requirements and market rules, LDCs build custom systems with which suppliers must interface, and ISOs and power exchanges are in varying states of readiness to support the basic functions of the new market. Any opportunities for standardization at the regional or national level undoubtedly would be welcomed by suppliers to reduce the administrative burden of creating unique interfaces for their entry into each utility franchise area. In essence, the new electricity market appears to suppliers to end and begin at virtually every state border, if not every utility franchise area boundary. Load shapes used as inputs to profiling are different, as are the computer system algorithms, loss factors, and reconciliation points. Because of these inconsistencies, ESPs have found it difficult and time consuming accounting for the differences between their monthly customer sales (measured by customer meter readings) and their power bills from generators (generally estimated using load profiling). Regional processing centers, using standard load curves and certified, high-quality assurance systems, may offer some relief. One such center has been created to serve several New England LDCs; however, the concept has been surprisingly slow to take off, due in part perhaps to the lack of expressed interest in the concept by ESPs operating in the region.

WHAT DOES THE CUSTOMER NEED TO KNOW ABOUT THE MARKET?

Most of the competitive infrastructure that must be put in place to enable customer choice of electricity supplier can be designed and implemented without direct involvement by the ultimate customer. Arguably, it is not productive to involve the customers directly in decisions about transaction processing systems and communication protocols when they will not be a party to these transactions, merely the subject of everyone else's attention and central to the transaction content. Utilities and regulatory bodies should consider the extent of customer education required merely to explain tariff changes and how to exercise choice of ESP, before communicating such nonessentials to customers.

Unbundling Rates and Bills

The New Hampshire retail access pilot program demonstrated the degree of confusion that can be created when customers are authorized to choose their ESP prior to being educated about the unbundling of their bills. Some pilot participants, upon seeing supplier prices in the two-cent range quoted in the media, believed that their chosen ESP would reduce their current price of nearly 15 cents per kilowatt-hour by 12.5 cents! How could one expect them to know better when unbundled bills had not yet been delivered by the utilities? So as a bare minimum, utilities should produce unbundled bills prior to the time when customers are invited to make a choice affecting only one part of the bill, when that part has never been culled out previously.

Expressing Their Preference

How customers obtain lists of qualified ESPs, obtain answers to their questions, and express their preferences must be a key focus of infrastructure; in these cases, standards of conduct will play a very meaningful role in determining how all of this plays out. The general model that has gained widespread acceptance is that the customer will be marketed by ESPs and will "enroll" with their ESP of choice, who in turn will send an electronic enrollment transaction to the LDC or third-party enrollment agent, which begins the set of transactions leading to a successful supplier switch on the customer's next scheduled meter read date.[4]

The challenge in customer enrollment is that three parties are involved—customer, ESP, and LDC—in ways that tend to force fundamental rethinking of the LDC's relationship with the customer, as was discussed earlier in this chapter. The design of enrollment is counterintuitive to many LDCs; the customer does not call the utility's call center to enroll and the customer is not routinely notified by the LDC that a change of supplier transaction is pending. Thus, the utility should take care not to design a process that makes it easy to become mired in disputes between customers and their competitive suppliers, or between two suppliers. As the enrollment agent working to maintain accurate records of which supplier is associated with which customer on a continuous basis, the LDC simply needs to know how to handle contingencies such as disputes. Quite simply, the transactions either should be all terminated or all executed, and the rules must be specific.

Utilities designing new business processes for customer choice should endeavor not to continue thinking about their role as central to the relationship with the customer. The utility generally will not be informed of the entire commercial arrangement or have access to the relevant contracts between customer and supplier, and intervention may be considered interference by the other parties, which could lead to unwanted legal involvement. Moreover, if the LDC has its own competitive affiliate, the LDC must exercise care not to show preference to its affiliate or even give the appearance of showing preference, and the safest way to insure this result is to remain uninvolved in the customer-supplier relationship. Finally, becoming involved in disputes will be costly from a resource standpoint, while also potentially compromising the LDC's neutrality with respect to its own competitive affiliate or affiliates.

What About Consumer Protection?

Much has been made of "slamming"—the practice of enrollment by a new supplier without the customer's knowledge—in early regulatory discussions about customer choice, as a result of previous experiences with telephone deregulation. While slamming has been virtually nonexistent in U.S. retail access pilot programs since 1996, requirements for consumer protection must be considered when confronting competitive infrastructure. Here are some of the questions that should be asked:

- Are all customers treated fairly, and consistently, by the processes of enrollment, billing, metering, and default service options? Do smaller customers have the same opportunities to save as large customers?

- Are charges for transactions reasonable and based on real, incremental costs imposed upon the LDC? Are customers and suppliers who request the services the ones who pay?

- Are the customers being given accurate information about the "ingredients" in their supplier's power based on its sources? Is "green power" really sourced from environmentally benign facilities, especially since customers are being asked to pay a premium price for such options?

- Are provisions present that limit the ability of the customer to make a choice? Are these appropriate limitations, e.g., a multiyear contract, offered at a preferred discount?

- Are low-income, elderly, and other lifeline customers adequately protected against a sudden loss of energy supply?

Most of these issues can be guarded against in the new process models; however, some ongoing oversight inevitably will be required and it should not be the responsibility of the LDC to identify abuses and initiate enforcement actions against suppliers.

WHAT DOES THE MARKET NEED TO KNOW ABOUT THE CUSTOMER?

ESPs and load aggregators want and need a great deal of information about customers to operate successfully in the competitive electricity market, particularly given the thin margins that have been observed to date. ESPs' abilities to target those market segments and individual customers that can be served profitably are a key to survival, and much information is held by the LDC that is valuable in this quest.

When the LDC is approached by an ESP for customer information, its typical response is, "What information do you need, for what period of time, and in what format?" Unfortunately, to date the suppliers' answer has often been, "Whatever information is available." This lack of specificity is troublesome for the LDC, since it may be facing literally thousands of customer information requests from dozens of suppliers. What is needed are agreed-upon guidelines for sharing customer information.

Common Sense Guidelines for Information Sharing

To facilitate the exchange of properly authorized customer information, the following issues must be addressed:

- Shift the responsibility for obtaining customer authorization to the supplier along with associated record retention requirements.

- Accept the reality that billing histories, load interval data, and class average load shapes, among other information series, are required for the market to work properly. Design efficient information clearinghouses to distribute this information as needed.

- Investigate electronic means for transfer of customer information.

- Define what is above and beyond existing releases of information to the customer and assess reasonable fees to recover the incremental costs.

- Enable timely information reporting and avoid bottlenecks that might otherwise occur during active recruiting periods.

Premium Services and Incremental Costs

Many of these activities are new and high-volume in nature, as early pilot experiences have demonstrated compellingly. While some of the demand for new information can be mitigated by improving bill stub information, data such as interval loads for large customers who are on the LDC's computers must be provided flexibly and quickly when properly requested by market participants, since restricting access to customers making direct requests will be viewed as nonresponsive behavior by other stakeholders. Producing 1,000 individual customer load reports for a large aggregator of health and educational facilities, for example, will have associated costs, as will the need to update information monthly or quarterly. These costs should be tracked and recovery from the supplier or customer may reasonably be sought by LDCs. How these costs are identified and defended as incremental to what is already covered in the LDC's cost-based rates, however, may be challenging.

Best practice in this area appears to be an approach defining basic information services that routinely have been provided to customers in the past, and creation of criteria for identifying incremental requirements as "premium services." These premium services should be clearly presented to the state utility commission or other ruling authority with reasonable fees for their provision to customers and suppliers alike on a nondiscriminatory basis. Experience suggests that suppliers are willing to pay the LDC for such services and will support such an approach to the extent that the fees are reasonable and the LDC's affiliate is not afforded special treatment.

UNANSWERED CHALLENGES

While a great deal of progress has been made in discovering the new requirements for processes and systems that support the competitive electricity market, major uncertainties remain. These uncertainties fall into several categories, which will be discussed in turn.

To what extent will utilities be able to design processes that work efficiently and are cost-effective to implement? The current range of costs for individual utilities to implement the mechanics of customer choice may be as high as $100 million or more for an individual utility, encompassing computer systems, more sophisticated metering, and enhanced telecommunications. However, these costs can be driven downward by many tens of millions of dollars through careful process engineering and skillful consensus-building with other stakeholders in the new marketplace. The degree to which cost-effective solutions will be pursued depends on utilities, their regulators, and suppliers alike. A collective failure to seek optimal solutions to these new challenges would have to be considered grossly irresponsible planning.

How many customers will choose to switch? Economic studies and opinions generally point to a minimum savings of 15 percent for significant numbers of customers to choose an alternative supplier. In fact, in some locations fewer than 25 percent of all telephone customers have switched long-distance carriers even once in the 10 years since long-distance service was made competitive; that is, 75 percent have remained with AT&T throughout the entire decade. The challenges in the electricity sector will be to:

- provide market leveling discounts to customers in ways that do not fundamentally disrupt the desired market processes;
- educate customers so thoroughly that "fear of the unknown" is taken away as the reason they choose to stay put; and
- implement highly efficient and consistent nuts-and-bolts systems and therefore minimize administrative overheads on suppliers who wish to operate in multiple jurisdictions, thereby creating wider customer options.

How will ESPs choose to compete? Under the current piecemeal approach, many suppliers are becoming niche players, focusing on those customer segments and product offerings that they hope will produce acceptable returns. It is already clear that much must change to enhance the ability of suppliers to pursue a large share of the overall market; otherwise, it is unlikely that robust economic savings for more than a few customers will be created for a very long time. The infrastructure challenges for suppliers will include :

- minimizing start-up and market entry costs;
- minimizing cross-jurisdictional costs, e.g., requiring unique interfaces to be built in each area due to lack of standardization;
- properly reflecting the supplier's rights to deal directly with customers without interference from the LDC or other parties when marketing and negotiating commercial arrangements; and
- installing provisions that are "aggregator-friendly." Indeed, the key to overcoming the high acquisition costs of smaller customers may well be load aggregation and new, unconventional market entrants can be expected.

What technology will be needed and how will technology innovations be financed? Clearly, the competitive electricity market will require more information on customers and their usage requirements, delivered more quickly and in ways which support product innovation in the marketplace. Similar patterns have been witnessed in other industries. One example is the revolutionary marketing and inventory management impacts of bar coding at the grocery check-out. Some of the technology challenges that need to be faced for the energy industry include:

- How will the tension between the need to implement advanced metering technology (such as network-based meter reading solutions) and the contest for control of metering and billing be resolved?

- Will regulators continue to view the 75 cents or so per month that it currently costs utilities to read a customer meter as the avoidable cost when new, more expensive technology will enable vastly improved service offerings and, in fact, will help redefine customer service itself?

- What systems should be regionalized and how should the increased risk of more distinct agents becoming involved in operational processing be balanced against the benefits of regionalization?

The final question posed in this chapter is the hardest to contemplate: **When will the full geographic market in the U.S. replace the piecemeal, jurisdictional market which thus far has been the defining characteristic of industry restructuring?** The reader is left to speculate on the answer to this key question, although early experiences with customer choice make it apparent that the status quo approach probably will not result in the full economic objectives of customer choice being realized any time soon.

CONCLUSION

The transition to a competitive electricity supply market in the United States has begun, with initial activity centered in California, New England, and Pennsylvania. Other states, including some with average electricity prices well below the national average, have passed legislation or opened regulatory dockets that make some form of national market for electricity all but inevitable. However, over the near term, the new market is likely to be characterized by unnecessarily high implementation costs, customer uncertainty, and administrative inefficiencies. Diligent pursuit of a competitive infrastructure that works well and predictably will help to insure that the current piecemeal approach will be supplanted by a robust, efficient market capable of delivering the economic benefits sought by its architects.

NOTES

1. Controversial issues include proposals for making metering and billing competitively provided services. Opinions are often highly polarized and negotiations between opposing parties have not been successful, in most instances.

2. Several jurisdictions have experimented with multiple suppliers per customer meter; however, this feature adds significant complexity to the business process models and introduces the possibility of inappropriate gaming where the local utility is compelled to offer some form of default generation service.

3. This is done in the U.K., where the regional electric companies bill ESPs a 'use-of-wires' charge and are guaranteed full payment, regardless of customer payments.

4. There is general agreement that the best treatment is to limit the number of times a customer can switch to a new supplier to once per month.

CHAPTER 19

Information: The Key to Unlocking Value from Customer Choice

Melanie Mauldin

Imagine a car dealership trying to sell their cars for twice the price charged by all other dealers. You'd expect that dealership to have to slash their prices, or go out of business soon. Why? The question sounds silly because the answer is so obvious, but this is because we are used to purchasing cars in competitive market. In this competitive market, consumers have access to information on dealers' costs, other dealer's prices, and they understand how a given purchase price would affect their remaining disposable income.

In fact, in any market that is workably competitive, consumers have access to certain types of information. Namely, consumers know, or can find out:

- the prices suppliers are selling the product or service for,
- how price and volume purchased affect the total cost of the purchase, and
- how much of the good or service they will or have purchased.

The latter two points—that consumers need to be able to understand how price and volume affect the total purchase price, and that they need to understand how much of the good or service they are purchasing—also sound pretty obvious. Consumers clearly know that if they go grocery shopping and mangos are $1 each, then three will cost them $3. However, when purchasing less tangible services, such as cellular phone calling time or long-distance phone service, users may need to learn over time how much it generally costs them to call a certain person at a certain time.

Similarly, suppliers need certain information to have a "fair chance" at competing and profiting in a market. Specifically, they need access to:

- information from which to estimate incremental and total costs of providing the good or service to customers;
- sufficient information on sales to use in forecasting future sales; and
- measurements of sales to particular customers.

This information helps suppliers price their product, bill customers for their purchases, and better manage their production. Of course, some suppliers will use this information better than others—just as some consumers will use market information to get a better deal than others. The point is simply that this basic information needs to be accessible to all market participants for the market to become and remain competitive.

Will customers and suppliers have access to needed market information as the competitive retail electric industry develops? The next section looks more closely at the information both groups will need, and whether and how they can currently obtain this information. This is followed by a discussion of the technologies and methods that can help fill any remaining information gaps. Next is a look at regulatory decisions potentially affecting the availability of this information. The final section looks at the strategic decisions utilities will have to make in determining whether to help fill remaining informational gaps, and provides assessments of the advantages and disadvantages of alternatives open to these decision makers.

ACCESS TO INFORMATION IN THE EVOLVING RETAIL ELECTRIC INDUSTRY

Availability of the information outlined above is important for the developing competitive retail electric market in two ways. Consumers will need to understand available pricing and usage in order to determine which suppliers are offering them the best deals. Suppliers will need information on their customers' usage to profitably provide service. With unequal access to information, suppliers with the information monopoly—rather than suppliers with better offerings—may survive while others fail. This could decrease customer benefits from competition.

What Do Customers Know?

Will customers know the pricing options available to them?

To select the provider and pricing option that best fits their needs, consumers will need to know the pricing options available to them. Experience in competitive retail electric markets and in pilot programs indicates that ESPs will devote significant resources to educating potentially profitable, large-volume customer segments on pricing options. In many cases, sales staff visit these customers and explain their offerings and the ways customers would benefit from them.

For smaller customers, the evidence is less clear. In various pilots and in the California market, ESPs have sent out mailings and/or advertised on television, radio, and so forth, to educate participants on their options. In New England, ESPs advertised set prices for the electric portion of their bill. In California, ESPs are primarily advertising prices relative to the utility distribution company (UDC) prices. In California, at least, there appeared initially to be substantial confusion among consumers as to what the discounts apply to. Many customers did not realize that advertised discounts applied only to the energy portion of their bills. Many others did not realize that they would receive a 10 percent discount from their UDC automatically, even if they did not switch energy suppliers.

Based on other markets, it appears that when market rules are set up that allow sufficient benefits to customers, and opportunities to suppliers, the market will ensure that customers are aware of price offerings. The lack of clear information on pricing for smaller customers in California may be more of a signal that the ESPs have not yet developed profitable, attractive offerings for these customers than a sign of a lasting information gap.

Do Customers Understand How Prices and Usage Affect Their Bills? Do They Understand How Much Energy They Purchase?

Historically, many customers—particularly smaller ones—probably have not understood exactly how their usage affected their bills. One reason for this is the relatively complex nature of electric rates: customer charges may be a significant portion of the bill, and all customers' bills vary, even with the same comfort or production level, with changes in season, weather, and equipment performance. In addition, block rates, demand charges, and time-of-use rates also can confuse the issue.

Retail rates in competitive markets may evolve to have different features than the ones common in regulated utility rates. For example, some ESPs may offer flat rates for the electric commodity, while UDCs will retain charges for transmission, distribution, transition charges, public good programs, etc. Other ESPs may offer prices that vary by time of day. These prices will allow customers more control in reducing their bills, and will provide ESPs with a way of better managing capacity and peak period purchases. Such rates could include hourly pricing (reflecting pool prices), time-of-use (TOU) pricing, or "super peak" pricing, which could include flat energy charges except during capacity constrained hours. Rates during these hours either could be a higher, flat price, or they could be based on pool prices. The super peak hours could be set a day in advance. Such an option also could contain a limit on the number of super peak hours that could be called, and could charge either a higher, flat price for these hours, or the pool price at these hours.

As new pricing develops, how can customers understand how this pricing, and their usage, affects their energy bills? One way consumers learn to "translate" this into expected bills is through experience. For instance, a new owner of a cellular phone may at first be hesitant to use it, for fear that the bill will be too high. After receiving a low bill, and not making calls he or she would really like to have made for a few months, the customer may then start freely using the phone. At the end of the month, if the bill is beyond what the customer can afford, usage most likely will be curtailed again. After a few months, the customer will better understand the usage/pricing trade-off, and will learn when to make calls, how long to talk, and so forth, while staying within his or her budget.

Electric customers on predetermined pricing plans (flat or TOU pricing) can follow the same process to eventually understand how usage affects their bills. As noted earlier, however, the electric consumer may have a more difficult time finalizing this "translation," because changes in weather or appliance/equipment performance can affect the bill. In addition, much of the electric customer's usage appears less discretionary than cell phone usage. While a customer only pays when he talks on the phone, a residential electric customers pays for refrigeration, for example, even when the lights are off, the heat is turned down, and no one is at home. When pricing options incorporate real-time prices, rather than pre-set prices, the relationship between usage and bills is likely to become too complex for customers to learn from experience.

So, what tools are available to help customers better understand how their usage will affect their electric bills? Interested residential and "mass market" commercial customers can obtain energy audit-type information on the World Wide Web. One example is Energy Interactive's On-line Home Energy Audit, which divides up energy dollars based on end use. The company offers similar on-line software for small commercial customers; businesses enter business type, region of the country, hours of operation, and so forth, to get more accurate estimates of their load profile and detailed energy usage. While this can help customers understand how appliance usage influences total electricity consumption, it cannot directly answer the question of how changes in their behavior will affect their individual bill—particularly if they are on a complex pricing option.

Large commercial and industrial customers typically understand this relationship much better than smaller customers do. One reason for this is that their bills for energy usage are significantly larger, and can comprise a significant portion of operating costs. Some of these customers have energy management systems to help control factors that affect their bills. A small number of these customers even have access to their hourly usage profiles in real- or near-real time. These are generally customers on real-time pricing (RTP) rates. Some utilities, such as Georgia Power, have software that allows RTP customers to not only view hourly usage and prices, but to perform "what if" scenarios to determine how changes and shifts in usage would affect bills.

How can smaller customers, or customers that do not yet fully understand the relationship between usage, prices, and bills, improve their understanding of this? Evidence (though scant) seems to indicate that access to real-time or near real-time information helps. This is one reason RTP customers with the software mentioned above can readily control their energy usage, and make operational decisions based on price information. According to a white paper by Plexus Research,[1] studies of prepaid metering consistently have shown that customers can and do use the real-time data showing the relationship between remaining credit and energy usage to manage consumption. The report explains that consumers can see how the displayed credit value declines faster when lights are left on than otherwise, and this motivates them to turn off lights before leaving home. According to the report, pilot tests of prepay meters (which generally use lower income consumers) have shown that savings of 15 percent or more are common. Another survey found that consumers with access to more detailed usage data than their monthly meter read reduced their energy usage by more than 10 percent.[2]

Commercial customers also benefit from more real-time and/or more "granular" usage data, such as hourly load profiles. This information can be used to compare usage across plants or branches, which can help identify equipment inefficiencies, as well as behavioral differences (if one branch manager does not turn down the cooling setting after the office is closed, for example). In the country's fledgling competitive markets, a number of companies are emerging that can provide commercial and industrial customers with information on their load profiles.

In summary, information already is available to help interested consumers understand how usage and (relatively simple) pricing options affect their bills. Evidence indicates that more frequent metering data could improve customer understanding of this relationship, even for customers on simple rates, such as flat kilowatt-hour rates. To the extent that pricing reflects real-time cost differences, availability of frequent and detailed (hourly) metering information may be the only way that customers can fully understand this relationship and control their bills. This ability to understand this relationship and control usage not only will help customers see greater benefits from the market, it will also improve both the short-run and long-run efficiency of the market, by decreasing the need for uneconomic capacity.

Supplier Access to Customer Information

Do suppliers have information to efficiently package services and purchase the commodity?
As noted earlier, energy service providers (ESPs) need to know the usage patterns of their customers to develop nonhourly price offerings. ESPs will see maximum

potential benefit if such information is available on a customer-specific basis, for this would allow them to customize offerings (including pricing and energy management services) for larger customers. For smaller customers, ESPs will benefit to the extent that they can receive this type of information for groups of similar customers (single family homes with over 2,500 square feet, for example), as opposed to receiving a single average load factor for each utility rate class.

To receive load profile information on larger customers, ESPs will need utilities either to adopt metering systems that can provide this data (with customer consent), or they will need regulators to allow themselves or potential customers to put in advanced metering options, and to grant them the ability to read these meters.

For smaller customers, ESPs may have to rely on statistical load profile data, at least initially. However, as margins in the retail energy business are expected to be small—less than 2 percent—ESPs will find significant advantages to knowing actual, rather than typical, usage patterns when developing offerings. The extent to which this information will be available will depend in part on the interest level of smaller customers, and the profitability of providing these customers retail energy service. Assuming there is a market for this information, ESPs and customers either will have to rely on utilities with sophisticated metering systems to provide this data, or will need regulations that allow them to directly acquire expanded metering information.

For the most part, hourly or time TOU data collected monthly will be sufficient for developing product offerings. ESPs also will need to be able to forecast usage by time of day to optimize power purchases. Again, obtaining hourly usage information on customers at the end of each month may be sufficient for this purpose in many cases.

Energy providers also will need to match power purchases with power delivery. For this information, ESPs may need access to near real-time usage information, at least for their larger customers, whose purchases may constitute the bulk of their volumes. This near-real-time information can help them avoid imbalance charges for not matching purchases with deliveries in specified geographic areas. Several ESPs have expressed an interest in receiving hourly information on their largest customers to help them avoid these penalties, which can be substantial.

In summary, it appears that actual (as opposed to estimated) load profile information will have considerable benefits to ESPs. The ability to access this information frequently—perhaps even hourly for some customers—will also be important in their business.

Do suppliers have a way to measure sales for billing purposes?
If customers are billed a flat rate based on total monthly usage (e.g., kilowatt-hour and potentially peak kilowatts), traditional monthly metering should

be sufficient for billing purposes. However, to the extent that ESPs offer additional pricing options to help attract customers—and to help customers derive greater benefits from the competitive market—more sophisticated metering will be needed.[3] Regulators will determine how this information is available; utilities may be required to install meters capable of providing this information, or ESPs may be allowed to install such meters.

Summary

The monthly usage information currently collected by utilities may be sufficient to allow retail electric markets to open to competition. However, there will be significant benefits both to consumers and suppliers in this competitive market from receiving more detailed data than monthly usage, and from receiving this data more frequently.

SUPPLYING INFORMATION IN A COMPETITIVE MARKET

Several methods of obtaining detailed usage information have been discussed for use in deregulating markets. This section provides a brief overview of these technologies, and also discusses the use of estimated ("statistical") load profiles in place of actual usage data.

"Metering-only" Options

One method of obtaining hourly usage information that has been discussed by California utilities is to rely on solid-state meters, or meter "add ons" such as interval recorders, that can store hourly information. This hourly information then could be picked up monthly, on a utility meter reader's regular route. In theory, an ESP also could pick up this data monthly by visiting the meter site, but with geographically disperse customers, the cost of doing so most likely would be prohibitive.

Providing monthly data in this manner has an advantage to the UDC in that it can continue its standard meter reading process. As noted, this advantage is not shared by ESPs, who probably cannot have these meters read cost-effectively without either employing one of the communication technologies discussed below, or relying on the utility for meter reads. (Whether relying on the utility for meter reads is cost-effective will depend on the UDC, and the pricing it is offering ESPs for this service.)

This method of acquiring detailed usage data has several disadvantages. Perhaps the most important for the development of the competitive retail electric market is the fact that this method does not allow frequent updates on usage. Retail customers cannot receive near-term feedback on how their energy management practices are affecting their usage. ESPs cannot monitor customer usage, and compare it to contracted power purchases on a daily (or hourly) basis. As mentioned earlier, this can result in ESPs incurring significant penalty charges.

The method has an additional, related, disadvantage: ESPs cannot have access to real-time data. This means, for instance, that the ESP will not necessarily be aware of a power outage that is affecting one of its customers. The ESP will not be able to check usage/bills for customers calling with questions or complaints without either visiting the meter site, paying the UDC a premium to visit the site in a timely fashion, or relying on the UDC to do this on the regular meter reading schedule.

Similarly, the availability of monthly only data means either that customers cannot switch suppliers between meter reads, or that they (or their ESP) must pay the UDC a premium for a special read during this time to allow the switch. All of these issues can be addressed easily if communication devices are installed in these meters.

The utility distribution company—as well as end-users and ESPs—also lose benefits from selecting this option, rather than one that provides more frequent data. Systems that can deliver frequent and/or near-real-time access can eliminate the need for special meter reads, for customers moving out or for contested bills, for example; this can lower meter reading costs. In addition, the UDC can use the more frequent data to improve power purchases (for customers they are serving as a provider of last resort), and improving distribution planning and operations. These benefits are explained in more detail later in this chapter.

Electronics to provide and store hourly information on a monthly basis, such as solid state meters, are becoming more widely available. Interval data recorders have been available for load research purposes for a number of years, and a variety of solid state meters are now available for commercial and industrial customers. Recently, meter manufacturers began to announce the availability of solid state meters for residential customers. (Landis & Gyr announced the first residential solid state meter in early 1998.) Because the up-front cost of these meter-only options can vary considerably for different customer types, it is difficult to determine how these costs compare to the cost of some of the meter reading systems discussed below. However, the operational costs may be higher than with automated systems, and the informational benefits to the retail market may be considerably fewer than those possible with the communication systems discussed below.

There are a number of technologies that can provide utilities, ESPs, or customers with frequent, detailed usage data. These are discussed below.

Telephone-based Automatic Meter Reading Systems

One of the oldest methods for collecting detailed data on a frequent basis is to rely on a phone-based automatic meter reading (AMR) system. These systems include devices that can record hourly usage from existing meters, and communications devices that use phone lines to call data into a central system. These systems have been used for a number of years for load research, and for monitoring curtailments for customers on these rates. Utilities also use these systems for collecting hourly usage information for real-time pricing programs.

These phone-based AMR systems provide a cost-justifiable way of collecting infrequent to moderately frequent detailed data for relatively large customers. (In California's competitive market, evidence to date indicates that these systems cost roughly $300 to $500 per large customer to install, and have ongoing costs of $18 to $35 per month.) They are well suited to monitoring usage for a small number of customers. Most systems cover a few hundred customers; the largest cover 10,000 to 15,000 customers. These systems are probably the most cost-effective solution for monitoring the usage of a few hundred customers. As the number of customers covered increases, they become less attractive relative to some of the other options discussed below. (See Table 19–1.)

Phone-based AMR systems also are well suited for geographically disperse customers, because they use existing infrastructure (phone lines), and therefore avoid the costs of building other networks. However, most phone systems are not capable of easily managing several thousand end points. They also typically have fairly significant computer hardware requirements, and computing needs become a more difficult problem as the number of meters increases.

In addition, these systems can have high installation costs. Because phone lines are not always located near meters, installers may have to trench lines to connect meters to phones.

These systems can use either a "dedicated" phone line, which provides continuous access to meters, or they can share an existing phone line. Systems using dedicated phone lines can have relatively high monthly operating costs, since an additional phone line must be rented. Shared line systems have lower operating costs, and can work well when data is needed daily. With these systems, data transfer "hangs up" if the customer picks up the phone to make a call, so that they are not inconvenienced. If real-time or hourly access to data is needed, dedicated lines typically are used to ensure that data can get back to the AMR operator.

Wireless Network Meter Reading Systems

Since 1996, the use of radio-based, or wireless, networks for meter reading has increased. These network meter reading, or NMR, systems (including CellNet's and others) add a radio transmitter to existing or new "advanced" meters. These transmitters send usage information to a pole-top radio unit, which then processes the data and sends it back to the system head end (software). Most installed systems currently bring back usage information on each customer daily, but these systems can bring back interval, or hourly,

TABLE 19–1	Comparison of Alternate Technologies for Obtaining Usage Information for Competitive Retail Markets
Existing Technologies	
Telephone-based Meter Reading Systems	
Advantages	Cost-effective for large customers, geographically dispersed customers, and for small numbers of customers.
Disadvantages	Installation costs can be high. Not well suited for automating several thousand or more meters.
Wireless Network Meter Reading Systems	
Advantages	Economies of scale make systems cost-effective for thousands to millions of meters. Can provide distribution company benefits including outage detection, and data for distribution system planning.
Disadvantages	Not cost-effective for a small number of customers or for geographically dispersed customers unless a network already exists in the area.
Developing Technologies	
"Drop in" Radio	
Advantages	Projected to be cost-effective for a small number of customers and for geographically dispersed customers.
Disadvantages	Not expected to be as low cost as existing systems for utility-wide metering. Not yet commercially available.
Satellite	
Advantages	Attractive for reading meters in remote areas.
Disadvantages	Not cost-effective for reading a large utility's entire meter base.
Broadband	
Advantages	Technology promises to allow delivery of a large number of services, including Internet access, pay-TV, etc.
Disadvantages	Cost-effective solutions for meter reading not yet developed; may be several years to a decade until commercially available. Likely to remain the most expensive option for meter reading.

data more frequently if desired. These systems allow allow real-time access to individual meters. Currently, over a million meters in the U.S. are being read by wireless network systems.

When used to bring back information on a large number of meters, (e.g. tens of thousands of meters or several hundred thousand meters), these systems deliver significant economies of scale. In these cases, wireless systems are available with all-in costs of less than half the per-meter costs of shared line telephone-based AMR systems. (When compared to the costs of phone-based systems using dedicated lines, cost savings are even greater.) In fact, some wireless NMR companies quote an all-in cost of equipment, installation, maintenance, and data management for providing daily metering information for residential customers at as little as $1 per meter per month.

A second advantage of wireless network meter reading systems is that when they are installed utility-wide, these systems can provide distribution companies (as well as end-users and ESPs) with valuable information. For example, by providing near real-time access to usage information, these systems can be used to determine whether a particular customer is receiving power. This information can help distribution companies more efficiently manage power outage restoration programs. In doing so, they ultimately can help reduce outage minutes. In addition, distribution companies can use detailed data such as aggregated hourly usage data for planning investments (or deferring investments) in their distribution system.

As noted, these systems can provide frequent, detailed data, on-line access to customer data, and additional benefits to the distribution company if installed systemwide, or to cover a large number of meters. In small volumes, however, the economies of these systems are lost. Therefore, while wide-scale deployment of such systems may make "advanced" metering cost-effective for all users, in a competitive market, with only a few subscribers signing up at a time, these systems do not cost significantly less than phone systems.

Wireless systems are not cost-effective for serving geographically disperse customers, unless these customers are located under an existing network. For this reason, wireless providers often supplement their systems with phone-based AMR. This allows low-cost wireless coverage of all customers in a metropolitan area, and additional phone coverage of a few large commercial and industrial customers located elsewhere in the utility's service territory.

Developing Technologies

Other technologies being developed to bring back detailed usage data frequently include "drop in" radio devices, satellite-based AMR, and broadband AMR. Drop in radio solutions, as currently being discussed, would generally have a paging device in the meter that would use a paging network to relay data. This would provide the same advantage as phone-based AMR, in that these units would use existing infrastructure. It appears that these units would be cost-effective for monitoring relatively small numbers of units, even in geographically disperse areas, but would not be as low cost as existing wireless systems as a way of automating meter reading for an entire utility.

Satellite meter reading will rely on low earth orbiting satellites to provide meter reads as frequently as hourly. This solution appears quite attractive for reading meters in remote areas. Satellite providers are not currently positioning this technology as a way of reading a large utility's entire meter base.

Broadband meter reading, which would use fiber-optic cables to the home, has the promise of being able to bring back far more data than just meter reads. It would allow service providers to offer additional services, including pay TV. At present, however, broadband technologies are not cost-effective for metering, and may not be so for possibly a decade.

In 1996, Nortel and Norweb announced they had developed a power line-based broadband technology that could provide high-speed Internet access as well as theoretically read utility meters. An electric company's investment for this type of system could be substantially less than for a fiber-based system, since it uses existing power lines. Current emphasis on the system appears to be on its Internet capabilities, rather than its meter reading ones, so it is unclear when a full meter reading system utilizing this technology will be available. In addition, the trial system has recently experienced technological snags.[4] Finally, differences in electric company infrastructure in the U.K. and the U.S. means that significant development work may be needed before the technology is ready for the U.S. market. The system uses data collectors at transformers, because transformers typically block power line carrier transmissions. In England, utilities average almost 100 houses per transformer, while in the U.S., we average five to eight homes per transformer. Thus, if the technology were to be adapted to the U.S. market, it appears as if significantly more data collectors would be needed, and the system would be considerably more expensive.

Use of Statistical Load Profiles

To avoid the need to install new metering technology, some in the industry have suggested the use of "typical" load profiles, rather than actual hourly usage. These estimated or "statistical" load profiles will be used in California, for example, for customers selecting open access who use less than 50 kW, but who do not choose to purchase hourly metering.[5] The idea behind this policy is that it allows small customers to participate in the open market immediately, without having to acquire hourly metering. After a transition period, the use of estimated load profiles may be discontinued.

Industry participants in a number of other states have discussed following a similar policy. They suggest limiting actual hourly usage collection to larger customers, and using load profiling for smaller customers. There are two advantages of the policy. First, it allows small customers to join the competitive market without potential metering delays. Second, it allows smaller customers to avoid the cost of hourly metering. While this cost can be as low as the $1 per month figure quoted earlier, it could be significantly higher in areas where customers have to rely on phone-based AMR.

The use of these profiles allows the "settlement process" to occur, but does not directly affect a customers' bill. In the settlement process, contractual deliveries to a given ESP are compared with the ESP's estimated takes (by hour) to calculate the difference between power input for the ESP and power taken out by the ESP. ESPs are billed (or credited) based on the difference, but their customers without time-differentiated metering typically still are billed based on total usage.

The use of these load profiles does not meet the competitive market's needs for information. For one thing, they do not provide the customer or the ESP with information on actual usage. Therefore, ESPs relying on these profiles must accept greater risk of imbalance charges when actual usage does not meet purchased power or when scheduled deliveries by geographic area do not meet customer usage. In addition, they have greater risk of incorrectly forecasting usage, which means they may not be able to optimize purchases.

In addition, if actual usage over time is not measured, it does not make sense to offer time-differentiated rates. Customers could not be billed on their actual usage by time of day, and they could not see bill effects from load shifting. Thus, a significant potential benefit to customers—and a potentially large benefit to society from deregulation—cannot occur without the use of a customer's actual load pattern.[6]

In short, the use of estimated load profiles allows the wholesale market to operate, but it does not address the informational needs of the retail market.

OVERVIEW OF REGULATORY DECISIONS AFFECTING METERING CHOICES

To date, the primary regulatory decision regarding metering has been whether metering and metering services will be "unbundled" from the distribution company—i.e., whether these services will be open to competition rather than remaining solely with the distribution companies.

While most states do not yet have definitive rulings on metering, it appears from what has been stated so far that:

- **Metering will be unbundled initially in several states.**
  Currently, California has unbundled metering, and allows certified

companies to provide physical metering services such as meter installations, and meter data management services, such as collecting and posting validated meter reads. New Hampshire has stated that metering will be unbundled for larger customers. Nevada's PUC recently identified metering as one of the areas to be turned over to competition when the market opens. In Pennsylvania, PECO's settlement will unbundle metering intially. At the time of this writing, Arizona has not reached a conclusive agreement on metering, but indications are that it will be unbundled.

- **Most states that have looked at the unbundling issue have decided to let the distribution companies retain metering during the "transition period."** Based on public statements, it appears that most of these states favor the idea of a competitive metering market, but believe competition in this market should be brought in only after the retail market has had time to transition to a competitive market.

- **Only one state (Rhode Island) has stated that metering will not be unbundled.**

Table 19–2 summarizes the decisions made to date by each state.

Advantages to Leaving Metering with the Utility

Proponents of leaving metering with the utilities have several arguments in favor of doing so. Based on the experience in California—which opened its metering markets to competition three months prior to its retail energy markets—some of these issues can be overcome relatively easily, while others may remain significant. This section provides a brief review of these issues, and also discusses whether and how they can be mitigated.

Issue 1: Slowing the Development of the Retail Market

As noted, most commissions that favor leaving metering (including meter ownership, meter installation and maintenance, and meter reading and data distribution) with the utility appear to be in favor of doing so temporarily. These

TABLE 19–2 State Commission Decisions on Unbundling Metering			
Unbundled Initially	Unbundled Later	May be Unbundled Later	Remain Bundled
CA NH NV AZ* PA (PECO & GPU)	ME	DE MD MA MN NJ NY PA AR	RI

*Initially, Arizona favored unbundling metering. The rules regarding deregulation are being challenged, and therefore it is not certain that metering will be unbundled initially.

Source: Table adapted and updated from "A White Paper on Direct Access Metering & Data Communication Requirements," (March 31, 1998), p. 74, prepared for NARUC by Plexus Research, Inc.

commissions seem to believe that opening the metering market at the same time as the retail commodity market could slow down deregulation and/or increase confusion in the market because of the complexity of the meter services market. Thus, an advantage of leaving metering with the utility, at least temporarily, could be that it does not slow the development of the retail market.[7]

Some commissions are discussing letting ESPs and/or large customers choose their own metering, even during the transition period. In these cases, it appears the utility would install the meters for the ESPs or customers, and also might continue reading these meters. Examples of states considering such an interim solution include Pennsylvania, Delaware, and New York. It is unclear whether customers with this metering option would be able to obtain frequent or real-time data, or whether they simply would be able to see more detailed usage data each month. This option could allow at least some customers to be able to take advantage of time-differentiated rates and other such options that their ESPs develop.

Based on the limited evidence available, it is difficult to determine how leaving metering temporarily bundled will affect the speed and development of the retail energy market.

Issue 2: Safety Concerns

Some involved in the metering debate bring up the issue of safety concerns with meters. Utilities offer significant training to their workers to ensure safety; removal of existing meters and installation of new ones could present hazards if the person performing the work does not know what they are doing. While safety concerns are a valid point, companies in competitive metering markets also can offer safe metering services. In California, working groups comprised of utilities, ESPs, meter manufactures, and others considering becoming meter service providers have developed detailed rules to help ensure safety in a competitive market. These rules require that the commission certify meter service providers. To qualify for certification, companies have to demonstrate that they have stringent training programs and experienced staff. These meter service providers follow strict rules, and are only allowed to install qualified meter types.

Issue 3: Stranded Metering Assets

Utilities often are concerned that if they lose the metering franchise, the investment they've made in metering assets may be lost. This could result in higher transition charges, or in losses to shareholders. While this, too, is a valid concern, there appear to be a number of options for mitigating these stranded costs short of preventing competition in the metering market. For instance, a state could conclude that the utility retains initial ownership of the meters (and earns a return on them). Customers and ESPs could have the right to purchase these meters at book value if they chose to upgrade to a higher functionality meter. Alternately, ESPs could add communications to existing meters to expand their functionality. In this case, the utility would retain meter ownership, but the ESP would own the communications.

Issue 4: Economies of Scale

One issue that could be raised in favor of keeping all metering with the distribution companies is that it is more economical for a single entity to read all meters. To date, this argument appears to be robust across different technologies. For instance, many utilities still read meters manually (with or without handheld devices to collect meter data, or with the use of a device operated from a van). These methods do not provide the competitive market the data it needs, because they cannot provide frequent access to usage information. Nonetheless, if such a utility were to lose meter reading to competitors, its average costs to read the remaining meters would increase significantly. (Meter readers would, by and large, still have to walk or drive the same routes, but would pick up fewer meter reads.) Thus, distribution companies that do not use any communications on meters see significant economies of scale in performing all meter reads themselves.

There are also economies of scale in providing the frequent, detailed usage data need by the competitive market. As noted earlier, phone-based solutions are currently the most economical method of providing communications for only a few customers in a service territory. However, as more and more meters are connected with communications, the economics shift in favor of network technologies, which generally are based on radio technology. For metering a large number of endpoints, the costs of installing, maintaining, and using these networks is significantly lower on a per-meter basis than phone-based metering (or other available metering communication technologies). Thus, one argument in favor of leaving metering—including meter reading—with the utility, is that, because of economies of scale, all customers will have easy, affordable access to frequent, detailed data.

These economies of scale are real, and can be significant. Note, however, that if metering is open to competition, the distribution company, or other providers, still could choose to install a network meter reading system to improve their competitiveness in offering meter reading. If the utility installed such a system, it also would help alleviate average cost increases brought about by losing metering customers to other providers.

Issue 5: Distribution Company Applications of Metering Data

Utility distribution companies can receive direct operational benefits from a ubiquitous network meter reading solution. For example, a distribution company could use a network to help it more quickly restore power after outages by having up-to-date information on what areas are affected, and their current power status. In addition, the utility could use the daily or hourly information available to help determine peak period loads for better distribution planning. If ESPs or others provide metering to only a few customers—e.g., larger customers that can afford telephone-based meter reading—the distribution company will not obtain these benefits. Of course, a distribution company in a competitive metering market could still choose to put a network system in place, and receive these additional operational benefits as well as the benefit of being able to cost-effectively provide metering to the market.

Advantages of Opening Metering to Competition

There are also a few basic arguments in favor of opening the metering market to competition.

Issue 1: A Direct Link to the Customer

ESPs often argue that metering needs to be open to competition because it provides them with a direct link to the customer. This provides them control over the metering information they receive, allowing them to offer special pricing and energy management services. In addition, it can allow them to answer customer questions on current energy usage, outage status, and so forth, in real time.

In theory, ESPs could access such data even if the distribution company provided metering—assuming that the utility had installed a network metering system capable of load profiling, real-time data access, and outage detection, and assuming the utility would provide this information to ESPs on their customers. Because few utilities currently have a system that would enable this, ESPs seem to be more comfortable with the concept of obtaining this data from competitive providers.

Issue 2: Bundling Metering Service Charges with Other Services

ESPs argue that they want to be able to bundle metering services with other services. They might bundle a charge for obtaining TOU metering data with their TOU rate, for example. This way, they could show a $3 monthly "TOU service fee" together with a $6 savings from the TOU rate. Or an ESP may need hourly metering to monitor the health of appliances or equipment. In this case, they might prefer to roll the cost of the metering into the fee for the service, since customers may not perceive a benefit for having their ESP receive detailed usage data, and therefore, they may not want to pay for it. However, they may value the services it enables, and not mind paying for these services. Allowing ESPs the ability to bundle metering services and service charges provides them more pricing and marketing flexibility.

If the distribution company provided metering services, ESPs are concerned that this would show up as a separate line item on the distribution company bill (or portion of the bill). Note, however, that it would be possible for the distribution company to bill ESPs for billing services directly, and allow ESPs to roll in these fees with other services.

Issue 3: Concern that the Market Won't Have Access to the Data it Needs

If a competitive market for usage data develops, ESPs and customers will be able to purchase the data they need to best run their business, and maximize benefits from retail energy competition. Competitive markets deliver services that customers are willing to pay for. Many fear that without such a competitive market—if utilities provide metering data—the market's needs for data will not be met. This is a serious, and a real, concern. Most utilities today offer only monthly metering information. (In contrast, in California, a few end-users are purchasing 15-minute usage data as often as every hour.) Unless utilities install some type of automated meter reading system, they will not be able to provide the data that ESPs and customers need to optimize purchases and operations. Even if utilities use advanced solid state meters that can store hourly information, they cannot bring this data back as frequently as some market participants will want, unless they have installed a communication system as well. And if utilities are safeguarded from competition in providing meter services, what would motivate them to invest in a state-of-the-art meter reading system?

As noted, this is a very real concern. ESPs' profits—even their survival—will depend on having frequent, detailed usage data, and on utilizing this to minimize imbalance penalties, minimize commodity purchase costs, and develop appropriate offerings. This is the primary reason that metering may be unbundled at some point in most states. Possible exceptions are states where large utilities already have network meter reading systems in place that could provide ESPs and customers with a wide range of data at frequent intervals.

Issue 4: Encouraging Innovation
One argument in favor of opening metering to competition is that competition spurs innovation, which could result in new offerings and new packaging of metering services. This argument is similar to Issue 3 above. Simply stated, it says that competitive markets will uncover market needs, and meet them.

The Devil Is in the Details

The arguments in favor of leaving metering with the utility distribution companies, and those for opening metering service to competition, have been presented. Which side wins? It seems as if—in theory—market needs could be met under either model. As usual, the devil is in the details. However regulators decide to go, they need to ensure that the system they set up allows ESPs and customers access to the data they need—which is significantly more data more often than utilities now collect. Furthermore, ESPs and customers need to be able to change their data requirements over time, and receive any additional data in a timely fashion. For example, if ESPs initially believe that daily usage information delivered daily will fill all their needs, but later learn that hourly data delivered hourly on their largest customers is needed to help them avoid imbalance penalties, the system needs to be able to provide this "upgraded" information to ESPs within no more than a month or so. If regulators require distribution utilities to install systems that can provide such information, it may be possible to develop processes that leave metering with the utility, and still address market needs and concerns. If, on the other hand, utilities cannot provide such information, regulators need to ensure that customers and ESPs can obtain this information from other sources.

In addition, regulators need to ensure that the system they set up encourages and rewards innovation—either by the distribution companies, or by ESPs and metering service companies, or by both. With systems that meet these criteria, it should be possible to meet market needs and address participant concerns.

STRATEGIC METERING OPTIONS FOR UTILITIES

For a utility, the key metering decision today is whether or not to automate meter reading to be able to provide the type of information the competitive market will need to operate efficiently. Utilities are in a unique position for offering these information services. Currently, only utilities have the ability to provide these services to their entire service territory (except in California, where any meter service company can provide these services). Utilities can take advantage of economies of scale that may be unavailable to other participants—for example, to ESPs, each of which probably will serve only a fraction of the area's customers. In addition, utilities are in a position to be able to use the data and system to better run the distribution company—and therefore potentially obtaining more value from the information than other companies would.

If a utility chooses not to automate its system—or a significant fraction of it—now, the utility could provide meter services (assuming regulations allow it to do so) in the same manner as other new metering service providers will. They could use telephone solutions, the drop-in radio solutions being developed, or they could purchase data from providers who have installed their own network meter reading systems.

The key difference in the decision to automate and the decision to wait is that, by waiting, the utility postpones an investment decision—perhaps avoids an investment—but sacrifices a potential competitive advantage in the metering business. As will be discussed, the utility may also sacrifice other benefits from having such a system.

Clearly, the advantages and disadvantages of automating meter reading will be influenced by the regulatory scenario facing the utility. This section reviews the advantages and disadvantages of investing in a meter reading system capable of meeting the needs of a competitive market under three different regulatory scenarios: metering will be unbundled, metering will remain bundled, and the future of the metering market is uncertain.

Scenario: Metering Will Be Unbundled

Decision: Should a utility invest proactively in a meter reading system flexible enough to meet the needs of a competitive market?

Advantages. There are several advantages to making such an investment now. These are:

- **Gain a profit opportunity once the metering market opens to competition.** A flexible, cost-effective meter reading system can provide a utility with an additional way of bringing in revenues and profits once the competitive metering market opens. The utility can sell data collected over the system to end users and/or to customer-approved ESPs. (California IOUs have developed tariffs to sell metering services to both these groups.[8])

- **Avoid average cost increases for meter reading.** In competitive metering markets, utilities are likely to lose some customers to alternate providers, even if they have installed a low-cost, flexible, meter reading system. As noted earlier, this loss of metered customers will tend to increase the average cost of reading the meters of remaining customers. A network meter reading system can help mitigate these cost increases. Some vendors sell meter reading services on a per-meter, per-month basis, so that loss of customers would leave average meter reading costs unchanged.

- **Use detailed data to improve utility marketing.** The utility can use this data to develop new services, offer innovative pricing options, and develop targeted marketing programs before its state opens to competition. These offerings can help retain customers once the retail energy market is deregulated.

- **Use detailed metering information to improve distribution operations, improve customer service, reduce costs, and increase profits.** Networks can help improve customer service and distribution operations by alerting the utility when a meter experiences an outage. Real-time reads can be used to verify when power has been restored to a site. Some advanced meters and data services can also deliver data on line voltages, reactive power, and power quality. Together these features can help a utility better manage the distribution system, and get customers back on line more quickly and efficiently in the event of an outage. Information on peak day (or peak hours) usage can help improve distribution planning; utilities with more data on load at various points given weather conditions, and so forth, potentially can avoid unnecessary capital investments in expanding portions of their systems. For utilities with PBR, decreasing distribution expenditures and outage minutes can increase profits directly. In addition, more refined usage information can help a distribution utility improve power purchases for those customers not selecting ESPs (those for whom the utility retains an obligation to serve). Note that if the utility did not automate, it probably still could purchase usage information for distribution planning, outage restoration, and power purchase planning decisions from companies providing metering. However, this data would probably only be available for a percentage of the utility's customers—those requesting advanced metering—and therefore would not be of as great a value for the utility.

Disadvantages. In this regulatory scenario, the primary disadvantage cited to investing now in network meter reading, or NMR, is the perceived risk of not recouping this investment—and therefore adding to the utility's stranded costs. In this section, we examine this perceived risk more closely.

- **Potential risk that competitive metering revenues and distribution company savings do not cover the cost of the system.** In California, the only state with a complete, approved plan for unbundling, the utility distribution companies have published tariffs stating their charges for meter reading and meter information delivery. The prices charged by some of the utilities are significantly greater than the prices they would pay for a low cost network meter reading system. This could be taken as an indication that competitive providers in other states will be able to cover the costs of such systems.

  In addition, the utility's distribution savings will help cover the cost of the system. While there is limited published information quantifying the distribution savings from such systems, a study on Ameren Corporation's system found that the direct benefits the utility is receiving from their network meter reading system exceeds the initial cost of the project.[9] These benefits accrue not only from meter reading cost savings, but also from operational savings and customer service center savings.[10]

- **Potential risk that investment in NMR systems increases a utility's stranded costs.** This concern is similar to the one above. If a utility covers the cost of the investment (and earns a return on it), it does not need to worry about stranded costs. One way to mitigate this risk is to outsource metering information services. The utility can pay a data service provider a monthly fee to collect frequent, detailed usage information on customers. The utility can contract and pay for only those services it wants at a given time, and still have the flexibility to upgrade to more advanced services later. The data service provider would retain ownership of the communications system that brings this information from the meters. This way, the utility can receive the data it needs for a competitive market, but does not have to invest in additional hardware. In addition, because the utility is not purchasing and operating a technology, but instead is being guaranteed that service will be delivered at agreed upon prices, the utility is shielded from concerns about potential changes in technology over time.

In summary, utilities proactively investing in an NMR system under a scenario where metering will be unbundled can expect to see the operational benefits that accrue to regulated utilities with such systems, as well as additional benefits. Published information indicates that regulated utilities already are finding that meter reading, operational, and customer service savings from these systems outweigh their initial cost, and that these utilities are working to gain further system benefits. For utilities with performance-based regulation, benefits from the improved distribution operations allowed by the NMR system will be even greater.

Utilities investing in such systems prior to competition also can use the information from these systems to improve their current marketing. In addition, utilities with NMR systems in areas with unbundled metering can gain an additional source of revenue by selling detailed usage data to end-users and ESPs, and they will minimize the cost increases that can occur when other metering companies are allowed to read the distribution company's meters.

For utilities considering such an investment, these benefits need to be weighed against the cost of the NMR system. Utilities considering such an investment should review existing benefit analyses, and will most likely want to meet with utilities that already have these systems to learn more about the benefits they deliver.

Scenario: Metering Will Remain Bundled

Decision: Should a utility invest proactively in a meter reading system flexible enough to meet the needs of the competitive retail energy market, given that it will be the only metering service provider?
In this scenario, the analysis on whether to purchase a network meter reading system is similar to the analysis that would be done even if no regulatory changes were taking place in the industry. The twist in the analysis is that utilities may be required to provide more frequent and more detailed data than they have had to in a fully regulated industry.

Advantages. Under this scenario, advantages of installing such a system include:

- **Reduced operating costs and improved customer service,** as mentioned elsewhere. For utilities under PBR, this can translate directly into increased profits.
- **Better power purchasing** for those customers not selecting ESPs or those customers using the utility as a "provider of last resort."
- **Avoided costs of developing statistical load profiles.** As noted earlier, if detailed usage data, such as hourly data, is not available, grid operators and ESPs may have to rely on statistical load profiles for estimating usage profiles of smaller customers. In California, utility distribution companies are responsible for developing load profiling for smaller customers (who first begin choosing alternate meter service and metering information providers in 1999). Estimates of the cost to develop and update load profiles in California and the U.K indicate that these costs are significant. Table 19–3 compares these costs to the cost of installing a high-volume network meter reading system that delivers hourly data

TABLE 19-3   Costs per Residential Customer for Providing Advanced Metering and Load Profiling

	Initial Non-Recurring Cost	Monthly Operating Cost	Monthly Load Research Cost
Metering	$100[a]	$1–$2	none
Profiling	$ 22–$24[b]	$1–$2	$0.10–$0.50[c]

[a]Cost estimate includes meter units and installation and communications infrastructure for data retrieval and management.

[b]Cost estimate includes deploying statistical load surveys and creating the data retrieval and analysis capability.

[c]Cost estimate includes updating profiles as customer population changes, and the cost of "dynamic load profiling," as required in California. Dynamic load profiling means that sample load profile data must be collected daily, to update the next day's forecast load profiles.

Source: Table adapted from "A White Paper on Direct Access Metering & Data Communication Requirements," (March 31, 1998), page 43, prepared for NARUC by Plexus Research, Inc.

daily. (The table's figures pertain specifically to residential metering, but many small commercial customers can use the same technologies.) With such a metering system, utilities can avoid the cost of developing and managing this load profiling.

- **Ability to cost-effectively offer advanced metering services.** If utilities keep the metering franchise, they may be required to provide "advanced" metering, such as TOU or hourly metering, provided monthly or daily, to end users and ESPs. An installed network meter reading system could allow the utility to provide this service cost-effectively.

- **Well-positioned in the event that metering is open to competition at some point.**

Disadvantages. There is only one significant potential disadvantage to investing in such a system.

- **Potential risk that the utility will spend more on the system than it receives in value from it.** As noted earlier, there is limited published information quantifying the benefits of large-scale network meter readings systems. Ameren recently completed a cost-benefit study on their NMR system (which currently covers more than 650,000 meters, and still is being expanded). This study concludes that "the direct benefits [the utility] is receiving from its use of the . . . NMR system exceed initial project costs."[11] The study found that these savings are comprised of labor and other meter reading costs, operational savings, and customer service center savings. A utility considering such a system should call on utilities with these systems installed to learn more about the costs versus benefits of such systems.

Scenario: The Future of the Metering Market Is Uncertain

Decision: Should a utility invest proactively in a meter reading system flexible enough to meet the needs of the competitive retail energy market, rather than waiting until the utility commission determines whether metering will be open to competition, or remain with the utility?

Advantages of Proactive Investment. If the utility eventually invests in NMR, there are several advantages of doing so sooner rather than later. For one, the utility will have more time to integrate the data from the system into their operations and customer service activities, and to reap the benefits of these distribution company savings and improvements. The marketing group—which is likely to spin off into the retail subsidiary—will have more time to work with the usage data, learn how to mine it for targeted marketing, and so forth. (Even though the utility affiliate probably will not be able to bring this data with them when they spin off, the lessons they learned about how to work with data, what data they want, and so forth, will be extremely valuable.)

Advantage of Waiting. One advantage to deferring the decision to automate is that the utility defers the costs of installing the system. These deferred costs need to be weighed against the deferred benefits discussed above. If the utility defers the decision and eventually decides not to install an NMR system, it can avoid the costs of installing such a system altogether. Whether this is in the utility's best interest depends of course on how the costs of the system compare to the benefits.

CONCLUSIONS

The competitive retail electric market will require customer usage information far beyond the monthly meter reads currently collected by most utilities. This information will help consumers evaluate ESPs' offerings and maximize their benefits from the competitive market. It will help suppliers optimize purchase decisions, and develop better offerings and marketing programs.

To meet these needs, utilities, ESPs, or metering service companies will need to adopt technologies that can provide detailed usage information frequently. These technologies are available and in use today; other offerings undoubtedly will develop as the market's demand for information accelerates.

Regulatory policies regarding metering and the provision of detailed metering information are just beginning to be developed. Only a few states have issued decisions on whether metering will be open to competition, and California is the only state that has defined the rules of the competitive metering market in detail. In theory, any regulatory decision that allows the market to cost-effectively obtain the information it needs to function efficiently could constitute sound policy from an economic point of view.

The uncertainty regarding the future of the metering market provides utilities with a unique opportunity to prepare to provide metering information services. Of course, utilities are learning that "opportunity"—one of the new buzz words of the industry—does not come without some risk. The risks of investing in technology that can meet the competitive market's usage information needs is the potential that this investment will not earn a sufficient return over time. Indications from regulated utilities are that these systems are covering their costs in operational savings alone. Indications from California are that market prices for information services will cover the cost of automated meter reading systems. Still, each utility will need to evaluate for itself the costs and benefits of such an investment, and may want to mitigate the risk of stranded costs by outsourcing the collection of frequent, detailed usage data.

The potential benefits from investing in systems flexible enough to provide competitive metering information are significant. Utilities can gain an additional source of revenue, they can improve distribution operations, and—particularly for those companies under PBR—can increase profits. In addition, as long as the utility (rather than its subsidiaries) is allowed to develop new marketing programs, this information can help deliver services that customers will value—and that may help retain them in the long run.

In the midst of the industry's uncertainty about how retail markets will develop, what rules of the game regulators will set, and how utilities will position themselves in the new markets, one thing is certain. The electric industry's journey into the information age is just beginning. The astute use of information will drive successful companies into the new competitive era. While it may be just the tip of the iceberg, detailed information on customer usage will be a critical part of the information needed to succeed.

NOTES

1. Plexus Research, Inc., "A White Paper on Direct Access Metering & Data Communication Requirements," (March 31, 1998, p. 25) prepared for National Association of Regulatory Utility Commissioners. The paper is available on NARUC's home page at www.naruc.org.

2. Richard D. Katzev and Theodore R. Johnson, *Promoting Energy Conservation: An Analysis of Behavioral Research* (Westview Press, 1987).

3. For a discussion of such alternate pricing options, see Chapter 13, "Creating Economic Value Through Risk-Based Pricing," by Stefan Brown, Douglas Caves, and Ahmad Faruqui in this book; and Melanie Mauldin, "Retail Risk Management: Pricing Electricity to Manage Customer Risk," *The Electricity Journal*, June 1997.

4. Mark Ward, "The Light Programme," *New Scientist*, 30 May 1998.

5. The California Public Utility Commission's current plans indicate that this size requirement may be reduced to 20 kW in April 1999.

6. Many observers point out that as long as it is possible to obtain and use hourly load profiles, customers who can benefit from doing so will be likely to do so. As long as customers have this freedom, many of the informational disadvantages of statistical load profiles are decreased. A remaining disadvantage is that it may be more difficult and/or expensive for customers to know or be identified as those that would benefit from the metering, if they are relying only on statistical load profiles.

7. Note, however, that leaving metering with the utility may slow the development of innovative price offerings, which could be a significant way customers realize savings from competition over time.

8. The California PUC has stated that utility distribution companies in California can invest in network meter reading systems to sell meter reading services to end-users and ESPs; the utilities are required to use shareholder dollars for this investment.

9. Wiesehan, Ray M., and Wayne Rechnitz, "Union Electric Unlocks the Power of NMR," *Transmission & Distribution World*, December 1997.

10. The study found that less than one-third of the direct benefits came from direct labor costs associated with eliminating the scheduled monthly meter read. The other direct savings were from such areas as decreasing special meter reads, reducing responses to customer equipment problems, and increasing meter accuracy. The article also discusses a number of additional benefits the utility believes it will realize in the future.

11. Wiesehan, Ray, and Wayne Rechnitz, "The CellNet Network Meter Reading System at Union Electric Company: A Cost-Benefit Analysis," September 14, 1997, p. 2.

SECTION VI

WHAT ARE THE KEY MARKET STRUCTURE ISSUES?

CHAPTER **20**

Give All Customers the Right to Choose, Immediately

Kenneth L. Lay

It wasn't too long ago that economists weren't much welcome in discussions over how best to organize some of our economy's most vital industries— such as airlines, natural gas, financial services, telecommunications, electricity, and water. But economists are nothing if not dogged, and with their help, we've seen industry after industry move away from monopolies and intrusive regulation to markets and competition. Consumers have seen equally dramatic results. The shift away from government planning has brought cost savings and innovations that have surpassed the expectations of industry restructuring's most ardent supporters.

It wasn't so long ago that those of us on the side of markets, competition, and choice were a lonely group, accused of being extreme, and worse, impractical. But today things are different. Governments around the world are looking to the private sector to deliver essential infrastructure. An increasing number of state legislatures and public utility commissions have sided with markets, and it's only a matter of time before Congress does the same. Even groups like the Edison Electric Institute claim to support competition and choice. Economists who opposed a common wisdom bent on guarding monopolies and who once were considered "fringe" are now mainstream.

This chapter is based on a speech given to the Western Economic Association Conference in Lake Tahoe, California, on June 30, 1998.

There's a reason for that. At Enron, we have a saying: "Facts are friendly." And the facts show beyond a doubt that when competition and choice replace monopolies and government planning, consumers and the economy are better off. It is difficult to overstate the importance of this accomplishment, and the vital role that economists have played in achieving it. It is an accomplishment our profession can be proud of.

CHANGING TIMES

Change hasn't been easy, and it won't get any easier. Understandably, the incumbent utilities that have benefited from 100 years of monopoly will not relinquish control willingly. But the change will come. It's only a matter of time. The electric industry, like the natural gas, airlines, and telecommunications industries before it, is on the threshold of a new era. We at Enron embrace the change. In fact, we're doing everything we can to hasten its arrival. So I want to congratulate and thank those whose work has helped to bring consumers the benefits of choice and competition that they have been denied for far too long.

Ideas and theories are important, but it is actions that matter in the end. In the battle of ideas that took place over electricity in states like California, New Hampshire, Nevada, Pennsylvania, and Arizona, markets, competition, and choice won the day. Competition and choice won the day when Congress moved to restructure the telecommunications, surface freight, airlines, and natural gas industries. But the real test has always been implementation. That test currently is under way in California.

First I want to congratulate the California Public Utilities Commission for starting the march toward consumer choice and competition in electricity in 1994. And I commend the California Legislature for providing the leadership to finish the job in 1996 with the passage of AB 1890, the legislation that made competition and choice law in California.

Are the rules that the California Public Utilities Commission initially established to govern competition and choice in California the very best that they can be? We don't think so. Is there room for improvement in AB 1890? In our view, there is. But more importantly, were the actions of the California PUC and legislature a step in the right direction? Are the state's consumers, its economy, and California's electric industry better off as a result? You bet. And for that reason, California should be applauded.

We're very excited about California. Enron currently has over 30,000 residential and business customers. We have seven major offices and 10 sales offices around the state and over 300 employees, including 200 serving the newly deregulated marketplace. We're currently bringing energy savings and innovation to the University of California and California State University systems, Kaiser Permanente, Pacific Bell, and Lockheed Martin. Thanks to California's decision to give consumers choice, we're helping to bring real energy savings and innovation to California's families, schools, health care system, and businesses.

That's why we staunchly opposed the initiative that some consumer groups placed on California's November 1998 ballot. The initiative represented a significant step backwards. It wouldn't have done much more than take the benefits of choice and competition out of the pockets of consumers and hand them over to lawyers and consultants in endless battles over the legality of the initiative's provisions. Is more work needed to ensure that Californians get real choice and real competition? Absolutely. And we intend to roll up our sleeves and work constructively with the California Public Utilities Commission and the legislature to get the job done. But the initiative was not the answer, and that's why we opposed it. The right thing is to continue to encourage and create opportunities for companies like Enron to revitalize an industry that has kept choice from consumers and performed poorly for too long, and at the same time to improve the current market structure and rules created by the legislature and the public utilities commission.

Before focusing specifically on California, let me share some general observations about the state of consumer choice and competition in the U.S. electric industry now that we have some experience under our belts.

Two issues stand out in particular. First, since electric industry restructuring began several years ago, many have taken their eyes off the ball. There's been a tendency to lose sight of why we decided to end 100 years of monopoly control in the first place. It's about consumers. Consumers, and the benefits that come with the freedom to make energy choices for themselves, must remain the focus. Unfortunately, consumers increasingly have become an afterthought. Consumers have taken a back seat to utility shareholders and managers, and to the interest groups who have relied on legislative and regulatory forums to secure funding for pet projects through utility rates. Clearly, these groups should be treated fairly in the transition, but we can't allow special narrow interests to ride roughshod over the interests of consumers, or restructuring will fail. I understand that those who benefit under the status quo might welcome failure. But if we do not succeed, consumers and our economy will lose, and the losses will be significant. So we must refocus our efforts on the consumer, and we must do it quickly.

The second issue is closely related to the first. If we require consumers to pay for every penny of the incumbents' "stranded costs," as the quid pro quo for choice and competition, we cannot also allow the incumbent monopolies to write the rules for competition. If allowed to write the rules, the incumbent monopolies will eliminate competition before it ever begins. We've already seen how this can happen, for example, by:

- guaranteeing the utility market share at no cost by leaving it with the role of default service provider;
- allowing the monopoly to use revenues from captive customers to fund unregulated affiliates and allowing the affiliates to leverage the incumbent's ratepayer-financed brand equity;
- allowing the monopoly to retain control over transmission and generation assets and use those assets to exercise market power;
- allowing the monopoly to devise arcane regulatory schemes to "estimate" stranded costs rather than rely on the market to value assets, and to keep us in the dark about the size of stranded costs;
- allowing the monopoly to retain control over other competitive market segments like metering, billing and distribution services.

If incumbent monopolies are handed 100 percent of stranded costs and permitted to write monopoly-friendly rules like these, consumers will continue to suffer from the same high prices, poor service, and lack of innovation that they have been subjected to for decades.

It took a lot of hard work to debunk the notion that we are better off with a policy that requires all of us to get our electric service from a monopoly. Now we have to work just as hard to make sure that the transition to choice and competition doesn't leave everyone worse off by trading in a regulated monopoly for an unregulated one.

THE CALIFORNIA EXPERIMENT

Let me turn to California and share how we see the electricity market shaping up over the next few years.

Enron entered the California market committed to serving all consumers, from the largest business customers to the smallest household, and our investment in marketing, sales, and back office operations speaks volumes about that commitment. We believed we could provide incentives for customers to switch and do so economically. Unfortunately, market conditions in California have forced us to refocus our efforts. What happened?

First and foremost, California's market structure and rules leave little, if any, room to offer residential consumers a reason to switch during the transition and give our shareholders an acceptable return on their investment. For example, the amount of stranded costs that consumers ultimately must pay the incumbents for bad investments is a big unknown. The failure to fix the amount injects a high level of risk into business planning. It's hard to set prices when one of the biggest components of your cost structure is so ill-defined. Real choice and real competition requires that the level of stranded costs be fixed, and fixed right at the start.

The 10 percent rate cut mandated by AB 1890 for residential consumers is another reason we've refocused. Based on press accounts and conversations with AB 1890's principal authors, we understand that the California Legislature believed there were a couple of good reasons for mandating the rate cut.

First, the authors feared that competitors would ignore residential customers and focus on large customers. They apparently saw the 10 percent rate cut as a way to guarantee savings for residential customers whether or not competition bloomed for small customers. Second, the authors hoped that the rate cut would protect small consumers. They figured that the market would have a difficult time competing with the 10 percent cut, which would make the residential sector unattractive enough that unscrupulous competitors would look elsewhere. So, as we understand it, AB 1890's authors never intended for competition to reach California's residential customers during the four-year transition and they structured the market with that goal in mind.

The legislation may have hedged these perceived risks, but in our view it overlooked another important risk. By focusing solely on a short-term rate decrease, California traded off competition's biggest benefit—innovation. All things being equal, California's residential consumers are better off paying 10 percent less for their electricity, but by handing the residential market over to the incumbent utilities in exchange for a 10 percent rate decrease, consumers will forego the benefits of innovation and perhaps even greater price cuts. That's because competition is the engine that drives innovation, and competition isn't going to be there—at least during the transition. In our view, we'll reap the greatest benefit by unleashing rigorous competition—and innovation—for all customers.

What is most troubling about the current market structure is the fact that it guarantees market share for the incumbent. Competitive markets just don't provide guarantees. But the way things are today in California—and in other states that have opted for consumer choice—the utility gets the privilege of serving any customer who fails to make a choice. The same thing happened when, for competition's sake, AT&T was forced to divest its long-distance operations. AT&T enjoyed what amounted to an exclusive franchise on every customer who did not choose to take service from another provider. And because AT&T's costs were significantly higher than its competitors, consumers who were stuck with AT&T as the default provider ended up paying billions of dollars in higher telephone bills.

We think there's a better way. Give all customers the right to choose immediately, but don't continue the utility's exclusive franchise for customers who neglect to make a choice. They will benefit more if we aggregate them and let competing suppliers bid to serve them. If no supplier can beat the incumbent's rate, then the incumbent will continue to serve them and no one is any worse off. If a competing service provider can beat the monopoly's rate, then the aggregated customers will be better off. We plan to press hard for what we think is clearly a superior alternative for consumers and will work hard with CPUC President Richard Bilas and the legislature to implement this sort of competitive framework for default service. It doesn't make sense to condemn California's electricity consumers to the same high prices AT&T's customers fell prey to.

These are some of the key features of California's market structure that led us to refocus our business strategy. It shows the importance of establishing a market structure that is friendly to competition at the outset if we want to maximize competition, innovation and savings for all consumers.

These shortcomings aside, California did the right thing by passing AB 1890 and deserves our congratulations. The best thing for California to do now is work diligently and constructively to improve as quickly as possible on what has been a good start. Passing the initiative that recently qualified for the ballot would be the worse thing to do.

Let me close with a few comments about our decision to suspend our residential efforts in California, and in other markets where the structure and rules preclude us from participating in a way that is attractive to consumers and our shareholders. Much has been made of that decision; too much, in my view. The more important question is: Are consumers better off when they're captive to a monopoly or when they're free to choose for themselves? The answer is clear, for one simple reason: the mere threat that a customer might choose someone else provides a powerful incentive to keep costs and prices low.

In California, as in other states, utility prices, like taxes, rose steadily for decades. But when the CPUC released its landmark choice proposal in 1994 and the utilities saw competition on the horizon, they started to trim costs, lower prices, and focus on serving customers better. In a very short time, the threat of competition accomplished something that California's regulators have struggled to achieve for a long time. So California made the right choice—it abandoned monopolies for markets—and we're going around the country urging other states to do the same. Just as importantly, the deals Enron has already done in California will help California's universities, hospitals, and telecommunications firms lower costs and provide better, more cost-effective service to Californians.

Enron is excited about the opportunities that exist today and the opportunities opening up every day. We're pleased about the success we've had, and we're confident that that success will continue. We will continue to assess our opportunities month by month, market by market; our obligation to our shareholders demands it. From time to time, circumstances will cause us to shift our focus and make mid-course corrections. But if an opportunity exists, we'll be there, like we are in California. We're here to stay. The opportunities are increasing every day, and we intend to seize them.

CHAPTER 21

Regulatory Issues in Customer Choice: The Case of Rhode Island

James J. Malachowski

It is a good thing my home state of Rhode Island does not have a complex about being the smallest state in the union and oftentimes forgotten about on the national scene. Rhode Island's founders left Massachusetts (upon request, by the way) in part to get away from oppressive government and establish a more independent union. This independence serves us well and continues to manifest itself throughout our history. Rhode Islanders burned a British ship in our bay before the Boston Tea Party, we were the first state to declare independence from England, and we were the last state to ratify the Constitution. This independence (and lack of recognition) now extends into the electric energy sector.

In July of 1996, the Rhode Island General Assembly passed a comprehensive utility restructuring act, which brought retail choice to large industrial customers starting on July 1, 1997. All remaining customers were eligible for choice on January 1, 1998. Despite all the attention focused on that large state located on the opposite coast, Rhode Island was the first state in the country to have retail choice for electric service. The Rhode Island Public Utilities Commission (PUC) does not lay claim on this initiative. Restructuring was mandated by an act of the General Assembly, and the accolades (or criticism) rightfully belong with the legislative body. The new law is very prescriptive and dictates the answers to all the major issues in the restructuring debate. The PUC has been working very hard to implement the law as well as working to develop a competitive wholesale market on a regional basis within New England. There is, therefore, some experience behind this discussion of choice.

The history of this case, which brought Rhode Island to this leading position, started with a general dissatisfaction with electric rates. Rhode Island suffers from the same dynamics that all of New England shares, which results in relatively high electric rates. These circumstances include the fact that despite the region's beauty and diversity, there is a lack of indigenous natural resources from which to generate electricity. Other factors include high labor costs, high local property taxes, and in some states a gross receipts tax on utility service is imposed. These competitive disadvantages were exacerbated by moves in the late 1970s to mid-1980s to implement rate design measures that were aimed at lowering residential rates. This was implemented by rate design decisions that force commercial and industrial customers to pay more than they should, thus cross-subsidizing the residential class. Apparently, at the time it was thought this was the "politically correct" thing to do (although the term "politically correct" was not in common usage at that time).

When the pressures came to lower rates for industry, the PUC responded by first engaging in rate rebalancing—removing the cross-subsidies. Development and implementation of a series of special rate discount plans followed. These initiatives included time-of-use rates, interruptible rates, "surplus power" rates packaged as economic development rates, and finally, special contracts. These special rates were very effective in lowering costs for industrial customers, so much so that the removal of these special rates became a problem in implementing market-based rates.[1] Nevertheless, the momentum for major change and a massive restructuring of the electric sector built to a point of no return. State government declared that the old system had failed. The indictment: unacceptably high rates. A new model had to be chosen. The new model is one that employs market-based mechanisms and competitive forces to drive efficiencies. The key behind this new model is choice for consumers, the ability to choose the supplier of the commodity.

When dealing with the utility sector, one must fully recognize the uniqueness of the sector. Utility service is one of the few remaining services or products that is delivered on demand to the home or business 24 hours a day, every day. The distribution systems necessary to fulfill this on-demand delivery function are expensive to build and maintain. Distribution costs are one of the major reasons the "old model"—a monopoly—was chosen. No new entrant can be reasonably expected to rewire every home or business in order to provide service. The most reasonable way to provide consumer choice is to mandate that incumbent monopoly utilities allow any new entrant the right to transport the commodity (the electrons, the electricity, the service) over the existing electric distribution system. The incumbent, now referred to as the wire company, will be paid a "rent" for the use of its distribution system. The physics of delivery, the accounting transactions, reliability, and system integrity should be transparent to the end-use consumer. The new model allows for choice of the commodity supplier, but customers are still captive of the distribution provider.

Electric restructuring brings with it a number of major policy, economic, and political decisions. These range from stranded investment recovery to divestiture of generation to standard offer service to distribution charges and more. This discussion begins after all the issues are decided. It centers on considerations relative to end-use consumer choice. Three areas will be highlighted: consumer information, aggregation, and information disclosure.

CONSUMER INFORMATION

A major obstacle in moving through the implementation phase of electric restructuring is the confusion and lack of understanding of the end-use consumers who have been given choice. One must not underestimate the natural resistance to change particularly when one endeavors in a somewhat technical or specialized field. There is significant segmentation within end-use consumers. The various classes may have an entirely different orientation, knowledge, and interest in electricity choice. Major industrial customers with substantial volumes and significant total bills may lust to enter a competitive marketplace. Major firms also may have a staff person or persons whose job it will be to understand the dynamics of restructuring, analyze the options, and recommend a choice. Importantly, these customers have the ability to allocate resources to this new area of decisionmaking.

The other end of the spectrum consists of individual households who, in today's busy world, barely have time to pay the bills and keep up with the routine household bookkeeping of balancing the checkbook. These customers feel they do not have the resources or time to allocate to this new decision-making chore, they doubt the savings will be material and, in fact, they may not have many real options offered to them. People in this class will be quick to perform a rough type of cost benefit analysis. It will be more of a "time" benefit analysis and will be done with minimal empirical information. If residential customers don't think the amount to be saved is worth the time necessary to analyze the situation and make a choice, they simply will not engage.

Experience in Rhode Island showed a surprising lack of understanding as to the basic underlying structure of this industry change. The basic principle, as outlined above describing the "wire company" and rent payment for use of distribution, is not understood by many. Large industrial customers who have followed, and in some cases participated in, the restructuring debate understand the basic principle. However, with the exception of this class, few others understand, including small industrials, the commercial class, and residential consumers.

Quite quickly many pointed to the PUC as the entity that would be responsible for the critically needed consumer information campaigns. In fact, the burden and expectations are even greater whereas this initiative is often referred to as "consumer education." This implies that commissions must go beyond the goal of providing information and, instead, actually "educate."

PUCs, as governmental economic regulators, would do well to avoid the use of the words "consumer education" and insist on "consumer information." The semantics here are important. Commissions are fine economic regulators. They maintain significant expertise in economics, finance, and accounting. They can dissect rate cases and analyze sophisticated and complex utility filings. This, of course, has been their mission. PUCs generally have been given the resources and developed expertise in these financial areas. They have not been created, however, to be "educational" institutions. By default, simply due to the lack of any other option, they are being assigned this informational/educational responsibility. The effort is further hindered by a general lack or severe limit to the resources allocated to PUCs to carry out this function.

Another interesting dynamic that directly affects the public's perceptions, expectations, and viewpoints relative to electric restructuring is the information that the government presents and distributes during the restructuring debate. In general, debates about utility service, its cost, and details have taken place at government regulatory commissions. These are quasijudicial tribunals whose decisionmakers serve, on average, a six-year term of office. There is also in most cases a staggered term of office. Most state utility commissioners are appointed. However, even in states where utility regulators are chosen by elections, the term of office and staggered terms generally insulate commissioners. PUCs are bound by rules of evidence and often by sunshine laws. The agencies are very process-oriented with appeal to the court as a right of any aggrieved party. Electric sector restructuring, however, requires activity by both utility commissions and state General Assemblies. These are two distinct bodies with two different processes.

General Assemblies have looser processes. Testimony is not cross-examined and the "record" is less formal. Ex parte communications are allowed and there often is less structure. General Assemblies are much better equipped and have more experience in working out consensus or compromise decisions. Members, however, are extremely sensitive to the public's demand for positive results. In fact, this demand has an even finer point. The public wants immediate positive results. In this environment, it is not surprising that legislative proposals often are heavily promoted by those in favor of pushing for the bill. In some cases the arguments proffered tend to overpromote the benefits.

This was the case in Rhode Island. The statements made and materials distributed during the announcement of the introduction of the electricity restructuring bill were sensational. Claims were made that rates would fall 15 percent to 20 percent immediately upon passage. Further, the restructuring bill was declared to be the most important legislation for the state's economy in 50 years. Passage of this legislation would make the state a magnet for business and job growth would be spectacular. Two full years after passage and one-and-a-half years after choice became available, rates are not any lower.

In Massachusetts, the General Assembly required in its restructuring law that rates be lowered immediately by 10 percent; government dictated what the price of the standard offer would be for each of its required seven years. The rate reductions are not 'real,' because even though the provider will see a loss of revenue, the law provides for each company's competitive transition charge will be adjusted in the future, thereby allowing them to collect—and the customer to pay—additional charges. It is also ironic that government price fixing is part of the state's move to a free market.

There is little wonder that consumers are confused and skeptical. Despite the obstacles and regardless of the name given these efforts, it is critically important to inform or educate consumers. PUC's must engage in this activity. In doing so, agencies should employ sophisticated marketing and consumer opinion tools. Consumer surveys, focus groups, and other types of research activities should be employed. Diversity of the target audience and techniques to reach various segments should be analyzed. Language issues and message delivery mechanisms are all factors to consider. A successful campaign will require a multifaceted approach and repetition. A one-time or one-shot effort will not suffice. This work is entirely different from economic regulation. Like regulation, however, it requires special expertise and an allocation of resources.

AGGREGATION

A skeptical commentator asked during the electric restructuring debate in Rhode Island, "How are rates going to be any lower if we simply are adding middle men?" One of the functions a monopolistic electric utility in the old model provided was aggregation. In fact, it was total aggregation for all customers. This function has provided value for end-use consumers. Functional separation of a vertically organized electric company into generation, transmission and distribution (wires company) and requiring divestiture eliminates the aggregation function.[2] In part the answer to the question above is that we are not "adding" a middleman but rather simply replacing an existing one we are losing otherwise. Aggregation has provided value for end-use consumers.

There is no question that those with buying power will have an advantage in the marketplace. Even prior to the passage of restructuring legislation in Rhode Island and Massachusetts, there were a number of entities actively and aggressively seeking to sign large industrial firms as customers to their aggregation enterprise. There was also a competing effort by a not-for-profit trade association whose membership was made up of large industrials, which was formed to track energy-related issues. Chambers of Commerce, trade associations, the League of Cities and Towns are just some of the types of organizations that are actively attempting to aggregate customers.

There are two important factors relative to electricity usage: volume (How much is used?) and load factor (When does the usage occur?).[3] The most attractive customers are those with both high volumes and high load factors.

An interesting illustrative exercise is to depict the entire end-use marketplace by these two factors, placing the highest volume, most attractive load factor customer on top. The next "best" customer would be placed immediately below this customer and so on until all customers were represented. The result would be a rough pyramid-shaped form with the most attractive but fewest customers at the top and the greatest number but least attractive (low volume, poor load factor) customers forming the base. (See Figure 21–1.) The customer pyramid then can be grouped into three sections:

- Triangle 1—High-volume, high load factor customers
- Quadrant 2—Moderate-volume, moderate load factor customers
- Base 3—Low-volume, poor load factor customers

The purpose of this exercise is to assist in describing the considerations consumers must be aware of in contemplating aggregation.

Customer A, the most attractive customer, has two basic choices. The customer can go it alone or sign on with an aggregator. When the customer is so attractive to suppliers, they must be careful that the aggregation does not dilute their own buying power.

Customer B, the least attractive customer in Triangle 1, should be very anxious to join an aggregation with any other group of customers from Triangle 1.

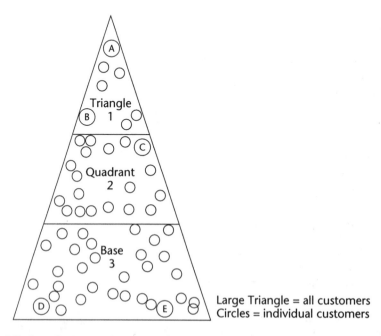

Figure 21–1   *Customer Segmentation*

Customer C is in a more interesting situation. This customer could decide to go it alone but it would be an attractive candidate for an aggregation of customers from Quadrant 2. The best option for Customer C, however, is to attempt to enter into an aggregation consisting of customers from Triangle 1. This type of "trading up" should be the general strategy of all buyers. The question is, will the Triangle 1 customer aggregation allow Customer C to enter? The aggregation may need the increased volumes that Customer C has to offer despite the poorer load factor Customer C brings to the group. Just as in any business relationship or consideration of membership in any group, one must strive to understand fully the details of whom you are dealing with.

Customers D and E have low volumes and poor load factors. They are residential customers. Opportunities for aggregation will be limited, particularly in the years immediately following restructuring. The market will develop first for high-volume customers. In considering a residential customer aggregation, it is likely that residential customers will be attracted to those entities attempting to aggregate where the customer feels some type of connection and has some level of trust. These entities may include employers, alumni associations, other utility sector service providers, or even other retail service providers when the household is already a customer.

It is anticipated that there will be a hesitance within the residential class to enter the competitive marketplace. There will be less price elasticity for these customers compared to the business sector. It appears that standard offer provisions in restructuring are designed solely as a transitional type mechanism.[4] These programs are intended to move residents to the market a few years after choice is available. This is done by government setting the prices for standard offer service. The prices start below the expected market price in the early years but escalate at a significant rate each year thereafter. The prices that have been set for the mid to late years of standard offer service are expected to be substantially higher than market prices. There is enormous irony here in that the government, in moving to a market-based approach and competitive forces, is doing so by engaging in one of the most onerous intrusions on markets—price fixing. The steep schedule of increases to the annual pricing of the standard offer service is a cause for concern. At some point the market price will intersect with the standard offer rate and later on the standard offer price will be higher than the market. At some point the spread between the market price and standard offer will be so great that residential customers, whether they like it or not, will be compelled to leave standard offer service. In both Rhode Island and Massachusetts, once a customer leaves standard offer service, they can never go back to that service. The concern is that if at this point the residential market has not developed or matured, residential customers will be in some type of market limbo, having high standard offer prices but no market alternative. The leap of faith is that market forces will prevail and sellers or aggregators will step in and meet this market need.

The state government's approach to standard offer service in both Rhode Island and Massachusetts has been very surprising. The model chosen forces people to go to the market whether they want to or not. Government is forcing people to do something they may not want to do. The force is the fact that eventually residential customers will pay more money for electric service than they are paying now. 'Do something or you will lose money,' is the message being sent. In this very complex and confusing area, government is saying they know what is best. People must go to the market where they will be very much on their own.

There is another concept, however—the existence of an entity that would aggregate for all residential customers under a type of "safe harbor" concept. Rather than the government dictating a continued staircased-type of predetermined price increases, standard offer service could be provided at a price determined by an annual bid or auction. The price of the service, therefore, would be determined by a market mechanism. Residential customers would be free, of course, to go to the market for service but, importantly, those that did not would not be harmed. Residential customers could stay in the "safe harbor" and still receive the benefits of competition.

The "safe harbor" model represents an entirely different approach, which gives residents a greater choice and more protection. If a customer does not want to engage, they have the choice to "stay put" and still receive savings. If the market is still developing or ill formed, residential customers will not be harmed. The aggregation function that the former integrated utility monopoly performed still would be performed for residential customers. If, on the other hand, a residential customer has an aptitude or appetite for markets, they have that choice. Finally, if the market for residential customers becomes robust with many service providers and aggregators, the "safe harbor" mechanism will dissolve because of the market forces. The approach here is one that offers more protections and greater choices. It is also one that is kinder and gentler and more representative of government's traditional role or response.

INFORMATION DISCLOSURE

Product differentiation has been a common marketing technique for quite a number of years. Efforts to distinguish a product from a competitor's are sophisticated and apply to products whether they are generic or not. Real or perceived differences are exploited to gain market share. When buying gasoline for the car, you can decide to "put a tiger in the tank" or buy the gasoline "that cleans your engine." Power versus cleanliness from the same basic product. Another example is, you can buy "regular" salt or the kind that "sticks to your food." Products that are seemingly the same can be differentiated, and so it is with electricity. You may not be able to distinguish one electron from another, but there are real and major differences in how the electricity may have been generated. The type of fuel used to generate the electricity varies greatly. Options range from coal, oil, natural gas, and nuclear fuel rods to water, wind, solar or chemical reaction and more.

There is also a major distinction in the type and quantity of emissions released into the atmosphere from the different types of generation techniques. Noncombustion generation has environmental and societal impacts. Thermonuclear generation plants produce "nuclear waste," and this country still does not have a long-term waste site under development. Hydroelectric power sources also have impacts. The Cree and Inuit native tribes of Canada have argued repeatedly that the flooding of land necessary for the hydro installations affects the very culture of their people. One of the issues that arises in electric restructuring, therefore, is what level of product information should be disclosed to the buying public. Will consumers be interested to know the fuel source and emission characteristics of the electricity they are buying? Will consumers' purchasing decisions be affected by this information? Will customers pay more for electricity generated a certain way?

These types of questions are new to the electric sector. There seems to be compelling reasons to provide information. Impacts on the environment are part of the consciousness of the public today. The environmental movement has been very successful in sensitizing people to the impacts of individual decisions or actions on the environment. Environmental quality depends on all of our individual and collective actions. Energy usage directly affects the environment. Information is power, so disclosure is important. In attempting to move in this direction, one is confronted with significant logistical problems in structuring a workable and verifiable information disclosure policy.

Seldom does an electrical energy supplier own just one generation unit. Diversity of fuel source and generation type were goals of every power company. Electric generation plants fall into three broad categories: base load plants that run most of the time; intermediate plants that run during times of increased demand; and peaking plants that only run when usage is at the highest. These designations stem from both economic and physical considerations relative to the power plant. There is also the matter of dispatch, which is when a plant actually runs due to availability or mandate. Add to these complexities the new wholesale market, which will dispatch plants based upon bid prices as opposed to cost. Determining which power plant is selling to which customer is, indeed, complicated.

In addition to fuel source and emission characteristics, clear and concise information pertaining to electricity price and key contract terms also must be made available to the buying public. Proper information in a form that lends itself to easy comparison will enhance the efficiency of the marketplace. Suppliers will be able to differentiate themselves and consumers will be able to direct electricity dollars to preferred attributes.

Despite the complexities, it is possible to construct and implement a workable and verifiable information disclosure policy. Two alternative approaches have been developed—a tagging system and a tracking system.

Under the tagging system, a generator applies for a tag independent of energy sales. A key feature of this system is the creation of a secondary market. Load-serving entities would acquire sufficient tags of specific characteristics to cover the total amount of their sales to end-use customers. Because tags represent actual monetary investment in particular generation characteristics, proponents argue that the generation mix will evolve over time, to reflect the sum total of consumer preferences expressed through the various tag-based transactions. It is also argued that to the extent a premium is paid for tags of any type of attribute, the system allows generators to receive these dollars.

The other alternative is a tracking system. This system is based upon historical data and tracks source characteristics, along with the bid and settlement process. This system is administered by the independent system operator or other entity that is responsible for the operation of the power pool and the wholesale power market. The tracking system traces the contract path of the electrons. The contract path concept is consistent with what has been used in the past for transmission billing purposes. Proponents of a tracking system claim it avoids the credibility problem associated with the tag system, which exists because the attribute tag is separated from the commodity power. End-use consumers may not feel they are actually purchasing the generation attributes they desire if they are simply purchasing a "tag."

Information disclosure has become an important and heatedly debated topic in electric restructuring. This topic, however, has to be approached somewhat differently and delicately in regard to jurisdiction. The most effective information disclosure system is one that is implemented on a regional basis. The regions should be at least as large as the affected power pool. Individual states within the same pool must be cognizant of the impacts that disclosure plans that are different in form or substance will bring. If states adopt different plans, they will be hindering the very effort they are trying to promote. Further, disclosure plans include specific requirements imposed on a supplier's advertising and promotional material. There would be an obvious burden on suppliers if they had to comply with a different disclosure rule in neighboring states.

CONCLUSION

The role of a state utility regulator clearly has changed. The mission has expanded to include new and important components. There has not been, however, any relief from, or elimination of, existing responsibilities. A monopoly wires company will continue to exist. The rates this utility charges for access to and use of its distribution system still must be analyzed and set. One of the major functions of an economic regulator dealing with for-profit firms is to identify and assign risk. This role will continue. New responsibilities and considerations are now included in the task. Regulators are responsible for the creation and development of markets. A new and major criteria in decision-making is how will the decision affect the market. Initially, the markets will be ill-formed and go through a development period. Regulators will not only make decisions affecting the development, but they will also have the responsibility to arbitrate disagreements between or among market participants. The uniqueness of utility service due to the continued monopoly on distribution dictates this role for regulators. As described above, regulators will have responsibility for consumer education/information programs and information disclosure policies. The effort to bring choice to consumers requires the creation and maintenance of efficient markets.

Changing an industry sector that has been in its current form for over 50 years will take great effort and close attention.

NOTES

1. In some cases, these special rates are providing industrial customers a rate for the commodity, which is lower than the market price. Therefore, moves to market-based rates result in rate increase for some.

2. The cost of aggregating what is "saved" by the vertically integrated company has not been considered in restructuring accounting. It probably should be deducted from stranded cost.

3. An individual customer's usage of electricity varies during the course of a day, week or perhaps from season to season. Load factor is a term used to describe a customer's average usage compared to their peak usage. The closer the average usage is to the peak, the higher the load factor becomes. Customers with high load factors are more attractive to suppliers because usage, in terms of kilowatt-hours, is closer to the capacity needed to serve.

4. In an effort to ease the transition to competition, some state legislatures, including Rhode Island's, have required incumbent utilities to provide default electric service for customers who do not want to participate in the marketplace. This service is known as "standard offer" service and usually is required for a limited number of years after choice becomes available.

CHAPTER 22

The Role of Power Exchanges in Competitive Electricity Markets

Becky A. Kilbourne

Exchanges bring two fundamental efficiencies to commodity markets: price discovery and risk management. Exchanges provide a market, or gathering place, for buyers and sellers to trade, thus dramatically reducing the costs of finding a counterparty to the buy or sell transaction. Because exchanges provide a centralized trading forum with price discovery, spreads are narrowed so that neither buyers nor sellers experience significant economic rents due to asymmetric information. Furthermore, exchanges allow buyers and sellers to trade independently of one another and significantly diminish trading risk.

Today's futures exchanges, such as the Chicago Board of Trade and the Minneapolis Grain Exchange, evolved from forums for trading physical commodities such as livestock and grain. Today they offer standardized contracts, such as futures and options, that provide fungibility to commodity markets and significant flexibility to buyers and sellers in physical markets. These instruments evolved as commodity exchanges became reliable and trusted trading forums providing credible price discovery. They enhanced the efficiency in markets by providing buyers and sellers the means to hedge risks, underlying their physical trades.

Power exchanges, which deal primarily with the physical market for electricity, are an infant industry compared to today's futures exchanges. Their existence will create the foundation for the trading of electricity as a commodity and the development of financial tools to manage price risk. It is widely predicted that as retail markets open to competitive electricity generation the electricity derivatives market will be greater than that for the combined trading of all other commodities.

This chapter explores and contrasts the development of electricity "power exchanges" around the world. It also describes the history and development of U.S. commodity exchanges, which now are trading primarily futures and other derivatives on underlying commodities, in order to draw parallels and establish the functionality of power exchanges in the commoditization of electricity. The existence of power exchanges can create the depth, liquidity, and price discovery in an open market necessary to truly provide consumer choice.

AN OVERVIEW OF RESTRUCTURED MARKETS

Restructuring in the U.S.

In the United States, the emerging restructuring of electricity markets involves allowing electric generation to be competitively procured in retail markets. Heretofore, U.S. electricity had been supplied by large, vertically integrated, for the most part investor-owned, utility companies. There were significant economies of scale associated with electric generation. In the interest of not creating duplicative distribution and transmission systems, these companies were granted a franchise to serve in specified geographic areas. In exchange for this guaranteed market (sometimes referred to as an obligation to serve), utilities were regulated at the state level for provision of retail electric service.

More recently, new technology and changing laws regarding fuel use have introduced lower cost, less-capital-intensive generation. The economies of scale in electric generation that once existed no longer exist. Because of this factor, and because of increasing deregulation in other industries (telecommunications, banking, trucking, railroads, and natural gas supply) there is substantial market pressure to open up electric generation to competitive procurement at the retail level.

In 1996, the Federal Energy Regulatory Commission (FERC) issued Orders 888 and 889 to disaggregate the generation and transmission functions to begin deregulation of the electricity industry in wholesale markets throughout the U.S. Since that time a few states have put in place laws to deregulate and open retail markets to competitive generation. Almost all states are evaluating and debating models to open their markets by the turn of the century.

California is the first state to open its entire regulated retail market. Legislation was enacted in California in September 1996, calling for the new electricity market to open just 15 months later, on January 1, 1998. Shortly before that deadline, the opening of the market was postponed three months, until March 31, 1998. In contrast to most other open markets, the California market was completely open from day one—it is not restricted to certain customers or electricity providers.

Worldwide Restructuring

Some markets in other countries have been restructured similarly to provide retail access. The British government, for example, was the first to outline a plan to privatize the electricity industry in 1988. As originally envisioned, wholesale competition in the U.K. would be between the two major utilities and existing alternative electricity suppliers in Scotland and France. However, 10 years later, retail competition all the way to the residential customer level is still being phased in for the U.K.

Since then, other countries that have opened retail markets have been able to take advantage of the lessons learned in the U.K. and from each other, helping speed up deregulation to some degree. Besides the U.K., Norway, Sweden, Finland, Argentina, New Zealand and Australia have opened their markets to retail competition.

Let's take a look at how the various markets around the world compare to one another. Table 22–1 is a chart that shows the progression of opening retail electricity markets around the world.

The Structure(s) of Restructured Markets

Just as each country has a different form of government and different public utility infrastructures—some started out as nationalized utilities, for example—each deregulated retail electric market in the world is structured differently and continues to go through different phases of restructuring. Similarly, in the U.S., each state or region has different infrastructure and history that is fashioning different forms of restructuring.

The many different approaches to restructuring have to do with whether the original utility structure is totally or partially a government-owned monopoly; to what extent there is vertical integration (generation, transmission, distribution, and retail customer service in one entity); whether it is regulated or operated on a state-by-state or national basis; and what the consequent driver is for restructuring. In the U.K., Argentina, and Victoria, Australia for example, restructuring began with the break-up and/or privatization of a government-owned, vertically integrated electric system. These markets immediately established power exchanges for trading, sometimes optional (Argentina) and sometimes mandatory (U.K. and Victoria).

In other markets, restructuring began with the lifting of the franchise from retail service providers without necessarily forming a pool or competitive wholesale generation market. In New Zealand for example, the retail market was opened for competing retail service providers prior to the breakup of the government-owned generation monopoly and prior to the formation of a centralized pool.

TABLE 22–1   Worldwide Progression of Opening Retail Electricity Markets

Nation	Year Opened	Market Size: Annual Load (TWh)	Market Structure*	Market Evolution
Argentina	1992	65 (Northern and Central Argentina)	Direct Retail Access with Optional Pool (CAMMESA)	1992—Privatization and creation of wholesale markets begin —Direct access for 2 MW and above —Formation of CAMMESA Current—Direct access for 100 kW and above
Australia (Victoria)	1993	37.1	Centralized Pool with Retail Competition Centralized Pool with Retail Competition and Access to Pool	1993—Victoria breaks up state utility, establishes power exchange (VPX) 1994—VicPool begins operation; participation is mandatory for wholesale providers; retail access to suppliers in pool—5MW and above 1995—Retail access to suppliers in pool—1MW and above 1996—Retail access to suppliers in pool—750MWh/yr. 1997—Preliminary stage of National Electricity Market begins with interface between power exchanges for Victoria and New South Wales 1998—Consolidated market and pool (National Electricity Market) for 4 states of Victoria, New South Wales, Queensland, and South Australia begins in late 1998; Any customer who uses 160 MWh/yr. or more can choose supplier from pool or have access to pool 2001—Retail market scheduled to open for all customers
Australia (New South Wales)	1996	54.1	Centralized Pool with Retail Competition	1996—NSW Wholesale State Market (SEM) established; participation is mandatory for wholesale suppliers; retail access to suppliers in pool for customers using 40 GWh/yr. or greater 1997—Preliminary stage of National Electricity Market begins with interface between power exchanges for Victoria and New South Wales —Retail access to suppliers in pool for customers using 750 MWh/yr. or greater 1998—National Electricity Market with 4 states of Victoria, New South Wales, Queensland, and South Australia begins in May 1998 —Retail access to suppliers in pool for customers using 750 MWh/yr. or greater 1999—Retail market scheduled to or greater open for all customers

TABLE 22-1 *continued*

Nation	Year Opened	Market Size: Annual Load (TWh)	Market Structure*	Market Evolution
England and Wales	1994–96	280.6 (96/97)	Centralized Pool	1990—Privatization begins; access for 1 MW and above 1994—Access begins for 100 kW and above Fall 1998—Access for all retail customers expected
New Zealand	1993	31	Retail Competition	1993—Retail franchise removed for customers with less than 5 GWh demand/yr. —EMCO formed to begin development of pool 1994—Retail franchise removed for all customers (access provided to different retailers)
			Centralized Pool with Retail Competition	1996—Government split monopoly generator into two separate companies (gov't owned) EMCO (10/96) begins operation for wholesale and retail markets
Norway, Sweden, and Finland	1991 1996 1997	150 Norway 136 Sweden	Direct Retail Access Direct Retail Access with Optional Pool	1991—Norway grants third-party open access to T&D networks 1996—Sweden grants third-party open access —Finland allows access for 500 kW and above —Joint Norway-Sweden power exchange (Nord Pool) opens —Finland launches El-Ex (electronic power exchange) 1997—Finland allows access to all customers
USA (California)	1998	208	Competitive Bilateral Wholesale Market Retail Competition with Optional Pool	1992—Open wholesale access transmission of electricity required in U.S. 1996—California legislation requires competition in retail electricity market in 1998 1998—Retail market opens

*Direct Retail Access: Customers given access to purchase supplies directly from generators
Centralized Pool: Market structure incorporates a mandatory pool or market for wholesale and/or retail buying and selling of electricity
Optional Pool: Market structure incorporates an optional pool for wholesale and/or retail buying and selling of electricity
Retail Competition: Customers have access to other retail providers or wholesalers but not to generators
Competitive Bilateral Wholesale Market: Wholesale market open to competition without organized pool or spot market

In most markets opened so far, retail access has been phased in. The U.K., Finland, Argentina, and Australia opened to larger customers first. In contrast, New Zealand phased in retail access from smaller to larger customers. The concern here was that as soon as access was made available to larger customers, there would be rapid and pervasive losses to existing providers. Norway and Sweden, like California, were opened to all customers the same day.

Centralized versus Decentralized Markets

In some deregulated retail markets, all buyers and sellers are required to bid their electric supplies through a pool or centralized energy exchange. This was the first market model used to restructure electricity markets in the U.K. However, in other countries, including Norway, retail markets initially were opened to bilateral trading. Subsequently, a pool came into existence to reduce trading risks and to provide an open forum for buyers and sellers to find each other.

Certain markets have been structured around the notion of a mandatory pool from their inception, while others, such as New Zealand, formed mandatory power exchanges subsequent to allowing some form of retail service competition. In a few cases, retail customers have had direct access to the pool. In others—such as in Australia prior to October 1998—customers could switch only between retail service providers that traded through the pool.

In some places, the pool operations are consolidated within the transmission operators' functions; in other places they are separate. In Australia, where different approaches were taken in different states, restructuring of the electric industry has taken a new turn toward a consistent national structure.

California's market structure is unlike any other market in the world. It is composed of an optional pool or power exchange (the California Power Exchange, or PX), an independent and separate system operator (the California Independent System Operator) and other entities known as Scheduling Coordinators. Both the PX and other scheduling coordinators provide balanced schedules to the ISO on an hourly basis.

Price Volatility in Restructured Markets

Generally, each new electricity market goes through certain predictable stages when it first opens. Increasing price volatility, as measured by degree of change in price over a given period of time, is one phase. During the initial "shakeout phase," dispatch methods—once controlled only by individual utility companies—can change dramatically as new players enter the market. This change in dispatch can lead to new congestion patterns that, in turn, can produce increased price volatility and uncertainty. There are new power brokers, new purchasers, and new generators. Even utility companies must bid for their power. These new market entrants may bid and schedule their power differently from the long-standing traditional economic-dispatch methods used by utilities. All participants must now deal with new risks, such as credit and performance risks that increase price volatility and uncertainty. As discussed later, this volatility and uncertainty has the potential to be more significant in competitive bilateral markets that do not have an exchange function.

By definition, restructured markets are more susceptible to gaming and chaos. Again, markets without an exchange function are more vulnerable to dislocations resulting from market power exercised by larger players with existing trading relationships and more information. By contrast, in most new centralized markets with an exchange function, some form of market monitoring function has been established to prevent destructive gaming. Although monitoring sometimes results in changes to the bidding rules, at other times market forces produce healthy self-correcting behavior. But even as patterns become more predictable, a change in bidding rules, laws, or technology can trigger a new shakeout in a competitive electricity market.

To understand price volatility in restructured markets we can look to information provided by exchanges in the U.K., Nord Pool, and Victoria. (By definition there is no auditable price discovery in decentralized markets.)

The U.K., in its early years (1994–95) experienced less price volatility because of existing vesting contracts that were in place. However, they experienced significantly increasing "uplift costs" associated with ancillary services. Rules were changed over time to correct for market anomalies and the potential for price manipulation.

In Scandinavia, Nord Pool continues to experience significant price volatility —sometimes as much as 400 percent—because of its dependence on hydroelectric generation. In 1996, a cold, dry year, prices were much higher and volatility was great. This situation created concern in Norway that suppliers were manipulating the market. However, higher rainfall levels in 1997 produced a significant drop in prices and volatility, and reduced those concerns. Victoria, Australia, on the other hand, experienced a decrease in average prices over time, but prices were still very volatile, with 1,000-percent price fluctuations. The graphs below demonstrate price volatility experienced in Victoria from December 1996 through January 1998.

The price spikes in January 1997 (see Figure 22–1) resulted from the loss of generation sources. The price spike in November 1997 was caused by transmission congestion. The price of generation in that month hit a high of $3,210 per megawatt-hour. The majority of the time, however, the price of a megawatt-hour was lower than $50/MWh. Over the course of a year, the price of a megawatt-hour in Victoria exceeded $50 only 1.8 percent of the time, as the price duration curve shows in Figure 22–2.

THE ROLE OF EXCHANGES IN COMMODITY MARKETS

History and Evolution of Commodity Exchanges

In understanding the role of exchanges in commodity markets, it is helpful to understand the history and evolution of exchanges.[1] Dating to the

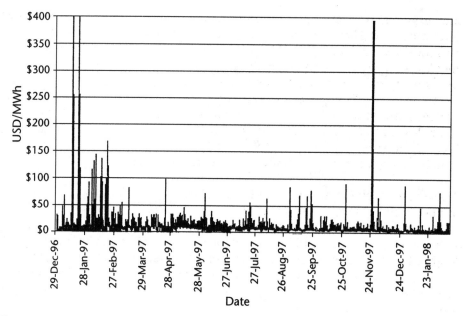

Figure 22–1   *Victoria Electricity Market Half-Hourly Prices*

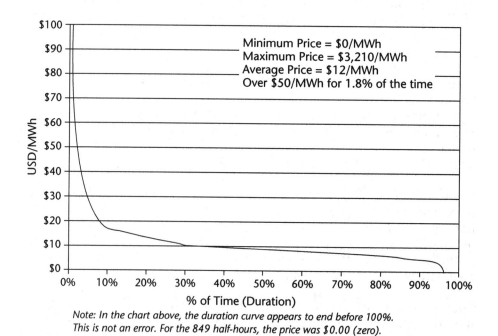

Note: In the chart above, the duration curve appears to end before 100%.
This is not an error. For the 849 half-hours, the price was $0.00 (zero).

Figure 22–2   *Victoria Electricity Market Prices, 1 December 1996 to*
                   14 February 1998

ancient Greek and Roman markets, formalized trading practices began with a fixed time and place for trading, a central marketplace, common barter and currency systems, as well as a practice of contracting for future delivery. In medieval England, a code known as the Law Merchant established standards of conduct, which performed a basis for common practices in the use of contracts, bills of sale, freight, and warehouse receipts, letters of credit, transfer of deeds and other bills of exchange. There were formalized rules and procedures for arbitration and dispute resolution. These rules were submitted to local authorities and administered by the trade associations themselves, which were the precursors to today's modern exchanges.

These early exchanges were created to facilitate the trading of products such as grain and livestock. They arose out of a need for individual buyers and sellers to trade in a centralized market where there were many other buyers and sellers and where they could easily discover the going price for their products. Without a centralized market, buyers and sellers had to find one another and rely on bilateral negotiations with little information as to whether they were receiving or paying a fair price. The exchange function in these days provided a marketplace whereby trading took place and buyers and sellers could negotiate price and delivery under a common set of rules and procedures.

Similarly, in the U.S., in the early 1800s it was common for farmers to bring grain and livestock to regional markets at a given time each year. They often found the supply of meat and grain far exceeded the immediate short-term needs of packers and millers. However, the glut of commodities at harvest time was only part of the problem. There were also years of crop failure and extreme shortages.

Even in years of abundant yield and high demand, supplies were quickly exhausted in urban areas, prices soared, and people literally could go hungry several months after the fall harvest and marketing of grain and livestock. The rural population, on the other hand, had crops they could not sell, and therefore, did not have the income to pay for needed manufactured products, such as tools, building materials, and textiles.

In the mid-1800s, in response to intolerable market conditions, farmers and merchants began to enter into contracts for forward delivery. Forward contracting is a cash transaction in which a buyer and seller agree upon price, quality, quantity, and a future delivery date for some commodity. Since nothing in the contract is standardized, each contract term must be negotiated between the buyer and the seller.

The exchanges that developed to support the forward trading in these commodities greatly reduced the time and resources expended for buyers and sellers to find each other. Further, organized exchanges did much to improve transportation and storage for commodities such that in conjunction with forward contracts there could be a more continuous supply of goods.

But forward contracts had their drawbacks. They were not standardized according to quality or delivery time and merchants and traders often did not fulfill these forward commitments. Then in 1865, the Chicago Board of Trade (CBOT) took a step to formalize grain trading by developing standardized agreements called futures contracts. Futures contracts, in contrast to forward contracts, are based on identical terms as to quality, quantity, and time and place of delivery for the commodity being traded.

At the same time, a margining system was initiated by CBOT to eliminate the problem of buyers and sellers not fulfilling their contracts. A margining system requires traders to deposit funds with the exchange to guarantee contract performance. Following these monumental steps, most of the basic principles of futures trading, as we know them today, were in place.

The late 1800s were critical to the scope and efficiency of futures trading. Trading practices were formalized, contracts were standardized and rules of conduct and clearing and settlement procedures were established. Trading became even more efficient as speculators entered the picture. Speculators assume price risk based on their forecasts of market prices. They are speculating that they have superior information about the market. They make profits by buying and selling commodities futures contracts and "trading out" of their positions prior to contract expiration. This added liquidity to the market and helped minimize price fluctuations.

All during history, exchanges continued to enhance efficiency in the market. The creation of standardized instruments and price discovery provided deep and liquid markets whereby buyers and sellers did not have to incur costs to search for a counterparty or discover price. The trading rules and credit management functions of the exchange significantly reduced risk to both buyers and sellers. Markets in general had continuous supplies of commodities at low cost. Today, because of the growth in exchanges and futures markets—from agricultural products to precious metals to financial securities—market participants can hedge and cross hedge nearly every price risk they face. Similarly, as described later in this chapter, the existence of exchanges in electricity markets has great potential to enhance efficiency in energy and other markets.

How Exchanges Function in Commodity Markets

Exchanges arise to provide liquid markets for commodities at fair prices. Initially this is done by providing forward contracts around generally traded products. Exchanges thrive by continuously providing more efficient markets to buy and sell commodities. Over time, the exchanges have evolved by providing additional risk management tools such as options on futures contracts, index products (Standard & Poor's 500, Municipal Bond Contracts) and other derivatives.

Exchanges in commodity markets have the following primary roles: (1) create a centralized mechanism to facilitate trading between buyers and sellers,

(2) provide reliable prices, not only for financial participants but for other market participants and related or interested businesses, (3) create a reliable clearing function, (4) manage counterparty risk, and (5) encourage the participation of market makers, brokers and other enterprises or individuals who assume price risk for buyers, sellers, and other market participants.

A Reliable Indicative Price

An exchange is a free market where the many factors that influence supply and demand converge and are then translated into a single figure—the price. An exchange therefore is a barometer for prices, continuously registering the impact of many forces on a commodity's price. This characteristic allows hedgers and speculators to gauge value. In addition, millions of people all over the world use the price information generated by an exchange as a benchmark in making financial, investment, and marketing decisions.

A reliable, indicative price for a standardized unit of trade (bushels, barrels, and so forth) for a specified delivery period (such as spot or 18-month contracts) at a specified quality and a determined place of delivery or receipt is essential to the development of futures, options, and other derivatives markets. Financial traders must have confidence that the exchange prices reflect a deep and liquid market in order to participate in derivative markets and assume price risk. Buyers and sellers want to understand and compare prices for over-the-counter trades (or bilateral contracts) against prices for trades through the exchange.

Through futures contract prices, exchanges also provide suppliers with valuable information about longer-run expected prices. When futures contract prices exceed long-run marginal costs, this signals productive capacity additions. Spot and shorter-term futures contract prices signal how to use marginal capacity relative to purchasing on the spot market.

The Clearing Function

The clearing function of an exchange ensures the financial integrity of the marketplace. Clearing allows buyers and sellers to trade independently of each other, thus contributing to the efficiency of the marketplace. The clearing function itself does not involve delivery or sale of the commodity; instead it provides the mechanism that settles trades so sellers can make delivery to qualified buyers. Indeed, the existence of a responsible clearing function can determine the growth and success of an exchange.

The role of an exchange to provide a reliable clearing function is central to a market operating consistently and within the context of rules that all participants understand. As a neutral player, an exchange will clear the market and settle trades fairly so that no market participant gains unfairly at the expense of others. Since the exchange handles credit clearances and margin calls by a specified set of rules, all participants are protected from default risks that they otherwise may bear without this clearing function.

Exchanges Operate to Reduce Counterparty Risk

Exchanges that have deep and liquid markets, that is, multiple buyers and sellers with large volumes being traded, significantly reduce the risk of failed delivery (for buyers) and nonpayment (for sellers.) This is because the pool, comprised of many buyers and sellers rather than a single entity, becomes the counterparty. This risk management benefit is particularly valuable to smaller participants who might not otherwise financially survive a large failure in a bilateral arrangement. In addition, to the extent an exchange has many buyers and sellers, the underlying commodity price cannot be easily influenced by any one trader. This offers stability and lower prices as compared to thinner, more volatile bilateral markets with wide price spreads. In capital-intensive electricity markets, as described later, the existence of a power exchange facilitates participation by many players—large and small—and promotes new entry into the market by providing a liquid market and easy trading forum.

Market Makers Assume Price Risk

The function of an exchange to offer a standardized trading product (quantity, quality, time, and place of delivery), a deep and liquid market and reliable price discovery, create the environment for market makers to assume price risk. Initially market makers enter commodity markets as individual speculators. As markets evolve and exchanges become more credible and gain more depth and liquidity, brokerages enter the market to trade on behalf of many speculators and hedgers. Ultimately, as a commodity market matures and an exchange is widely used, other enterprises, such as proprietary trading firms that use sophisticated computer trading models to hedge prices in multiple markets, use the exchange to participate in the market. This results in even further narrowing of spreads and more efficiency in the market.

Market makers and speculators do not desire physical delivery of the product, but instead thrive on their ability to forecast prices and make profits through options and futures. They assume price risk in the market by buying and selling options, futures and other instruments based on their projections of market prices. They make money when their price forecasts are right and lose money when they are wrong. What they do for commodity buyers and sellers (and hedgers) is remove price risk. This is because buyers and sellers (and hedgers) use the financial instruments to lock in their desired price for the commodity, irrespective of where the market price goes. The entrance of market makers and speculators in commodity markets, that neither consume nor produce, adds more depth and liquidity to markets and enhances market efficiency.

THE ROLE OF POWER EXCHANGES IN ELECTRICITY MARKETS

Commodity-based trading of electricity is relatively new compared to the trading of agricultural products and precious metals. Significant financial markets for options and futures in electricity do not exist today. While the New York Mercantile Exchange (NYMEX) has been trading electricity futures at the California-Oregon border and Palo Verde, Arizona, since March 1996, volumes are small compared to the underlying physical market for electricity.[2] As retail electric markets open to competition—and there are many new entrants, both buyers and sellers—financial tools for price risk management will be a natural evolutionary step. As in other commodity markets, a centralized electricity exchange that provides a trading forum and price discovery for standardized forward transactions can facilitate this evolution.

Most restructured markets around the world to date operate around the concept of some form of a power exchange. In many countries the power exchange is a mandatory pool where all supply and demand must be bid into the pool. Only in Norway and California do the market designs accommodate bilateral or multilateral trading as an alternative to bidding through the pool.[3] In order to understand the role of exchanges in electricity markets it is useful to explore the various market models that have been evaluated. In California, shortly after the California Public Utility Commission's 1994 ruling that provided a blueprint on how the state should restructure (also known as the "blue book"), an extensive debate took place as to what role, if any, an exchange should play in California's restructured marketplace.

The Great Debate in California

In California the concept of a mandatory "pool" model similar to that in the U.K. was met with significant resistance. California regulators initiated an open and public process to design the restructuring of the retail electricity market. A great debate ensued, particularly because many of the participants were successful incumbents from the recently deregulated gas markets where a less centralized approach was used. This debate centered around criticisms of a "PoolCo" or centralized pool and advancement of a less structured approach using bilateral contracts to effect retail competition in California.

Critics of PoolCo were adamant that a centralized, mandatory pool administered by the ISO was monopolistic by nature and inconsistent with free markets.[4] They were concerned that because the pool would be administered by the ISO, the commercial function of the market would suffer at the expense of the ISO's mandate to assure system "reliability." They felt an ISO as a regulated entity would have little incentive to run an efficient market and that the market itself would be limited to a single, uncustomized spot price. This would eliminate price signals around the variety that exists in the physical electricity product. Other criticisms were noted regarding the complicated and costly systems that would be required, as well as the method proposed to determine the market-clearing price—a single-price auction—resulting in merit order dispatch of generation.

However, there were plenty of criticisms lodged against the "Bilateral Market Model" as well. Most notably, there didn't seem to be a concise vision as to how the bilateral model would work.[5] Most arguments for the bilateral model were couched as criticisms of the PoolCo model. It was clear that the bilateral model was not a centralized market. It would operate on the premise that wholesale access could be provided to retail markets, perhaps under existing rules that had been enacted by the FERC to provide more open access in wholesale electric markets. Presumably a competitive supplier could negotiate a bilateral contract with a retail customer and use open access tariffs filed with FERC to deliver the electricity.

This model was criticized not only for its vagueness, but because it would thrive on chaos and result in "aggressive rent seeking behavior"[6] that would make markets less efficient. It was criticized as being inconsistent with the physical market realities. Because electricity cannot be directly delivered from a particular seller to a particular buyer, it was argued that it would require expensive metering to measure congestion management, ancillary services, load following and other control area services provided by parties outside of the bilateral contract. Further, it was argued that the bilateral model would exploit consumers because there would be no clearly visible market price and unchecked discriminatory pricing practices would result. The fact that large players and incumbents would have the advantage because of their access to information would limit, rather than expand, entry by new players.

The California Model: The Best of Both Worlds
Resolution of restructuring was stalled by the great debate. The investor-owned utilities (IOUs) in California were divided, with those in the south favoring PoolCo and those in the north favoring the bilateral model. The California Public Utilities Commission (CPUC) had ruled to advance the PoolCo model, but it was on a split 3–2 vote. After debating the pros and cons of the PoolCo versus bilateral market for over a year, and because there was much resistance, particularly by organized industrial customers, a negotiated Memorandum of Understanding (MOU) was reached between utilities and consumer representatives in December 1995.

The MOU resulted in a CPUC compromise order that would create a PX separate from the ISO and allow for the coexistence of a bilateral marketplace. It required that in order to separate generation from distribution functions, the regulated IOUs would be required to buy and sell their retail energy through the PX during a four-year transition period. This would remove effectively any possibility that the utilities, through their bidding processes, could exercise undue market power and influence the market-clearing price calculated by the PX. Finally, the compromise order and resulting legislation (AB1890) in September 1996 set the date for full and open access by *all* retail markets to be January 1, 1998.

Unique New Channel Structure in California
Thus, a new model was conceived in California that is not PoolCo and not the bilateral model, but instead a unique channel structure whereby "scheduling coordinators" or "SCs" (of which the PX is one) bring supply and demand to-

gether in balanced schedules which are submitted to the ISO for delivery. This means that marketers have great latitude to manage both sides of an energy trade—that is both supply and demand—rather than being a price taker in a centralized pool for energy. And, the PX would be available as a spot market that could be used by all SCs to balance their schedules. In this way, the variety in the physical electricity product can be fully captured by SCs as they develop their supply portfolio, and this portfolio can also be managed against the particular demand characteristics of their customers.

Some hypothesize that this structure, unlike most others around the world, will accelerate market development and the creation of energy efficiency measures and other services for retail customers. At the same time, it means that there will be a deep and liquid spot market, a market-clearing price determined by the PX that serves as a benchmark for consumers, and a mechanism to determine stranded asset recovery by utilities. Most competitive energy providers in California already are indexing their prices to the price "at the PX." Further, because of its initial depth (IOUs are required to bid all load and resources through the PX during the transition period), the PX can cost-effectively meet needs for load following and provide a mechanism for all SCs to balance their schedules at a representative market-clearing price.

The Institutional Role of the California Power Exchange in the Competitive Retail Market

In California, the PX was created by state law to provide "an efficient competitive auction for electrical energy, open on a nondiscriminatory basis to all buyers and suppliers." It is a nonprofit public benefit corporation governed by a diverse group of stakeholders. Its structure and legislated design does not allow for private ownership interests. It must be available to provide services to all qualifying market participants. It was designed by law to be an integral part of deregulation in the State of California with the express purpose of reducing electricity costs to California consumers. Its role is to provide a visible and credible market-clearing price and disseminate market information, implement and adapt trading rules and practices, monitor the market for abuse of market power and schedule physical delivery of power, as required by market participants.

Unlike models where the spot market is embodied within the structure of the ISO, the PX's main focus is the commercial trading of energy. It's core functions include bidding via an electronic data interchange, scheduling energy delivery through the ISO, settling and clearing the market, and billing (and paying) participants for energy transactions and services provided. The PX administers a credit policy and cash settlement so that participant risks are diminished. It performs a market monitoring function that ensures fair trading. It provides training and information to participants to facilitate their success in the new market. And, it provides the price discovery that is unparalleled by any other electricity trading or information forum.

The PX thus provides a solid and reliable underlying market for development of options, futures and other financial instruments. It does this because of its visible and credible price, and by maintaining a deep and liquid market.

Functions of Power Exchanges

Centralized Trading

As in other commodity markets, a power exchange that has a deep and liquid market (many participants and significant supplies) reduces the search costs of its participants. They do not have to search for a counterparty, and they can be assured of receiving a representative market price for their trade. In hourly electricity markets, this can be a very valuable service. In California, each scheduling coordinator must submit to the ISO a balanced schedule, where supply matches demand for each and every hour. Since the PX has depth and liquidity (it represents 90 percent of the transactions in the California market with participants from all over the western U.S.) the PX provides an easy mechanism for all SC's to balance hourly schedules at a representative market price for energy.

Reliable Indicative Price Discovery

Historically, in electricity markets, price discovery has been provided by market survey. However, market surveys cannot provide the same information that is available from a power exchange for several reasons. A power exchange is a free market where the many factors that influence supply and demand of electricity converge through an auction or other bidding process to establish a price in a specific forward physical market for electricity. Prices are based on actual volumes traded of standardized trading units. Market surveys, on the other hand, capture only information *volunteered* by market participants. Because trading can take so many forms (firm, nonfirm, weekly, monthly, block, and so forth), it is impossible to capture standardized information through price surveys.

Clearing and Settlement

The role of a power exchange to provide a neutral and credible clearing function is central to a market operating consistently and within the context of established rules that all participants understand. As a neutral player, an exchange will clear the market and settle trades so that no market participant gains unfairly at the expense of others. Since the exchange handles credit clearances and manages credit risks by a specified set of rules, all participants are protected from default risks that they might otherwise bear without such a clearing function.

Counterparty Risk Management

Markets that open without a power exchange function may risk liquidity in spot markets, but even more importantly, counterparty risks could cause significant dislocations. Significant new risks arise because of hidden factors such as uncertainty around the underlying capability of a marketer to deliver.

Marketers may buy and sell the same energy over and over, often making future commitments as to price and/or quantity variables. These practices increase trading risk dramatically and the resultant "value-at-risk" is not disclosed as a liability on a firm's financial statements. Resultant counterparty "credit risk" is difficult to measure or understand. In the absence of a pool or exchange, if a large player fails, like a house of cards, their counterparties may fail and major price dislocations can result.

In Norway, for instance, when electric markets were first opened, there was no energy exchange or pool. Restructuring created more market volatility and many bilateral suppliers and marketers went out of business or otherwise defaulted, which caused increasing price volatility and instability. As a result, both buyers and sellers faced significant counterparty risk of failed delivery or failed payment. Nord Pool was created to create a pool for multiple buyers and sellers, thus establishing the pool as the counterparty and significantly reducing born risks for its participants.

Similarly, the price spikes in electric wholesale markets in the Midwest during the summer of 1998 demonstrated the volatility of bilateral markets reacting to failure of relatively small players. Power exchanges, through their credit management practices, also reduce credit risk that may otherwise be unknown or ignored. This is because the exchange manages credit judiciously so that "value-at-risk" is collateralized. Also, credit risk is reduced by the vary nature of the pool acting as the counterparty. This credit risk management function and counterparty risk mitigation provided by a pool accommodates new or smaller market participants who might not otherwise financially survive a single counterparty default in a bilateral market.

PRICE RISK MANAGEMENT IN RESTRUCTURED ELECTRICITY MARKETS

Worldwide Developments

Generally, in countries where the model for restructuring has been premised on the creation of a centralized power exchange, all electricity is supplied and purchased on a spot basis. As a result of uncertainty and price volatility associated with a spot market, various financial risk management tools or derivative markets have developed. In some cases, the power pool or exchange itself developed them. In other cases, they were developed by securities exchanges or traded "over-the-counter" between buyers and sellers of electricity and other financial intermediaries. In all cases, they were derived from, and are based on, the price discovery provided by optional or centralized power exchanges.

Early on, in the U.K., buyers and sellers wanted to lock in a price and thus "hedge" their price risk over longer periods of time or to mitigate price volatility. Contracts for differences (CFDs) were the first form of price risk management that arose in the U.K.

CFDs first require buyers to pay the spot price from the pool. Then any difference between the contract price and the spot price is paid to the seller if the contract price is greater than the spot price. Conversely, if the contract price is less than the spot price, the seller reimburses the buyer for the difference. These contracts are generally highly customized to meet whatever risks buyers and sellers agree to take on. They can allocate and separate different risks to buyers and sellers. While these contracts are not standardized and they trade bilaterally rather than through organized exchanges, they do effectively hedge risk in the physical market for electricity. And, counterparty risk in these bilateral transactions is limited to the hedging risk rather than the entire energy contract. In addition, and more recently, we see trading of futures, options, swaps, and other instruments in open electricity markets.

Futures contracts allow buyers and sellers to hedge price risk independent of physical delivery. That is to say that they can enter a futures contract to buy or sell a specified quantity at a specified delivery point over some specified future delivery period (usually a month ahead). These contracts are "marked-to-market," that is to say that they are settled day-to-day based on changes in projected electricity prices during the delivery period. They can be bought and sold at any time, thus allowing participants to "trade out" before a contract goes to physical delivery. To the extent that there are many buyers and sellers in these markets, including financial players who provide liquidity to the market, futures contracts can be an effective means for buyers and sellers to hedge price risk.

Nord Pool offers standardized futures contracts that are struck against the pool price. Standardized contracts that do not require an underlying physical trade encourage the participation of financial intermediaries that add liquidity to the market. Nord Pool's futures trading has increased steadily and, in 1996, surpassed sales in the underlying spot market. In 1998 futures were trading at a multiple of about nine times of power traded in the Nord Pool physical market.

Swaps, sometimes characterized as "fixed for floating swaps," are similar to CFDs except that there is a financial intermediary between buyer and seller that assumes the price risk for one or both parties. Swaps also are traded over-the-counter or bilaterally.

Over-the-counter trading of electricity swaps, CFDs, futures, and options has developed in many countries, including Scandinavia, Australia, and New Zealand. Over-the-counter instruments offer the advantages of greater customization without margin calls, but do not provide the same price transparency, anonymity, and reduced counterparty risk offered by exchange-traded products.

In the U.S., the electricity futures contracts developed by NYMEX, the Chicago Board of Trade, and the Minneapolis Grain Exchange are standardized around specified quantities for a month-ahead delivery period at specified delivery points. So far, electric futures trading has not been significant in the U.S. As retail markets open, the success of these contracts will depend largely upon the depth, liquidity, and credible price discovery at the specified delivery points. That is where a power exchange can make a difference.

How Power Exchanges Can Facilitate the Development of Risk Management Tools

A power exchange's ability to maintain and grow its depth and liquidity, and create reliable and representative market-clearing prices, is the foundation for the development of new financial instruments around standardized and evolving trading products. In some cases, such as Nord Pool, the power exchange itself expanded its existing clearing and settlement functions from physical to derivative markets by creating and trading futures and options. Other power exchanges exist to provide information services (including price discovery and relevant indices) that accommodate and supplement the needs of market participants and financial traders to understand, evaluate and create new standardized trading products, futures, options and other instruments to manage price risk.

Power exchanges have the databases that can be used to evaluate and engineer various financial instruments to manage around the many risks in electricity markets. There are an enormous amount of risk variables between electricity consumption and production functions, quite distinct from any other commodity market. These risks involve such factors as weather, transmission congestion, forced and planned generation outages, transmission outages, and underlying fuel costs, among other factors. Most businesses are interested in hedging those risks not endemic to their business. Power exchanges not only can provide a solid underlying market around which financial instruments are traded, but they have the data and information to provide basis differentials and spreads.

One critical role of any exchange is to take an active role in identifying participant needs and shaping products and services around those needs. Power exchanges that deliver the physical product of electricity are in a unique position to understand and enhance their participants' trading practices. This information has great value in the creation of financial products to hedge or manage electricity price risk.

The development of futures and options that can be traded easily will enhance the efficiency of the energy market relative to a world of long-term fixed bilateral contracts. As electricity markets open to competition, both buyers and sellers will face dramatically changing individual circumstances. Rather than having to go into complex negotiations with the other party, who may not have corollary interests, commonly traded futures or options contracts provide buyers and sellers significant flexibility at a lower cost, particularly in volatile new markets.

As electric markets evolve, new standardized underlying physical markets may also evolve that can be structured by a power exchange. Most power exchanges today provide a spot market for electricity for either hourly or half-hourly periods, usually for the day ahead. However, over time, different forward markets may also be desirable such as month-ahead, week-ahead, day-week, and night-week contracts. These standardized trading products will be based on the needs of market participants.

SUMMARY

Power exchanges have played an important and vital role in restructuring retail electricity markets worldwide. Some ask, what would happen in restructured electric markets without an institutionally established power exchange? The answer to that question depends upon how the market is structured, the level of credible price discovery available to consumers, the availability of cost-effective energy for load-shaping needs, and the nature of regulatory oversight to monitor market power. At this stage, it's impossible to create a "one size fits all" market structure. Each state or country will develop the market model that is best suited to their socioeconomic and political needs.

Some believe that the power exchange functions should be provided by an ISO as one entity. The counter argument is that a combined entity would be concerned principally with operating a reliable system, trading off efficient market operations when necessary to ensure system reliability. To the extent that an ISO's central function is ensuring system reliability and it is regulated as such, its primary focus would not be to expand and enhance services to facilitate the evolution of robust competition in the commercial energy marketplace. A separate power exchange model focuses the exchange on creating and operating a reliable commercial market for energy.

In the absence of an institutionally established exchange either within an ISO or independent of an ISO, the market would operate on the basis of bilateral arrangements. Some believe that a fundamentally bilateral marketplace would be chaotic and easily manipulated by large players. There would not be adequate price discovery, particularly valuable to consumers and new entrants. Also, without an exchange there may be limited flexibility for variable sizes of trades during different time periods. The hourly energy market may thus become more unstable with greater opportunities for economic rents.

Some believe that if a power exchange function facilitates markets, it can and will be created by the marketplace. An alternative view is that the incumbent bilateral marketplace is so powerful that a commercial exchange never would be able to gather enough depth and liquidity to take hold. In this alternative world, it is feared that an oligopolistic bilateral market thrives at the expense of consumers and smaller producers.

This chapter presents the case that the price discovery and credible physical market provided by power exchanges offer more flexibility to consumers. Because a power exchange provides the environment that facilitates the development of standardized risk hedging mechanisms through futures, options and other derivatives, consumers have considerably more flexibility in meeting their changing needs without having to rely on a single counterparty. They easily can change their physical position and trade in and out of futures contracts to manage price risk while being assured a "market price" with no hidden margins.

The most important role a power exchange can play in newly restructured markets is to be a neutral, independent market facilitator, focused on meeting the market's evolving needs. In this neutral capacity, a power exchange can provide credible price discovery to market participants, consumers, and financial markets. It will be a trading forum for all market participants, thereby equalizing the advantages that large incumbents already have through existing trading relationships. All market participants, then, will have a place to readily buy and sell power at market prices. Because a power exchange will have many participants and liquid markets, and since it manages credit risk and performs the settlement function, it can significantly add value by reducing trading risk, stabilizing prices, simplifying trading, and commoditizing electricity.

NOTES

1. For additional information, see "History of the Commodity Markets, at http://www.futuresbroker.com/history.htm.

2. New futures contracts also are being introduced by the Chicago Board of Trade and the Minneapolis Grain Exchange.

3. The U.K. reportedly is reevaluating its market structure, which may result in a decentralized market structure similar to California and Nord Pool.

4. Presentation by Robert A. Levin, "Centralized Pools: Description, Critique, and Recommendations; A Contribution to the Dialogue," April 1995.

5. Paul L. Joskow, "Restructuring to Promote Competition in Electricity: In General and Regarding the PoolCo vs. Bilateral Contracts Debate," January 1996.

6. Ibid.

CHAPTER 23

Creating Equal Opportunities for All Competitive Electric Service Providers

*Kenneth W. Costello**

Under customer choice, retail customers will have unprecedented opportunities to select among competitive energy service providers (ESPs) for electric energy, standby service, and so forth. These providers, who may have to satisfy certain certification requirements as a condition for entry, likely will include generation companies, marketers, brokers, load aggregators, and retail energy service companies. Some of them may have an affiliation with the local utility —whether it be the telephone company or the cable company— who will retain the legal right to be the monopoly provider of local delivery service.

The concern of state public utility commissions (PUCs), nonaffiliate providers, and some consumer groups is that the local utility or its parent company may leverage the monopoly power it enjoys in the delivery components of its operations to gain undue advantages in its lines of business that are susceptible to competitive forces. The local utility then may have the ability to foreclose or impede the development of competition in markets for unbundled electric services. In the jargon of economists, this situation is a vertical-control problem.[1]

---

*The views and opinions of the author do not necessarily reflect those of The National Regulatory Research Institute, the National Association of Regulatory Utility Commissioners, or their contributors. I thank Jaison Abel, Larry Blank, John Hilke, Robert Michaels, Wayne Olson, and Edwin Rosenberg for helpful comments on an earlier draft. The author, of course, is responsible for any remaining errors or omissions.

The leverage theory is premised on the idea that a firm with power in one market can exploit that power to attain or preserve power in a second market. This behavior can harm competitors and create monopoly power in a market that is naturally competitive. Many economists would argue that most attempts to leverage would be unsuccessful in expanding market power. The reason for this is that most markets are workably competitive and regulation is able to prevent monopoly profits. Certain situations, such as when a firm operates under rate-of-return regulation in one market, potentially can cause leveraging that produces an anticompetitive outcome in an unregulated market. Specifically, a regulated firm can evade a regulatory constraint by operating in an unregulated market. The idea is related to, although different from, traditional leverage theory.[2]

In order for retail customers to realize the greatest possible benefits from choice, regulatory rules must be designed and enforced to allow all service providers the same opportunities to compete via nondiscriminatory comparable open access to essential distribution facilities and support services. "Essential" can be defined as facilities whose cost of duplication by another entity would be uneconomically prohibitive. This implies that retail customers can obtain access to service providers who, under fair rules, can serve them most attractively. In the absence of this condition, retail customers likely will be worse off. "Fair rules," as discussed later, require that all firms play by the same rules, thereby producing winners on the basis of their performance in serving retail customers. This means that inputs such as advertising and capital requirements should be regarded as barriers to entry only if all firms are unable to gain access to them.

The primary objective of this chapter is to propose a regulatory strategy aimed at mitigating the vertical control problem in retail electricity markets. The chapter first will discuss various definitions of the concept of equal opportunities. It then will identify the underlying sources of nonequal opportunities among retail service providers, some of which are evident in today's electric industry market. Next, the chapter will outline the various positions of interest groups and others with regard to the requisite conditions for an equal-opportunity setting. These positions reflect varying perspectives on the sources and nature of the problem. A discussion of regulatory options along with recommendations for a comprehensive regulatory strategy concludes this chapter.

THE CONCEPT OF "EQUAL OPPORTUNITIES"

Efficient competition requires that all incumbent and prospective firms be given equal opportunities to compete for customers. Equal opportunities have different connotations among the different interest groups, as well as among economists. A local distribution company (LDC), for example, may interpret standard-of-conduct rules as overly restrictive, placing its affiliate at a disadvantage, while nonaffiliates may regard these rules as necessary to avoid what they perceive as inherent favoritism toward the LDC affiliate. Generally, however, all groups seem to agree that standards of conduct are required. What is considered fair by some may be viewed as unfair by others. In the context of competitive sports, fair rules are supposed to show no partiality toward any team or individual. They should result in outcomes that depend solely on the skills of the participants—that is, the best should always win. In the marketplace, fair rules should produce winners on the basis of their ability to satisfy consumers, nothing else. This means that new entrants should have the same opportunities as incumbents to succeed while, at the same time, incumbents are not unduly restricted in their market activities. Of course, most individuals may define fair rules in terms of their self-interests. Social policy should ignore this perception, as it would frequently cause harm to the aggregate population.

The realization of efficient competition requires the absence of what economists call "barriers to entry." Two schools of thought occupy the basic perceptions of barriers to entry.[3] The first, attributed to Joe Bain, includes as a barrier anything that enables an incumbent firm to charge monopoly prices without attracting new entry.[4] Obstacles to entry such as economies of scale, product differentiation, and absolute cost advantages are considered barriers that stifle competitive forces' ability to reduce the incumbent's prices and profits. The second school of thought, led by George Stigler, defines a barrier to entry as costs that must be incurred by an entrant that are not being, or have never been, incurred by incumbents.[5] This definition, compared to the first, imposes a higher threshold on what constitutes a legitimate barrier to entry. As long as entrants have access to the same technological and market opportunities that incumbents do, no barriers to entry exist. Sunk costs and advertising, for example, would not be regarded as barriers. The basic reason for this is that an incumbent had to incur these costs in the past to gain its current market status.

The definition of barriers to entry invariably drives the specification of the conditions required for an equal-opportunity marketplace.[6] Regardless of definition, when barriers to entry exist, an incumbent would have an "unfair" advantage over prospective entrants. An example would be an insurance requirement, such as a bond, imposed onerously upon marketers. When excessive restrictions are imposed on an incumbent, the outcome also is unfair, even though this problem falls outside the conventional perception of barriers to entry, which traditionally has been viewed as entry deterrence by incumbent firms. Prohibiting the use of a utility's logo or name by an affiliate arguably may be such a restriction. As a point of principle, entry restrictions and licensing requirements should apply equally to both a utility affiliate and other new entrants.

The Stiglerian definition of barriers to entry seems more appropriate in terms of developing a regulatory strategy to create equal opportunities among service providers. It also is more in line with the condition that for an entry barrier to be considered a problem requiring a governmental remedy, consumer well-being must decline.[7] Under Bain's definition, almost anything that delays entry can be interpreted as a barrier. One then can assign any cost differential between an incumbent and potential entrants as a barrier (e.g., the learning curve)—which is certainly a loose definition that can lead to mischief in terms of setting rules and policy.[8] Consequently, Bain's definition would tend to favor new entrants, even when economic efficiency is diminished. Overall, Bain's broad definition of barrier to entry fails to distinguish between good and bad barriers. Good barriers may result from those advantages enjoyed by an incumbent because of its efficiencies.

Under the Stiglerian definition of barriers to entry, a utility affiliate, or marketers, would have an unfair advantage when it does not have to incur costs that nonaffiliate competitors incur. Genuine examples of barriers to entry would be discriminatory transmission access fees favoring an affiliate, shifting cost to a regulated affiliate, and disclosure of customer information only to an affiliate.[9] If these actions are successful, less than the ideal number of nonaffiliated firms will compete in the market. This will result in inefficiency and likely hurt consumers. One example of fair rules would be the golden rule variety, where the utility owning essential facilities treats its affiliate's competitors the same as it treats its affiliate with respect to prices and terms and conditions.

SOURCES OF THE PROBLEM

When customer choice is available, if the local utility has an unregulated marketing affiliate, the utility will be operating in a mixed competitive-monopoly environment. In this setting, the utility or its parent company may have both the ability and the incentive to leverage its position in the regulated monopoly market to gain an advantage in unregulated markets. For example, if regulatory rules are not completely effective in preventing all discriminatory behavior, the utility may establish local distribution access and pricing rules that discriminate against nonaffiliate marketers. The most serious outcome of this leveraging is the diminution of competition and economic losses to retail customers.

The pertinent problem here is one of vertical control, where regulated utilities will continue to have a monopoly franchise in local distribution. One current trend is the formation of marketing affiliates by both electric and natural gas utilities, or their parent companies. Their interest in forming marketing affiliates may be temporary. Some utilities or their parent companies may believe there is currently a window of opportunity for them to earn economic profits until the market develops more competitively. Because of consumer inertia[10] and other incumbent advantages, a utility affiliate may be able to gain high market share, temporarily. The marketing function may be integral to a utility's or its parent company's strategy in a more competitive marketplace and also may be critical for a utility to maintain its status as the only service provider with local experience.

One must start with the premise that marketing affiliates are established to increase the profits of the parent company. They can do so in various ways, some of which stem from anticompetitive behavior while others are compatible with improving consumer welfare. Anticompetitive behavior could result from the affiliate selling services to the regulated utility at excessively high prices or from the affiliate buying services from the utility at excessively low prices, and from aggressive pricing by the affiliate with the intent of driving out competitors, and thereby earning monopoly rents in the long run. (Of course, the definition of "excessive" varies among analysts and others, casting doubt on whether a specific transfer price is actually anticompetitive.) An example of increasing profits while improving consumer welfare would be the benefits derived from economies of scope, where an affiliate of a utility is able to provide a service at lower cost, or less cost than its competitors.

Cynics argue that marketing affiliates exist only to evade regulation or to function as a hedge for the utility against the ill effects of increased competition. Establishing marketing affiliates to evade regulation in general or specific regulatory constraints can produce either good or bad outcomes: good in the sense that regulation prevented the utility from offering a variety of prices and services demanded by consumers, bad in the sense that the utility along with its marketing affiliate is able to engage in anticompetitive behavior that causes harm to both consumers and society as a whole. Supporters of marketing affiliates point to the benefits of integration economies and of an additional, potentially efficient entity in the marketplace. Integration economies would derive from the ability of the firm to use some inputs more intensively or productively and from the avoidance of market transaction costs.

State regulators have a long history of raising concerns over utility diversification but have implemented various mechanisms to address them, with varying degrees of success.[11] Particular attention has been given to the situation where an unregulated entity sells a service to an affiliated utility that owns and controls "essential facilities." This is the vertical-control problem alluded to earlier—a problem that in theory could lead to diminished customer choice, and consequently, inefficient competition in nonregulated markets.

The problem of vertical control causing discriminatory and anticompetitive behavior by the regulated utility can be largely attributable to two sources. The first is imperfect regulation. If regulators were able to detect all abuses, such as cost-shifting, inflated self-dealing prices, and disclosure of confidential customer information to an affiliate, this problem would not exist. A major problem for regulators is to identify and measure those costs that are commonly incurred to provide more than one kind of service. It is unreasonable to expect regulators to possess this information, which thereby allows the regulated firm the opportunity to engage in abuses hurting both competitors of utility affiliates and consumers.[12]

Rate-of-return (ROR) regulation also provides incentives, or at least no disincentives, for abuses. Under ROR regulation, when the regulated firm reported higher costs its prices would rise, assuming regulatory approval of these costs. Of course, the fact that the regulated firm can pass through misreported costs implies that regulators have imperfect information. Yet it is fundamentally the cost-plus nature of ROR regulation that makes it particularly susceptible to cost-shifting and other abuses.[13]

The second cause of a vertical-control problem is imperfect competition in retail markets. Given that local distribution service will remain a monopoly service for the foreseeable future, consumers cannot readily respond to a higher price for this service by cutting back demand. Consequently, an artificially high price induced by regulation can be charged more easily to final consumers. A parent company would tend to shift costs to markets that are the least competitive and where the company has the least incentive to minimize its costs—in other words, the local distribution market under ROR regulation.

An additional problem arises from the tendency of policymakers to overprotect new entrants or utility-unaffiliated firms, under the aegis of consumer protection. The outcome may turn out to be harmful to consumers since handicapping the incumbent has real costs and only hypothetical long-run benefits. Such actions would give new entrants an advantage over the utility affiliate until sustainable competition develops in the local market for the service. The rationale is that the benefits of assuring a more dynamic and competitive market in the long run would more than offset any efficiency losses that may ensue in the short run. The argument presumes that the incumbent holds an inherent market advantage that justifies handicapping it until new entrants have a fair chance to compete.

Favoritism toward new entrants is analogous to the "infant-industry" argument that has been applied to policymaking in international trade. According to this argument, during its infancy a domestic industry may require protection against foreign competition. The underlying premise is that the new domestic entrant in a developing country, for example, would face high costs during the initial period, but it could compete in the long run. It is argued that without immediate protection, the new entrant would find it extremely difficult to compete. Consequently, in the absence of the protection, the long-run benefits that the country would otherwise realize from more competition would be foregone.

Critics of the infant-industry argument offer persuasive counterpoints. First, once protection occurs it is difficult to terminate.[14] Those who benefit would strongly oppose any change in policy; they would expend significant resources, in the form of rent-maintenance costs, to argue that protection should continue because the industry has not "grown up." Second, policy-makers would find it difficult to know the appropriate time to end the protection. Opponents of termination would be expected to argue that they are still infant, even after several years, and therefore continued assistance would be required to stay in business. Third, policymakers lack the necessary information to quantify or estimate the size of the potential benefits from protection. How much more competitive would retail markets be, for example, with short-run protection to new entrants? What would be the benefits to retail consumers? Fourth, protection of new entrants represents an inferior way to cope with the problem of incumbents holding a market advantage if indeed they do. If genuine barriers to entry exist, they should be identified and remedied.

As a policy, protection for new entrants is as likely to inflict losses on society as it is to benefit society. Extended protection may keep inefficient firms in the market at the expense of a more efficient incumbent. Finally, the presumption is that new entrants cannot compete with an incumbent. In the context of this chapter, this translates into new entrants being unable to compete with a utility affiliate. If barriers to entry are eliminated or minimized, no reason exists for concern by policymakers. The dynamics of markets allow new entrants to compete successfully when they are more efficient, innovative, and customer-responsive than incumbents. This phenomenon has been observed across a broad range of industries with naturally competitive characteristics; computer and financial services are conspicuous examples. Expect this to hold for retail electricity markets also, as long as genuine barriers to entry are eliminated.

POSITIONS OF PARTIES

The debate over how to create equal opportunities for all service providers centers on the parties' perception of barriers to entry. All parties have an incentive to have a "home field advantage," whereas the job of regulators is to eliminate any tilt that favors one category of providers.

Not surprisingly, marketing companies unaffiliated with the local utility take the position that utility affiliates have a definite competitive advantage in four areas: (1) cost-shifting from an unregulated utility affiliate to the regulated utility, (2) transfer of the incumbent utility's name to an affiliate, (3) vital customer and other information distributed only to the utility affiliate, and (4) an education bias whereby consumers believe that a utility affiliate's service is more reliable. Nonaffiliated companies generally support mandatory structural separation of a utility's regulated activities from its nonregulated ones, strict standards of conduct, prohibition of transferring a utility's logo to an affiliate, and nondiscriminatory information access. Their sentiment is that explicit and well-defined regulatory or legislative rules are required to assure that retail customers reap the benefits of new competitive choices from industry restructuring activities. Independent marketers argue that if the local utility gets away with bestowing preferential treatment on its marketing affiliate, such as providing the affiliate with confidential commercially sensitive information, competition would be seriously stifled. This surely would result in less consumer benefit. Overall, independent marketers support strict command-and-control mechanisms to prevent favoritism by the local utility of its affiliate, thereby ensuring their perception of fair competition.

The electric utility industry favors limited restrictions on a utility affiliate. As the president of the Edison Electric Institute (EEI) said:

> Market rules should provide a basis for equitable competition. They should ensure, for example, that regulated parts of the business do not subsidize the unregulated, competitive parts. . . [W]e quickly spotted a disturbing trend in some proposals to go much further and offer rules that would actually handicap traditional utilities in competing. This would be to the detriment of the marketplace and consumers.[15]

An EEI executive sent a letter to the National Association of Regulatory Utility Commissioners (NARUC) that identified three major contentious issues: (1) structural separation, (2) the sharing of information with an affiliate, and (3) joint marketing or advertising.[16] It is no surprise that EEI opposes the requirement of structural separation and the prohibition against information sharing, joint marketing, and advertising.

In a critique of a paper advanced in 1997 by the NARUC Subcommittee on Accounts and the Staff Subcommittee on Strategic Issues, EEI reiterated its concern for regulatory rules that handicap utilities and their affiliates and that show undue favoritism toward competitors. As conveyed by some utility personnel, state PUCs are making rules that reflect a "make the world safe for Enron" mentality.

As a whole, consumer groups support strict regulation on the interaction between a utility and its affiliate. One group, the Electricity Consumers Resource Council (ELCON), has argued that:

> [R]egulatory oversight is required for diversified electric utilities that continue to own or control essential facilities and have captive customers. . . .ELCON is adamantly opposed to future diversifications by vertically-integrated utilities unless and until: (1) such diversification is accompanied by adequate regulatory protection, or (2) consumers can protect themselves by having real customer choice.[17]

Over the last few years, several state PUCs have ruled on the appropriate interrelationships between a utility and its marketing affiliate. Until recently, more of the rulings involved natural gas distributors than electric utilities. Typically, the rulings include a standard of conduct with the following components: (1) prohibition of utility preferential treatment of an affiliate, (2) structural separation of operational functions, (3) complaint procedures, (4) periodic reporting to identify problems, and (5) monetary penalties for violations.[18]

PUCs generally have allowed self-dealing transactions between a utility and its affiliate subject to strict rules that guard against possible abuse. Structural separation, by placing a "Chinese wall" between regulated and nonregulated activities, allows PUCs to better track the actual costs incurred by the regulated segment.

The much-talked-about December 1997 ruling by the California Public Utilities Commission requires nondiscriminatory access to local distribution information disclosure, allows the use of the utility's logo by an affiliate, and specifies separation standards and the degree to which a utility's nonregulated activities may be conducted by an affiliate.[19] The commission mandates that utilities and their affiliates be structured separately and maintain separate books and records. Utilities and their affiliates, for example, may not share office space or information systems, or employees, although in some instances utilities may transfer employees to affiliates. They may share office supplies, and support, payroll, and tax services.

REGULATORY OPTIONS

As argued in the previous discussion, an unfettered posture toward utility-affiliate interactions will fall short of achieving equal opportunities for the different service providers in retail electricity markets. ROR regulation engenders serious incentive problems and requires regulators to have certain information that is unlikely to be forthcoming. Specifically, it does not discourage such practices as cost-shifting and self-dealing abuses.

Regulatory rules and other prescriptions also will be required to prevent certain market abuses. For example, the utility would have the incentive to favor its affiliate by imposing barriers on competitors, independent of the operative price-regulation paradigm. (Even if the affiliate is structurally separated from the regulated utility, this problem exists since profits from the affiliate would accrue to the same parent company.) The expected outcome would be retail consumers being deprived of the most efficient and consumer-responsive service providers.

One task of state PUCs, therefore, should be to do something to assure equal opportunities. What specific actions commissions take as the electric power industry undergoes restructuring will have a major effect on the allocation, as well as the size, of benefits from a more competitive industry.

Regulatory alternatives can be separated into two categories, command-and-control and market-based. The first comes in the form of regulatory mandates that attempt to prohibit the utility from undertaking objectionable practices.

These mandates include standard-of-conduct or safe-harbor rules, prohibition of affiliated transactions and certain intracorporate information flows, and forced structural separation or divestiture. These approaches all represent what some industry observers would classify as heavy-handed regulation.[20] They encompass both behavioral rules and structural safeguards. Structural separation and divestiture would fall under the latter category. Behavioral rules require monitoring, policing, and enforcement, which could be quite difficult in view of the incentives utilities face under ROR regulation.[21]

Although structural separation helps prevent cost-shifting, it has the problem of losing any economies of scope or coordination that otherwise would occur when a single entity provides both marketing and distribution services. Structural separation also does not eliminate the incentive for the utility to engage in certain abuses relating to self-dealing and preferential local distribution access and information disclosure. Consequently, behavioral, or standard-of-conduct rules, would need to accompany a structural-separation mandate. Taking everything into account, the social attractiveness of structural separation hinges on (1) the ability and incentive of a regulated utility to pursue anticompetitive practices in unregulated markets, and (2) the degree of economies of scope between the regulated and unregulated lines of business.

Divestiture, which most state PUCs presently do not have the authority to order, is considered by some analysts to be the only sure way to give all service providers equal opportunities to sell in retail electricity markets.[22] It would eliminate the need for regulatory oversight. On the other hand, divestiture would eliminate any economies from vertical integration. It may create costs, for example, in the form of incomplete contracts and lost coordination. Divestiture is regarded by most antitrust experts as a last resort measure that, as the courts have articulated, may be justifiable when affiliation between two entities has harmed consumers, has violated antitrust principles, and has injured the public interest. Some electric utilities likely will voluntarily divest their assets either as a business strategy or as a way to avoid grief from regulators.[23] Indeed, some utilities already have divested their generation assets.

The market-based approach relies explicitly on incentives and competitive forces to prevent market abuses. It is premised on the notion that behavioral rules are costly, and perhaps impossible, to enforce as long as the utility has the incentive and the ability to evade them. As mentioned above, it is extremely difficult for regulators to detect violations and evaluate utility actions.

Market-based approaches include price-cap regulation and the promotion of competition and service unbundling in retail markets. Price caps, relative to ROR regulation, would be less information-intensive for regulators and would dull the incentive for certain abusive practices. By severing prices from a utility's reported costs, the utility or its parent company would have little or no incentive to shift costs and transfer excessive prices from self-dealing to the regulated utility. For example, under a pure price-cap mechanism, a utility simply could not pass through an inflated price for electricity purchased from its marketing affiliate. The underlying reason for this is that price changes do not necessarily correlate with the utility's change in reported cost.[24]

Expanding the scope and intensity of competition also would engender more equal opportunities among retail service providers. With customer choice, retail services would become unbundled, with some susceptible to competitive services. Prices for these services would be more transparent to consumers. Consequently, price inflation by way of affiliate abuse would erode the profits of utilities or their parent companies from competitive services.

Regulatory options may involve combining the command-and-control and market-based approaches to achieve an equal-opportunity market environment. The next section proposes a strategy that incorporates both approaches.

A PROPOSED REGULATORY STRATEGY

Any strategy should have as its objective the maximization of consumer welfare. This requires the condition that service providers be given equal opportunities in serving consumers. Equal opportunities, in turn, demand that no genuine barriers to entry be imposed upon any of the actual or prospective service providers. State PUCs and legislatures can best achieve this by eliminating government-induced barriers to entry and then making certain that market failures are not responsible for any remaining barriers. Because entry barriers are difficult to measure and quantify, their revelation requires a close look at the relevant market.

As argued earlier, genuine barriers are similar to the Stiglerian perception of barriers to entry. Accordingly, barriers would be limited to those costs that new entrants would have to incur but which never were incurred by an incumbent. As an example, customer information dispersed at zero cost from a utility to its affiliate but not disclosed to other marketers would constitute a genuine barrier to entry.

The controversial question of whether a utility should be able to transfer its logo to an affiliate is less clear.[25] One can argue that if the logo is transferred at zero cost to the affiliate, the affiliate would enjoy an unfair advantage relative to its competitors; that is, a Stiglerian barrier to entry exists. Competitors may have to expend large sums of money to advance their reputations to the level of the utility affiliate when it acquired the utility logo. On the other hand, if entities competing with the utility affiliate, such as Exxon or Enron, are allowed to use their own logo or that of an affiliate, the utility affiliate would be at a disadvantage if it could not use the utility's logo.[26] Consequently, it is not clear whether transfer of a utility's logo to an affiliate is a genuine barrier to entry. It seems, at the minimum, that if the affiliate compensates the utility for use of the logo, no barrier to entry would exist. (In theory, compensation should correspond to the market value of the logo to the affiliate.) Nonpayment to the utility, however, may violate this condition, depending in part on whether competitors are simply generic companies in the eyes of consumers or are companies having distinctive reputations and brand names themselves. Branding, by and in itself, falls outside the Stiglerian definition of a barrier to entry, since the firm with the brand has had to incur advertising, promotional, and other costs to acquire its current reputation.

The proposed hybrid regulatory strategy has four general components: (1) market-like incentives for a regulated utility, (2) limited behavioral rules that prohibit genuine (Stiglerian) barriers to entry, (3) regulatory threat or backstop of mandatory structural reorganization following evidence of anticompetitive practices, and (4) promotion of real competition in retail electricity markets.

As mentioned earlier, price caps could diminish greatly the incentive for cost shifting and abusive self-dealing. Arguably, price caps could also reduce the need for structural separation, whose major benefit would be to mitigate cost shifting.

While price caps can reduce certain distortions, they do not eliminate all motives for anticompetitive behavior. They do not address the incentive of a utility to discriminate against the competitors of its affiliate, for example. Consequently, some standard-of-conduct rules would need to be in place. Elements would include: (1) information disclosure requirements, (2) prohibition against unfair trade practices, (3) complaint procedures, (4) reporting requirements, and (5) penalties for violations.[27] These rules seek to eliminate genuine barriers to entry resulting in economic losses to consumers.

A backstop where regulators can resort to mandatory utility reorganization (e.g., structural separation or divestiture), if needed, makes up the third component of the strategy.[28] Evidence compiled after a period of operation may show that price-cap incentives, in conjunction with standard-of-conduct rules, fail to prevent discriminatory/abusive behavior. In that event, more drastic measures in the form of mandatory utility reorganization may become necessary. Reorganization may result in lost economies of integration, but they may be more than offset by the reduction in anticompetitive behavior.

The last component of the strategy, the promotion of "real competition" in retail markets—a term that implies that no individual firm possesses sufficient market power to engage in collusive and other abusive practices—is self-explanatory. Competition makes it more difficult or impossible for consumers to pay inflated prices for services. In a competitive environment, consumers could switch to another provider if discriminatory treatment raises prices charged by the utility affiliate. Services such as local distribution will remain regulated indefinitely, so competition by and in itself will not be able to avoid all anticompetitive possibilities.

CONCLUSION

A regulated utility's interactions with an unregulated affiliate should be a concern of regulators who hope to promote efficient competition and maximize benefits to retail electricity consumers. Under prevailing regulatory pricing practices and expected market structures over the next several years, electric utilities have both the ability and the incentive to engage in behavior contrary to the equal-opportunity criterion laid out in this chapter.

While state PUCs should take a proactive stance, designed to prevent anti-competitive behavior, they must avoid actions that give an unfair advantage to any group of service providers, including new entrants. This is a difficult task, in part requiring commissions to distinguish between bad and good (or important and unimportant) barriers to entry. Differences in the positions of the various interest groups, in a large way, reflect disagreement over the definition of socially harmful barriers to entry. As a basic principle in developing a strategy, regulators never should lose sight of the goal of antitrust enforcement to promote consumer welfare via competition, not to promote the interests or financial health of any particular category of competitors.

Command-and-control methods are the most popular at this time for addressing the problem of vertical control at the retail electricity level. This chapter has argued that more consideration should be given to market-based strategies, especially in nurturing competitive forces and in providing incentives for cost minimization by the regulated utility.

A preferred regulatory strategy would combine both the command-and-control and market-based options. Under this hybrid strategy, price caps for regulated service along with unbundling of retail services would be supplemented by standard-of-conduct rules limited to mitigating genuine barriers to entry, and, as a backstop, the threat of mandatory utility reorganization.

NOTES

1. In this chapter, vertical control means that an electric utility providing local distribution under regulated and monopolistic conditions can misuse its power in that market to attain market power for an affiliate selling electric energy and other competitive services.

2. See Timothy J. Brennan, "Why Regulated Firms Should Be Kept Out of Unregulated Markets: Understanding the Divestiture of AT&T," *The Antitrust Bulletin* 32 (Fall 1987): 741–93.

3. As one noted economist has expressed, "The discussion of barriers in economic litera-ture hardly reflects consensus . . . [The] differing definitions allow their authors to hold different opinions about specific sources of barriers." See Harold Demsetz, "Barriers to Entry," *American Economic Review* 72:1 (March 1982), 47. Demsetz criti-cizes the conventional definitions of a barrier to entry for focusing only on the differ-ential opportunities of incumbents and potential entrants. He uses the example of some legal barriers, such as taxi medallions, whose opportunity costs to incumbents and potential entrants are the same.

4. Joe S. Bain, *Barriers to New Competition* (Cambridge, MA: Harvard University Press, 1956).

5. George Stigler, *The Organization of Industry* (Homewood, IL: Richard D. Irwin, 1968). The phrase "or has never been" interprets Stigler's definition of a barrier to entry in terms of permitting a firm to sustain economic profits in the long run.

6. In most industries, several competitors exist and so entry conditions may not be es-sential to competition. In the electric power sector, however, where there was only one incumbent firm in a retail market, entry is indeed critical.

7. Christian C. von Weizsacker, "A Welfare Analysis of Barriers to Entry," *Bell Journal of Economics* 11 (Autumn 1980): 399–420. Similar to Bain, von Weizsacker defines a bar-rier to entry in terms of a particular outcome. A barrier can reduce consumer well-being by keeping out firms that can compete successfully on the basis of their relative efficiencies.

8. Using the game of golf as an analogy, under Bain's view of the world, new golfers on the PGA tour would be handicapped since they do not have the same experience as current golfers. Therefore, until they gain this experience, the other golfers will con-tinue to earn a disproportionate share of the prize money (unless the rookie is Tiger Woods). As a Bain-like policy, new golfers would be given a handicap, that is their scores would be reduced by a specified number of strokes, until they have gained the experience of the other golfers. Certainly, few people would favor such a practice; to the contrary, most people would consider it detrimental to professional golf and ap-propriate only for recreational golf. From a Stiglerian perspective, the only legitimate concerns are whether all golfers are playing by the same rules—e.g., all golfers re-ceive the same penalty when they hit their ball in the water, and are given equal op-portunities to practice and purchase the equipment required to play professional golf.

9. Cost-shifting is not always anticompetitive. Certainly, it has the effect of raising the prices of regulated services. Yet it could have no effect on the unregulated market—it simply may be a way for the parent company to increase its profits by cost manipula-tion, rather than predation or other strategies giving an affiliate an advantage over its competitors. See Jaison R. Abel, *An Economic Analysis of Marketing Affiliates in a Deregulated Electric Power Industry* (Columbus, OH: The National Regulatory Research Institute, 1998).

10. This inertia may reflect more than anything the transaction cost of switching to an unaffiliated service provider. Transaction costs may affect small consumers more, thereby reducing their willingness to switch providers.

11. See, for example, Robert E. Burns et al., *Regulating Electric Utilities with Subsidiaries* (Columbus, OH: The National Regulatory Research Institute, 1986); and Edwin A. Rosenberg et al., *Regional Telephone Holding Companies: Structures, Affiliate Transactions, and Regulatory Options* (Columbus, OH: The National Regulatory Research Institute, 1993).

12. In spite of accounting safeguards, it is presumed here that a utility still would be able to misallocate costs for regulated and unregulated services in an effort to maximize its profits.

13. So long as the utility has unexploited market power in the regulated sector (i.e., the price lies below the unregulated profit-maximizing price), it can earn excess profits in the unregulated market. For example, by paying an inflated price for services from an unregulated affiliate, the affiliate receives higher profits while the utility's profits remain unchanged, assuming dollar-for-dollar passthrough. Cost-shifting is another example where, assuming regulation prevents monopoly prices in regulated markets, the utility-affiliate aggregate profits would increase.

14. See, for example, Robert E. Baldwin, "The Case Against Infant-Industry Tariff Protection," *Journal of Political Economy* 77:3 (May/June 1969): 295–305; and Anne O. Krueger, "Trade Policy and Economic Development: How We Learn," *American Economic Review* 87:1 (March 1997): 1–22.

15. Thomas R. Kuhn, "The Electric Industry—Meeting the Competitive Challenge." Address to the New York Society of Security Analysts, Inc. New York, New York. February 11, 1998, 2–3. One can interpret the industry's position as advocating no regulatory restriction on a utility in participating in any business activity so long as the *necessary* standards of conduct are in place. Of course, it is expected that the industry's definition of "necessary" would be less stringent than that of independent marketers and other interest groups.

16. David K. Owens, Letter sent to Lou Ann Westerfield of the NARUC Subcommittee on Strategic Issues, May 15, 1997.

17. John A. Anderson, Letter sent to Lou Ann Westerfield of the NARUC Subcommittee on Strategic Issues, August 6, 1997.

18. See, Laura Murrell, "Workable LDC Affiliated Marketer Standards," *Natural Gas* (December 1996): 12–15.

19. See, *Opinion Adopting Standards of Conduct Governing Relationships Between Utilities and Their Affiliates*, Decision 97–12–088, 183 PUR4th 503, A. 97-04-012 (Cal. P.U.C.), December 16, 1997; and "California Affiliate Rules Back Away From Ban on Use of Utility Names, Logos," *Electric Utility Week* (December 22, 1997) 1, 10–11.

20. Command-and-control approaches require regulators to broaden or extend their activities. They characterize the regulatory *modus operandi*.

21. Enforcing rules on those who have an incentive to violate the rules would be difficult, especially in light of the difficulties associated with knowing whether the rules were actually broken (i.e., information asymmetry).

22. The argument used by the U.S. Department of Justice to break up AT&T was that regulation was incapable of stopping AT&T from (1) recovering any losses from undercutting competitors by shifting costs to its regulated monopolies, and (2) vertically foreclosing competitors by controlling its strategic bottleneck of the local telephone monopoly. The government believed that the combination of litigation and ROR regulation could not promote competition in long-distance and equipment markets or protect "captive" customers from inflated prices.

23. Regulators of "wires" services will find it difficult to regulate "wires" utilities that also sell commodity and other services. As regulators impose tougher restrictions and become increasingly frustrated over their inability to prevent market abuses, utilities may decide on their own to divest.

24. In practice, price caps may not produce the desired results. Pure price caps generally are not politically acceptable, requiring in many instances an earnings-sharing component that resembles the profit-constraining characteristic of ROR regulation. Further, price caps may not eliminate cross-subsidization or cost-shifting if the service baskets are excessively broad to include subservices that vary widely in the degree of competition faced by the regulated utility, or if the initial prices for different services were set improperly on the basis of cost.

25. Electric utilities in California have argued that prohibition against the transfer of a utility's logo to an affiliate violates their constitutional rights, namely their First Amendment right to commercial speech. The Federal Trade Commission has recently filed comments before the Texas Public Utility Commission raising the issue of logo transfer as a potential consumer-deception device.

26. Some states, including California and New York, have allowed the transfer of the utility's logo to an affiliate under the condition that consumers are provided certain upfront information from the utility, or that consumers will not receive favorable treatment from the utility when they deal with the affiliate.

27. Simultaneous disclosure of information provided to an affiliate can mitigate discrimination. This requirement is similar to what the Federal Trade Commission calls a "firewall," which is defined as a legal prohibition against specified divisions of vertically integrated firms communicating about proprietary information received by one of the divisions from outside parties. An example of a prohibition against unfair trade practices would be offering the affiliate favorable treatment in contracting for distribution service.

28. As mentioned earlier, most PUCs currently do not have the authority to order divestiture. State electric-industry restructuring legislation may give some regulatory agencies that authority in the future, however.

EPILOGUE

CHAPTER 24

What's in Our Future?

Ahmad Faruqui
J. Robert Malko

As we look at the future, it is clear that customer choice is here to stay. It is an idea whose time has come. To quote the immortal words of Victor Hugo, "No one, not even all the armies of the world, can stop an idea whose time has come."

It also is clear that different states will implement customer choice in a variety of different ways. As they begin the process of creating processes for instituting choice, regulatory commissions, incumbent utilities, and new entrants in these states will find it useful to review the diverse and complex information presented in this book. In addition, they will need to resolve the following complex and thorny issues that are unresolved at this time:

- **How much does it cost to create a competitive infrastructure?** In California, estimates in excess of one billion dollars are being reviewed by the state legislature as we go to press. One of the key legislators behind reform, Senator Steve Peace, expressed his disappointment with this staggering figure, and indicated that legislators were expecting to see a figure in the $100 million to $200 million range.

- **Should states mandate rate reductions of 10 percent—or of any amount—for their residential and small customers?** In a keynote address to a conference on Pricing Energy in a Competitive Energy Market, former California Public Utility Commission Chairman Daniel Fessler indicated that the California PUC did not intend to mandate any such reduction, because this would be contrary to the objective of letting markets set prices.[1] However, such a reduction was imposed by the legislature as part of a political process to get the bill approved.

- **Should states mandate the establishment of a non-bypassable charge on the distribution of electricity?** Such a "wires" charge is being used in California to fund energy efficiency, renewables, and other public purpose programs, as a necessary condition of implementing customer choice. In the same address mentioned above, Daniel Fessler indicated that such charges were introduced in California as a one-time activity, to deal with job losses stemming from cuts in the federal defense budget.

- **Will customers become confused and disenchanted by the choices being offered to them?** As English customers of natural gas were provided choice this year, a British Gas Trading executive stated that they were getting "10,000 calls a week" from customers who were confused by the offers in the marketplace. A customer in California indicated that she expected choices to proliferate, prices to go up, and reliability to go down.

- **Is it possible to make money selling electricity as a commodity?** In California, none of the major energy service providers (ESPs) are in the black thus far. One of them, Enron, has pulled out of the residential market.[2] In the U.K., margins in the commodity retail market have averaged under 1.5 percent.[3] In New Zealand, similar numbers have been reported.[4] However, there is evidence that one ESP in the mid-Atlantic region has broken into the black, after two years of operation.

- **How many customers will switch?** The competitive power markets in the U.K. and Australia show evidence of substantial switching behavior. As of 1997, in the population of large customers exceeding 1 megawatt of load, 57 percent in the U.K. and 60 percent to 70 percent in Victoria, Australia, had switched their supplier. The trend over time is toward increasing switching activity. In 1990/91, about 40 percent of the output for large customers had switched providers in the U.K. This figure rose to about 70 percent in 1996/97. Similar trends can be observed in the U.S. natural gas and long-distance telecommunication industries. In 1985, 31 percent of the industrial load had switched natural gas providers. This number rose to 81 percent in 1996. AT&T's share of revenues in the U.S. long-distance market fell from 79 percent in 1987 to 45 percent in 1997.[5]

- **Will switching become an addictive process?** A recent study that draws upon data in the long-distance telecommunications market indicates that about 44 percent of customers who have switched once will switch again.

It is too early to address these questions in this book. We hope to address them in future updates of this book.

NOTES

1. The conference was organized by the Electric Power Research Institute in Washington, D.C., in June 1998.

2. For details, see Chapter 20 in this book.

3. Alex Henney, K. Simmons, and J. Percival, "Is There Value in Value-Added Services?" (Electric Power Research Institute report, February 1996).

4. Personal conversation with Bruce Turner of Mercury Energy.

5. For details, see "What Drives Customer Choice in Competitive Power Markets" (Electric Power Research Institute report, October 1998).

Kenneth R. Bartkus is Associate Professor of Marketing in the College of Business at Utah State University. He holds a bachelor's degree in economics from California State University at Long Beach, an MBA degree in finance from Humboldt State University, and a Ph.D. in marketing from Texas Tech University.

Stefan M. Brown is vice president and senior economist at Economic Insight, Inc., in Portland, Oregon. He is a specialist in natural resource and environmental economics, and production economics. While at Christensen Associates he directed the development of EPRI's Product Mix Model, a retail product analysis tool, which analyzes the benefits and risks to merchants and consumers associated with a portfolio of electric products in a competitive market. He also developed optimal pricing equations for various retail electric products under different market structures. Dr. Brown holds a Ph.D. in agricultural economics from Purdue University.

Sheryl C. Cates is a Project Manager in Research Triangle Institute's Center for Economics Research in Research Triangle Park, North Carolina. Her work focuses on using conjoint analysis and other quantitative techniques for assessing customer needs and preferences for new technologies, programs, and services. Before joining RTI in 1986, she was a rate analyst at Duke Power Company in Charlotte, North Carolina. Ms. Cates has a B.A. degree in Business Administration from the University of North Carolina at Charlotte.

Douglas W. Caves, vice chairman of Christensen Associates, is an expert on issues facing industries going through deregulation. His early work centered on restructuring issues facing airline, railroad, and telephone companies. In addition, he has long involvement with the electric utility industry having directed Christensen Associates' electric utility work since 1976. Dr. Caves has made numerous improvements in the methodology for economic analysis of the industry, and has been active in the design and evaluation of a wide range of innovative pricing and service programs. He is well-known for his pioneering role in the development of real-time pricing and he has advised numerous companies on issues related to industry restructuring and deregulation. Dr. Caves also has been active in the field of economic evaluation and its use in the design of utility programs and regulation. His scholarly work has been published in a number of journals. He holds a Ph.D. in economics from the University of Wisconsin-Madison.

Eric P. Cody is President of the Retail Access Advisory Group, a division of NEES Global, Inc., which specializes in detailed implementation planning and business process development for full-scale retail access with customer choice. He also serves as Vice President of NEES Global. Mr. Cody was the executive directly responsible for detailed implementation planning and process development for NEES distribution companies moving to customer choice in Massachusetts, Rhode Island and New Hampshire prior to the creation of

NEES Global's utility consulting practice. Mr. Cody holds a BA degree (cum laude) from Amherst College and a Masters degree in City and Regional Planning from Harvard University, where he specialized in energy planning and policy analysis.

Kenneth W. Costello is Associate Director for the Electric and Gas Division at the National Regulatory Research Institute, a position he has held since July 1990. He previously worked for the Illinois Commerce Commission, the Argonne National Laboratory, Commonwealth Edison Company, and as an independent consultant. He is widely published on topics relating to energy industries and public utility regulation. Mr. Costello has provided training and consulting services to the governments of Argentina, Bolivia, Costa Rica, Egypt, Russia and Alberta, Canada. Mr. Costello received both a master's and a bachelor's degree from Marquette University and has completed some doctoral work in economics at the University of Chicago.

Robert W. Crandall is a Senior Fellow in the Economic Studies Program at the Brookings Institution. He holds a M.S. degree and a Ph.D. from Northwestern University. He has specialized in industrial organization, antitrust policy, and the economics of government regulation. His current research focuses on regulatory policy in the telecommunications sector, with particular emphasis on universal service and competition. He is the author of several books on telecommunications and regulation and numerous journal articles. Dr. Crandall has taught economics at Northwestern University, MIT, The University of Maryland, George Washington University, and the Stanford in Washington program. Prior to assuming his current position at Brookings, he was Acting Director, Deputy Director, and Assistant Director of the Council on Wage and Price Stability.

Kerry N. Diehl is Vice President of PNR and Associates, Inc., a subsidiary of INDETEC International. Prior to joining PNR he concurrently served DQE as General Manager of Marketing and Customer Service for Duquesne Light and as Vice President of Duquesne Enterprises with responsibility for regulated and unregulated energy sales, marketing, as well as telephone-based service and collection activities. He joined DQE from the cellular industry. Mr. Diehl has a background in strategic planning, mergers and acquisitions, and currently is concerned with the tactical, strategic and financial implications of targeted marketing. He has authored or co-authored five EPRI market research reports on consumers and small/medium businesses. Mr. Diehl holds an M.S. degree in Industrial Administration from the Graduate School of Industrial Administration, Carnegie-Mellon University and a B.S. degree in Chemical Engineering from Lehigh University.

Jerry Ellig is a Senior Research Fellow at the Mercatus Center and the Institute for Humane Studies at George Mason University. He is also a senior fellow at Citizens for a Sound Economy Foundation in Washington, D.C. He previously served as senior economist for the Joint Economic Committee of the U.S.

Congress, as assistant professor of economics at George Mason, and as a consultant to the President's Commission on Privatization. Dr. Ellig has published numerous articles on government regulation of business and business management. He holds a Ph.D. and M.A. degree in economics from George Mason University, and a B.A. degree in economics from Xavier University.

Joseph T. Ewing is Strategic Sourcing Manager, North American Energy for Procter & Gamble, a position he has held since 1993. He is responsible for approximately $200 million in energy purchases and energy strategy development for 50 North American sites. He has held various manufacturing and purchasing positions since joining Procter & Gamble in 1969. Mr. Ewing is on the Steering Committee of the Industrial Energy Users of Ohio. He is a Program Instructor for Effective Energy Sourcing and Delivery at the University of Wisconsin-Madison, Management Institute. He is also a member of the Process Gas Consumer (PGC) group. He earned a Bachelor's Degree in Accounting from the University of Cincinnati.

Ahmad Faruqui manages the retail and power markets area at EPRI. Over the past two decades, he has worked with more than 50 energy companies and government agencies on restructuring energy markets, retail business strategy, marketing tactics, pricing design, and demand forecasting. In his career, he has held senior management positions at several consulting firms, including A. T. Kearney, Barakat & Chamberlin, Battelle-Columbus Division, and Hagler Bailly. He has also worked at the California Energy Commission and the Applied Economics Research Center at the University of Karachi. He has authored or co-authored more than 100 articles on energy issues and is co-editor of the book, *Changing Structure of American Industry and Energy Use Patterns*. He graduated from the University of Karachi, Pakistan with a B.A. degree in economics and has a Ph.D. from the University of California, Davis.

Patricia B. Garber has over 15 years' experience in designing, managing, and utilizing market research for manufacturers, distributors, advertising agencies, utilities and other public agencies. In addition to the energy field, she has worked for computer hardware and software manufacturers, automobile manufacturers and distributors, financial/educational services, and consumer packaged goods manufacturers and their advertising agencies. As Manager of Retail Products Marketing at EPRI, Dr. Garber studies retail marketing strategy, customer loyalty, market segmentation, and new products. She has a Ph.D. in sociology from the University of California, Los Angeles.

Clark W. Gellings is Vice President of Client Relations and Vice President, Energy Utilization for the Electric Power Research Institute in Palo Alto, California. He joined EPRI in 1982 as a Program Manager and subsequently served as a Senior Program Manager and as a Director. Between 1992 and 1997, he served as Vice President of Customer Systems. Prior to joining EPRI, he spent 14 years with Public Service Electric and Gas Company. Mr. Gellings is a registered Professional Engineer, a Fellow in the Institute of Electrical and

Electronics Engineers (IEEE), a Fellow in the Illuminating Engineering Society (IES), a Vice President of the U.S. National Committee of CIGRE, and is active in a number of other organizations. He has degrees in Electrical Engineering, Mechanical Engineering, and Management Science.

H. Dean Jones II is Vice President, Gas and Power Origination for Williams Energy Services. His career in the energy industry began in the mid-1970s as an engineer for Exxon Chemical Company USA. He joined Texas Gas Transmission Corporation in 1980, where he held positions in gas process studies, tar sand engineering, and planning. He has served as Director of Marketing for Texas Gas Transmission, as General Manager of TXG Gas Marketing Company, and as President of TXG Gas Marketing. In 1992 he was named President and Chief Executive Officer of Transco Energy Marketing Company; the following year he became Senior Vice President of Transcontinental Gas Pipeline Company and served as Senior Vice President of Transco Gas Marketing Company in 1994–95. Mr. Jones holds a Bachelor of Science degree in chemical engineering from the University of Kentucky.

David E. Jones is the Chief Executive Officer of S4 Consulting. He specializes in marketing and strategy issues for clients in the energy industry, with a particular focus on key account management and pricing. Mr. Jones has more than 20 years experience in management consulting. Previously he was Vice President and managing director of the Atlanta office of Temple, Barker and Sloane (now Mercer Management Consulting) and was a faculty member of The Goizueta Graduate School of Business at Emory University.

Becky A. Kilbourne is Director of Market Services for the California Power Exchange Corporation (PX). Her responsibilities include planning, directing, and coordinating the activities of the marketing, communications, and related regulatory functions of the PX. They also include coordinating services to customers, helping to shape the business strategies of the PX to evolve with changing markets, developing and implementing business strategies to meet business objectives, managing corporate communications and directing complex negotiations with major customers and regulators. Ms. Kilbourne has 20 years' experience in the electric utility industry, including serving as Director of Marketing and Business Development for the Public Service Company of New Mexico just prior to joining the PX. Ms. Kilbourne has an M.B.A. degree in Finance and a B.B.A. degree in Accounting from the University of New Mexico; she is a Certified Management Accountant.

Michael King is Senior Vice President and Managing Director of the Generation Asset Valuation and Strategy Group at Hagler Bailly, with nearly 20 years' consulting experience in the energy industry. An economist by training, Mr. King currently specializes in valuation of generation assets, asset acquisition strategies, risk management, and bidding strategies for power in restructured markets.. He led the valuation of the NEES generation assets in an independent valuation for one of the final round bidders where he directed the application of option pricing methods to value investment opportunities. He has prepared expert reports to secure more than $2 billion of financing to support purchase of generation assets. He has developed methods for applying option pricing theory to hedge price and quantity risk associated with product offerings in the wholesale electric markets. He has played a key role in start-ups for two power marketing companies, and is currently developing bidding support systems that assist generators in maximizing the yields from bidding their generation into power exchanges. Previously, Mr. King served as a senior vice president of Synergic Resources Corporation where he was responsible for the firm's Latin America, Pacific Rim, and Western U.S. business. He has an M. A. in economics from the University of Wyoming.

Raymond W. Lawton is Director of the National Regulatory Research Institute and Adjunct Associate Professor at the School of Public Policy and Management at The Ohio State University. He received his Ph.D. from The Ohio State University. Dr. Lawton specializes in telecommunications competition, regulatory reform, modernization, management audits, water policy, and qualitative research methods. He regularly makes presentations at national regulatory forums and conferences and is a member of the National Association of Regulatory Utility Commissioners Staff Communications Committee. He was a faculty member at a U.S. Department of State-sponsored telecommunications tariffs seminar in Tashkent, Uzbeckistan for the Confederation of Independent States.

Kenneth L. Lay is chairman and chief executive officer of Enron Corporation. He was named to that position in February 1986, following the merger of Houston Natural Gas and InterNorth, Inc. in July 1985. Previously, he was president of Continental Resources Company (formerly Florida Gas Company) and executive vice president of the Continental Group, the parent company of Continental Resources Company, before joining Transco Energy Company in May 1981, as president, chief operating officer, and as a director. He joined Houston Natural Gas in June 1984 as chairman and chief executive officer. Mr. Lay holds both an undergraduate and a master's degree in economics from the University of Missouri, and a Ph.D. in economics from the University of Houston.

David C. Lineweber is Vice President, Market & Survey Analysis Group, for Hagler Bailly. Recent projects have focused on competitive positioning analyses, customer loyalty assessments, strategic customer satisfaction analyses; customer segmentation modeling, and new product and service development. Dr. Lineweber is also an experienced quantitative analyst, specifically interested in customer segmentation models and the development of customer satisfaction models that link to observable changes in customer behavior. Trained at Stanford University, Dr. Lineweber has taught undergraduate university courses in sociology, with a special focus on research methods.

James J. Malachowski is Chairman of the Public Utilities Commission and Chairman of the Rhode Island Energy Facility Siting Board. He is active in a number of professional organizations including serving as Chairman of the Board of Directors of the National Regulatory Research Institute. This organization is the official research arm of the National Association of Regulatory Utility Commissioners. Mr. Malachowski earned a B.S. in Business Administration from the University of Rhode Island and an M.B.A. degree from Providence College.

René H. Malès is the former president of IES Utilities, the Cedar Rapids-based electric and gas company serving Iowa customers. He now provides strategic planning services to the energy industries as president of Strategic Decisions Inc. Mr. Malès was a vice-president of EPRI where he was responsible for energy and environmental research. Also, he has been an executive of other firms. Mr. Malès holds an M.B.A. from Northwestern University, and a B.A. in mathematics from Ripon College. He has lectured in the graduate business programs at Northwestern and DePaul Universities. He has published articles on energy and environmental issues in a number of journals, contributed chapters to several books and has spoken frequently on these subjects.

J. Robert Malko is Professor of Finance in the College of Business at Utah State University. He serves as an Advisory Council Member of the Society of Utility and Regulatory Financial Analysts, and was president of the society from 1988-90. He is on the Board of Directors of the National Regulatory Research Institute at The Ohio State University and serves on the Advisory Council of the Center for Public Utilities at New Mexico State University. Dr. Malko has served as Chief Economist at the Public Science Commission of Wisconsin (1975–77 and 1981–86) and has served as Chairman and Vice Chairman of the Staff Subcommittee on Economics and Finance of the National Association of Regulatory Utility Commissioners. He also served as Program Manager of the Electric Utility Rate Design Study at the Electric Power Research Institute. Dr. Malko received a B.S. degree in Mathematics and Economics from Loyola College and M.S. and Ph.D. degrees in Economics from the Krannert Graduate School of Management at Purdue University.

Melanie Mauldin is Strategic Marketing Manager for CellNet Data Systems. Prior to that, she was a Project Director with Barakat & Chamberlin, a consulting firm that is now part of PG&E Energy Services. In that position, she consulted to utilities on pricing and product development issues. Dr. Mauldin is the co-author of *Winning Retail Strategies: Beyond Innovative Rate Design*, has published and presented numerous papers on electricity pricing, and has taught pricing and marketing courses for EPRI. Dr. Mauldin holds an undergraduate degree from Massachusetts Institute of Technology and a Ph.D. in economics from the University of California at Berkeley.

Robert J. Michaels is Professor of Economics at California State University, Fullerton. He holds an A.B. degree from the University of Chicago and a Ph.D from the University of California, Los Angeles, both in economics. He has served as Staff Economist at the Institute for Defense Analyses and Senior Advisor to Hagler Bailly of Arlington, Virginia. He writes and consults on deregulation and competition in the electricity and natural gas industries and has advised state commissions, electric utilities, power marketers, natural gas producers, pipelines, public interest groups, and governments on regulatory and antitrust matters. He has participated in electricity restructuring proceedings in California and other states, served as expert witness in electric utility mergers, testified before the House Subcommittee on Energy and Power, and advised the Government of New Zealand on electricity denationalization. Mr. Michaels has written extensively on stranded investment, wholesale pooling, antitrust analysis, and electrical restructuring and gas regulation.

Michael R. Peevey is the President and Chief Executive Officer of New Energy Ventures, L.L.C. Before cofounding New Energy Ventures in 1995, Mr. Peevey was the President and Director of Edison International and Southern California Edison Company from 1990 to 1993; he had been with Edison since 1983. Before joining Edison, he served 10 years as president of the California Council for Environmental and Economic Balance. He also worked with the California Labor Federation and the AFL-CIO, and he was an economist for the Department of Labor during the Kennedy and Johnson administrations. Mr. Peevey received his B.A. and M.A. degrees in Economics from the University of California, Berkley.

Michael W. Rufo is Director of Consulting of the Western Region for XEN-ERGY, Inc. Mr. Rufo has led a number of unique studies on retail competition since the mid-1990s, including a comprehensive analysis of the opening of the California market that included both large-scale customer surveys and in-depth regulatory analyses. Mr. Rufo also co-led (with Dr. Kenneth Train) the design and implementation of several stated and revealed preference models of customers' choices in a competitive electricity industry, which produced state-of-the-art results for the retail electric industry. Mr. Rufo has studied the critical infrastructure issues that underlie the transition from regulated to competitive energy markets. His work has included analyses of billing options, metering protocols, and load profiling methods. Mr. Rufo has an M.A. degree in Engineering and Policy from Washington University, St. Louis, and a B.A. degree in Environmental Studies and Planning from California State University, Sonoma.

Dr. Francine Sevel is editor of the NRRI *Quarterly Bulletin*, a research publication of the National Regulatory Research Institute (NRRI). At the NRRI, Dr. Sevel conducts research on consumer protection, consumer education, service quality, and public information issues facing public utility commissions. Dr. Sevel also provides onsite assistance to state public utility commissions regarding these topics. Dr. Sevel has a Ph.D. in education and a M.A. in journalism from

The Ohio State University (OSU); previously, she was an Assistant Professor at OSU. Dr. Sevel serves as a member of the NARUC Ad Hoc Committee on Consumer Affairs, the NARUC Staff Subcommittee on Consumer Affairs, and the FCC/NARUC Working Group on Consumer Education.

Dr. Karl E. Stahlkopf is Vice President of the Energy Delivery & Utilization Division at the Electric Power Research Institute in Palo Alto, California. He has overall responsibility for the direction of a $200 million R&D program, as well as responsibility for the Institute's marketing efforts in the Western Region of the U.S. Prior to joining EPRI, Dr. Stahlkopf was a Research Fellow at the University of California, Berkeley. Earlier he was a technical consultant on nuclear submarine propulsion on the staff of U.S. Chief of Naval Operations. He received a B.S. degree in Electrical Engineering and a B.S. degree in Naval Science from the University of Wisconsin in Madison, and M.S. and Ph.D. degrees in Engineering from the University of California, Berkeley.

Kenneth Train is Chair of the Center for Regulatory Policy at the University of California, Berkeley, where he teaches econometrics and industrial organization. He is also Vice President of National Economic Research Associates, Inc. He has published two books and numerous articles, has served as an expert witness in regulatory proceedings and litigation, and frequently consults to private firms and government agencies throughout the world. He has an A.B. degree from Harvard and a Ph.D. in economics from the University of California, Berkeley.

Daniel M. Violette is a Vice President in the Economics & Analytics group of PHB Hagler Bailly. With 20 years of utility and energy industry consulting experience, he has led numerous consulting engagements for electric and gas utilities. He has performed enterprise structure and performance analyses, organizational effectiveness studies, management reviews and audits, market positioning studies, asset management and yield studies as well as new product and service analyses, and resource planning. He has provided support to utilities in merger and acquisition analyses, rate cases, regulatory hearings, as well as in securities and environmental litigation. Dr. Violette also has undertaken major assignments in the telecommunications, iron and steel, and chemical industries. Dr. Violette served three elected terms as the President of the Association of Energy Services Professionals. He previously held senior positions at Electronic Data Systems (EDS) Management Consulting Services and at A.T. Kearney Management Consultants. Dr. Violette received his Ph.D. in Economics from the University of Colorado.

Robert E. Wayland is president of Robert E. Wayland & Associates, Inc., a Concord, Massachusetts consulting firm providing business and marketing strategy and management development services. He is the author (with Paul Cole) of *Customer Connections: New Strategies for Growth* (Harvard Business School Press, 1997), which provides a comprehensive framework for demand-side strategy. Mr. Wayland began his 24-year career in the energy sector as the Chief Economist of the Public Utilities Commission of Ohio where he was one of the architects of Ohio's "self-help" gas program, a forerunner of open access and customer choice. Before starting his own firm, he headed the energy practice at Temple, Barker and Sloane and its successor firm, Mercer Management Consulting.

Sue Winemiller is an Account Director at M/A/R/C Research, where she specializes in helping clients adapt customer wants and needs into clear, operational strategies. Her experience comes from a variety of industries, including utilities, high technology, financial services, and telecommunications. She specializes in the areas of Voice of the Customer; Quality Function Deployment; Customer Satisfaction Research; Process Improvement; Action Plan Development; Organizational Assessment; and Analytic Interpretation. Ms. Winemiller earned her B.S. and Masters degrees from Bowling Green State University.

David W. Wirick has been Associate Director for Administration and Special Projects at the National Regulatory Research Institute since 1986. He is the author of reports and articles on topics such as accounting for regulated utilities, the evaluation of water utility financial capacity, public utility commission strategy, the health effects of electromagnetic fields, the natural gas futures market, the gain-on-sale of utility assets, and alternative dispute resolution. He leads the NRRI's program to provide assistance to state public utility commissions engaged in transformation in response to changing utility and regulatory environments. He is a mediator and mediation trainer.

Lisa Wood is a Principal in the Market & Survey Analysis Group at Hagler Bailly, where she specializes in the design, implementation, and interpretation of quantitative market research. She has extensive experience in the areas of customer preferences for new product and service offerings, market share analysis, and willingness to pay and has directed a wide range of studies in the electric utility industry over the past 10 years. She has published in leading journals and spoken at numerous marketing and electric utility industry conferences. Dr. Wood previously held a senior position at Research Triangle Institute. Dr. Wood holds a doctorate from the Wharton School of Business at the University of Pennsylvania.

SUBJECT INDEX